FLORA ZAMBESIACA

Flora terrarum Zambesii aquis conjunctarum

VOLUME THREE: PART TWO

FLORA ZAMBESIACA

MOZAMBIQUE

MALAWI, ZAMBIA, ZIMBABWE

BOTSWANA

VOLUME THREE: PART TWO

Edited by
J.R. TIMBERLAKE, G.V. POPE, R.M. POLHILL & E.S. MARTINS

Published by the Royal Botanic Gardens, Kew,
for the Flora Zambesiaca Managing Committee
2007

First published in 2007 by
Royal Botanic Gardens, Kew
Richmond, Surrey, TW9 3AB, UK
www.org.uk

ISBN 978 1 84246 139 6

British Library Cataloguing in Publication Data
A catalogue record for this book is available from the British Library

Design and typesetting by Christine Beard,
Kew Publishing, Royal Botanic Gardens, Kew

Printed in the United Kingdom by Hobbs the Printers

This book is printed on paper that has been manufactured using wood from well managed forests certified in accordance with the rules of the Forest Stewardship Council.

For information on or to purchase all Kew titles please visit
www.kewbooks.com or e-mail publishing@kew.org

All proceeds go to support Kew's work in saving the world's plants for life

CONTENTS

ARRANGEMENT OF TRIBES IN VOLUME 3

LIST OF NEW NAMES PUBLISHED IN THIS PART

Page

61. LEGUMINOSAE
Subfamily CAESALPINIOIDEAE

by R.K. Brummitt, A.C. Chikuni, J.M. Lock and R.M. Polhill*

Caesalpinioideae DC., Prodr. **2**: 473 (1825) as Caesalpinieae.

Trees, shrubs, sometimes climbing, or rarely herbs, unarmed or often armed in the tribe Caesalpinieae. Leaves mostly alternate, usually pinnate, sometimes bipinnate, rarely unifoliolate or simple; stipules paired, often caducous. Flowers in spikes, racemes or panicles (compound racemes), sometimes reduced to fascicles, small to large, rarely very small, usually ± zygomorphic, mostly bisexual and 5-merous. Sepals usually imbricate, rarely valvate or open from an early stage, free or sometimes ± connate, sometimes raised on the edge of a bowl-shaped to tubular hypanthium (receptacle). Petals imbricate in bud, usually with the dorsal one within and overlapped by the adjacent lateral ones, free or sometimes connate below, usually 5, sometimes ± reduced in number and size or absent. Stamens 10 or fewer, free or shortly united; anthers various, but without an apical gland; pollen grains usually simple. Ovary free or when stipitate the stipe sometimes adnate to the hypanthium; ovules anatropous. Pods various. Seeds without an areole except in *Burkea, Senna* and *Tamarindus*, with an apical or subapical hilum; embryo generally with a straight radicle.

About 170 genera and 2300 species, widespread but most common in tropical South America, Africa and SE Asia. The circumscription of tribes and genera follows Lewis, Schrire, Mackinder & Lock, Legumes of the World (2005). The tribe Swartzieae is included in the Papilionoideae. The subfamily is paraphyletic with respect to Mimosoideae and Papilionoideae and includes the basal elements of the family, some of which are poorly discriminated in current molecular and general analyses. The following tribes provide an interim, pragmatic arrangement.

Tribe 1. **Cercideae** Bronn, De Formis Pl. Legum.: 131 (1822) as Cerceae.
Bauhinieae Benth. in Hooker's J. Bot. **2**: 74 (1840).

Trees or shrubs or with herbaceous stems from an underground tuber, sometimes climbing, sometimes with tendrils or intrastipular spines. Leaves alternate, usually simple, bilobed or entire, sometimes with 2 separate leaflets (usually with a small mucro at apex between the lobes), usually palmately nerved; stipules caducous or persistent; intrastipular trichomes present, sometimes forming a spine or glandular. Flowers in racemes, panicles or coymbs, bisexual or unisexual, slightly to greatly zygomorphic; bracts and bracteoles small to large, deciduous or persistent. Hypanthium nearly absent to greatly enlarged. Sepals free or joined, entire, valvately 2–5-lobed or spathaceous. Petals (2)5(6), subequal or generally unequal, erect to reflexed, rarely absent. Stamens 10(25), all perfect or with 2–9 reduced to staminodes; filaments free or partly joined; anthers dorsifixed, opening by longitudinal slits or by a central pore. Ovary 1–many-ovulate, stipitate; style filiform, stout or apparently absent; stigma small to peltate. Pods flattened or turgid, woody or thin-valved, dehiscent or indehiscent, continuous, filled or septate within. Seeds rounded, elliptic or pyriform, ± compressed; hilum circular or crescentric, with or without two funicular aril-lobes usually adnate to the testa and leaving a U-shaped or hairpin-shaped scar.

Now regarded as basal to the family, with 12 genera and 322–348 species, mainly tropical apart from the genus *Cercis* L. in north warm-temperate regions. Genera 1–4.

---

* Genera 1–4, 6, 10, 24 and 28–39 by R.K. Brummitt; genus 21 by A.C. Chikuni; genera 5, 7–9, 11–16, 18–20, 22, 23, 25–27 by J.M. Lock; genus 17 and generic keys by R.M. Polhill.

Tribe 2. **Detarieae** DC., Prodr. **2**: 521 (1825).
Amherstieae Benth. in Hooker's J. Bot. **2**: 73 (1840).
Cynometreae Benth. in Hooker's J. Bot. **2**: 74 (1840).
Macrolobieae Breteler in Adv. Leg. Syst. **7**: 59 (1995).

Trees, rarely shrubs or with annual stems from a woody rootstock, unarmed. Leaves pinnate, rarely 1-foliolate or simple; leaflets opposite or alternate, often glandular punctate or with specialised glands; stipules intrapetiolar, if negligible bud scales well developed. Flowers spiral or distichous in racemes or panicles, small to showy, regular to markedly zygomorphic, usually bisexual; bracteoles small to large, imbricate or petaloid and valvate. Hypanthium shallow to tubular. Sepals 0–5, variably developed. Petals 0–5, equal or abaxial ones rudimentary. Stamens 2–10 or rarely more, sometimes with staminodes; anthers sagittate to dorsifixed, opening by longitudinal slits. Ovary with stipe free or adnate to hypanthium or sessile. Pods usually dehiscent with twisting woody valves, but sometimes indehiscent. Seeds diverse, with thin or hard testa, sometimes with an aril. Genera 5–23.

About 82 genera and 750 species, mostly in the tropics. Genera 5–23 and the cultivated genus *Pterogyne*.

Tribe 3. **Cassieae** Bronn, De Formis Pl. Legum.: 130 (1822).

Trees, shrubs or herbs, generally unarmed. Leaves normally pinnate, sometimes with extrafloral nectaries on petiole and rachis; leaflets opposite to alternate without specialised glands; stipules lateral or absent. Flowers in racemes or panicles, regular or zygomorphic, generally showy, generally bisexual; bracts and bracteoles generally small. Hypanthium cupular and infilled with nectarial tissue or inconspicuous. Sepals imbricate to subvalvate. Petals 0–5, small to elaborate. Stamens variable in number, sometimes with staminodes; anthers dorsifixed or more often basifixed, usually opening by pores. Pods diverse, 2-valved and dehiscent to turgid and indehiscent. Seeds with a small subapical hilum, sometimes with closed areoles.

About 21 genera and 730–735 species, tropical and subtropical. Genera 24–27 (*Ceratonia* is now included in Caesalpinieae).

Tribe 4. **Caesalpinieae**

Trees, shrubs or rarely herbs, sometimes climbing, quite frequently armed with spines or prickles. Leaves opposite to generally alternate, pinnate to bipinnate, the axes sometimes with specialised glands; leaflets opposite to alternate, sometimes glandular punctate or glandular hairy; stipules lateral or lacking; stipels occasionally present. Flowers in spikes, racemes or panicles, small to showy, regular to zygomorphic, unisexual or generally bisexual; bracteoles small or lacking. Hypanthium usually cupular, sometimes very short or tubular. Sepals usually free to the hypanthium, sometimes joined into a tube, imbricate or valvate, not always covering the petals in bud. Petals (1)5(6), resembling sepals to markedly differentiated, rarely absent, the adaxial petal often further modified. Stamens (4)10(20), some or all sometimes staminodal; anthers sagittate to dorsifixed, opening by lateral or introrse slits. Ovary central or rarely basally adnate to hypanthium, 1–many-ovulate; style short or elongate; stigma bilobed, funnel-shaped, peltate or very small. Pods very diverse, 2-valved and dehiscent to indehiscent and variously turgid or winged. Seeds compressed to globose, with hard or thin testa, sometimes winged, with a small hilum, and genrally a small adjacent sunken discoloured patch (the lens of Papilionoideae), this occasionally extended as a line around the seed (*Erythrophleum*), rarely with areoles (*Burkea*).

About 56 genera and 420–450 species, widely dispersed, but mainly in the tropics and subtropics. Genera 28–39 and all the cultivated genera except *Pterogyne*.

The Caesalpinioideae are relatively protean in the family. Features which are consistent in genera or even tribes of the more advanced subfamilies often vary within genera, species or sometimes even on the same plant, e.g. pinnate and bipinnate leaves in *Gleditsia* or numbers of stamens in *Dialium*. In using easily visible features for the generic keys it is somewhat disconcerting to find that *Baikiaea* has alternate leaflets in F.T.E.A. but opposite ones in the F.Z. area: both are correct, just different species. Even a cursory scrutiny of the traditional tribal descriptions above reveals an immense diversity. The subfamily has tended to receive less attention than the other two and much still needs to be done to understand the morphology and its underlying biology.

Because there is so much morphological diversity, floristic accounts have tended to use only a small number of familiar features. Closer examination of, for example, the leaflets of Detarieae reveals numerous differences which are difficult to categorise and describe in simple language. Even the concept of a petiolule becomes difficult when the base of the two sides of the leaflet end at different points and in different planes and different degrees of attenuation, and then the base of the petiolule itself is also inserted at different angles and subtended by or fused with outgrowths of the rachis. Traditionally such differences are described simply as petiolule twisted or not. Wieringa in Wageningen Agric. Univ. Papers **99**(4): 32–50 (1999) has effectively used diagrams and photographs of live plants to illustrate some of the diversity of leaves in the group of Detarieae he was studying. He has also provided more insight into stipules that extend in front of the petiole and may become united to cover the younger parts rather than the leaf subtended. This tendency, which is characteristic of the Detarieae, has been referred to simply by saying the stipules are intrapetiolar, but again there is a lack of terminology for the very varied extents of the modification. He also criticises the imprecise use of the term panicle to describe the branched racemose inflorescences in the family. He considers that they seem always to be derived from contracted branch systems and thus do not conform to a purist teutonic concept of panicles. In any case what we for simplicity call panicles and what he calls compound racemes, show a range of contraction and further modification that is generally unrecorded.

The seeds of Caesalpinioideae lack the complex hilar valve that facilitates the drying out of hard seeds in Papilionoideae, but a few genera do have mechanisms that allow the testa to shrink without undue stress. The most obvious of these are known as areoles. In *Burkea* the areole is open, a horseshoe-shaped crack on the face of the seed, similar to that found in many Mimosoideae. In *Tamarindus* and *Senna* the areole is closed, with a continuous margin, comprising an area of the face demarcated by a distinct change in level or surface marking of the testa. Other genera may have wrinkles or transverse cracks but not clearly defined zones.

Molecular evidence, summarised in Lewis et al. (Legumes of the World, 2005), has brought new insights in recent years. Cercideae now seem to have an unequivocal basal position and the sister group seems to be the other families of Fabales, Polygalaceae, Surianaceae and Quillajaceae. Some unexpected associations of genera have emerged, notably *Gleditsia, Ceratonia* and *Umtiza,* previously placed in three different tribes now comprise a single early clade, characterised by considerable diversity in the flowers but all dioecious, an otherwise rare condition in the family.

A number of species, all Caesalpinieae except *Pterogyne,* are recorded in cultivation but do not yet appear to have become naturalised.

Acrocarpus fraxinifolius Arn., native to southern and SE Asia from India to Thailand, a fast-growing tree grown for ornament, shade and timber, is probably more widespread than the records suggest, although it requires a moister environment than pertains in most of the region. **Zimbabwe.** E: Mutare, Mutare (Umtali) Park, fl. 1.ix.1948, *Chase* 871 (K, SRGH); Chipinge, Gungunyana Forest Res., in grounds of forester's quarters, fl. vii.967, *Goldsmith* 81/67 (K, SRGH). **Malawi.** S: Zomba, Veterinary Office compound, imm.fl. 4.iii.1981, *Chapman & Patel* 5573 (K); Thyolo (Cholo), Masambanjati Agric. Station, fr. 26.x.1960, *Chapman* 1025 (K). Tree to 30 m with smooth grey bark, young growth pink; leaves bipinnate; pinnae 3–5 pairs; leaflets 4–9 pairs per pinna, 5–15 × 2–7 cm, ovate to ovate-oblong, acuminate; racemes dense, flowers deflexed; sepals green; petals red, 6–10 mm long; stamens yellowish orange, twice the length of the petals; pods brown when ripe, 8–15 × 1–2 cm, flat, oblong.

Ceratonia siliqua L., Carob, native of the Mediterranean region, has been grown in gardens. **Zimbabwe.** C: Harare (Salisbury) Park, Rhodes Avenue, Moffat St., fl. 10.xii.1972, *Biegel* 4454 (K, SRGH). E: Mutare (Umtali) Park, fl. 5.iii.1949, *Chase* 1305 (K, LISC, SRGH). **Mozambique.** M: Goba, near R. Maiuáua (Maiuana), cult., fl. 2.xi.1960, *Balsinhas* 166 (K, LISC, LMA). Evergreen dioecious tree; leaflets in 2–8 pairs, 3–6 × 2.5–4 cm, elliptic; flowers in racemes from axils and old nodes, small, without petals, stamens of male flowers on a broad disk, the tomentellous ovary in female flowers with an enlarged, sessile, grooved stigma; pods dark brown, leathery, 10–30 × 1.5–2.2 cm, laterally compressed, pulpy, indehiscent, many-seeded.

Gleditsia triacanthos L., Honey Locust, native of North America, has been grown in gardens and is liable to spread. **Zimbabwe.** C: Harare, Parktown Nurseries, fr. 31.x.1974, *Biegel* 4687 (K, SRGH); Shurugwi (Selukwe) Dist., Farm Depoto, at foot of Wanderer Valley, st. ii.1977, *Wells* s.n. (K, SRGH), "in *Brachystegia glaucescens* woodland; origin of this exotic is unknown, but thought to be spontaneous". **Mozambique.** M: Maputo, Jardim Tunduru (Jardim Vasco da Gama), imm.fr. 12.iv.1971, *Balsinhas* 1828 (K) & fl. 4.xi.1971, *Balsinhas* 2252 (LISC). Tree or shrub, dioecious, usually with straight or branched spines; leaves usually pinnate and bipinnate on the same shoot; leaflets 1–3.5 × 0.5–1.5 cm, oblong-elliptic to elliptic-lanceolate, crenulate; flowers greenish, small, in axillary racemes from short shoots; tepals all similar; stamens 3–10; pods 15–30 × 2.3–3.5 cm, often twisting with age, laterally compressed, indehiscent, with many seeds embedded in the pulp.

Intsia bijuga (Colebr.) O. Kuntze (*Macrolobium bijugum* Colebr.; *Afzelia bijuga* (Colebr.) A. Gray) [not in the key], a large timber tree of coastal forests around the Indian Ocean including Zanzibar and Madagascar, has been planted in Maputo and probably elsewhere. It is a tree to 20 m tall, with simply pinnate (1)2–3-jugate leaves, and white flowers in lax terminal panicles. The pods are leathery and glabrous, 7–25 × 3–6 cm. **Mozambique**. M: Mata de Marracuene, cultivated, fl. & imm.fr. 29.ii.1952, *Barbosa & Balsinhas* 4618 (LISC).

Libidibia coriaria (Jacq.) Schltdl. (*Caesalpinia coriaria* (Jacq.) Willd.; *Poinciana coriaria* Jacq.), a native of tropical America, is widely cultivated in the tropics and often naturalised. It is known from one garden in the Flora area — **Mozambique.** M: Maputo, Jardim Tunduru (Vasco da Gama), fr. 8.v.1952, *Pedro* 3939 (LMA) and imm.fr. 8.v.1985, *Groenendijk & Dungo* 1690 (K, LMU). It is a usually spreading, much-branched tree up to 10 m high, without prickles or spines; leaves with the petiole and rachis up to 12 cm; with 4–9 pairs of opposite or subopposite pinnae and a single terminal pinna inserted at the same point as the distal pair; pinnae with 10–30 pairs of narrowly oblong, subsessile, gland-dotted leaflets up to 2.5 mm broad; flowers small, in short, dense racemes or panicles, with cream, white or greenish-white petals, and blackish, coiled or twisted, indehiscent pods up to 6 × 2 cm, somewhat resembling those of *Dichrostachys cinerea* (subfam. Mimosoideae).

Moullava spicata (Dalzell) Nicolson (*Caesalpinia spicata* Dalzell; *Wagatea spicata* (Dalzell) Wight), native of India, has been recorded once from the Flora area. **Zambia.** C: National Council Scientific Research, Chilanga, fl. xii.1978, *Critchett* 11/79 (K). Climbing shrub with recurved prickles; leaves bipinnate; pinnae 4–7 pairs, 8–12 cm long; leaflets 5–8 pairs, 2–4 × 1–2 cm, obliquely oblong-elliptic; racemes subspicate (pedicels up to 2 mm long); hypanthium shortly cupular; sepals red, lower one slightly larger; petals yellow, 1–1.3 cm long; pod 5–8 × 1.5–2 cm, oblong, subtorulose, indehiscent, 2–4-seeded.

Pterogyne nitens Tul., native of South America, is said to have been grown in Zimbabwe (Biegel, Check-list Ornam. Pl. Rhod. Parks & Gard.: 90, 1977). Small tree; leaves pinnate; leaflets mostly alternate, 4–7 × 1.5–3 cm; flowers small, yellow, in racemes shorter than the leaves, with 5 petals and 10 stamens; fruits samaroid with a single seed at proximal end and a distal parallel-veined falcate wing.

Schizolobium parahyba (Vell.) Blake (*Cassia parahyba* Vell.; *Schizolobium excelsum* Vogel), native of Brazil, is planted as a street tree or in gardens; it is highly decorative, especially when young. **Zambia.** W: Kitwe, fl. 16.ix.1968, *Fanshawe* 8 (K). **Zimbabwe.** C: Harare, Mt. Pleasant, fl. 27.x.1981, *Biegel & Pope* 5875 (K, SRGH). **Malawi.** S: Zomba Botanic Garden, behind Zomba Power Station, fl. 13.x.1970, *Ndovi* s.n. in SRGH 208045 (K, SRGH). **Mozambique.** M: Maputo, Jardim Tunduru (Vasco da Gama), fl. 13.x.1971, *Balsinhas* 2243 (K, LISC). Erect tree to 30 m, usually branching into several equal parts 5–10 m above the base; leaves bipinnate, up to 60 cm or more long, with 15–20 pairs of pinnae each with 10–25 pairs of oblong obtuse leaflets 2–4 cm long; racemes 20–50 cm long; flowers golden yellow; fruits obovate, obtuse, 8–12 cm long, each with a single large winged seed.

Tara spinosa (Molina) Britton & Rose (*Caesalpinia spinosa* Molina), native of South America from Venezuela to Bolivia and Chile, now widely cultivated, often used in fences, has been recorded from one town in the Flora area. **Zimbabwe.** W: Bulawayo, small street tree, st. v.1977, *Webber* in SRGH 253302 (K, SRGH); Bulawayo, Fairbridge Way, street tree, fr. vii.1978, *Webber* in SRGH 261308 (K, SRGH). Tree up to 5 m; branches armed with short prickles; leaves bipinnate, usually with short prickles at the insertion of the pinnae and leaflets; pinnae 1–4 pairs; leaflets 4–7 pairs per pinna, 1.5–4.5 × 0.6–2 cm, oblong-elliptic; racemes 10–20 cm long, many-flowered, sparsely prickly; lower sepal larger and fimbriate-pectinate; petals up to 1 cm long; filaments a little shorter, yellow, pubescent below; pods pinkish brown to red, 5–9.5 cm long, oblong, thick, indehiscent.

*Key to genera based mainly on vegetative and floral characters**

1. Leaves simple, bilobed, palmately nerved 2
– Leaves with 2–many leaflets (if a single pair sometimes bridged across their basal junction and with a small appendage at the fork beneath and exceptionally in *Guibourtia* fused into a single leaflet, but not bilobed) 5
2. Stems trailing or climbing, herbaceous, from a large tuber, generally with tendrils; fertile stamens 2, accompanied by staminodes; calyx with 2 upper sepals partly or completely fused, the others free .... **3. Tylosema**
– Stems forming shrubs or small trees, seldom climbing but then woody and without tendrils; stamens and calyx as following couplet 3
3. Flowers usually unisexual and on separate plants; female flowers with the stigma sessile on the ovary, capitate, flattened-globose; male flowers with 10 fertile stamens; calyx turbinate with 4–5 short broad lobes .... **4. Piliostigma**
– Flowers hermaphrodite; style elongate; fertile stamens 1–10; calyx spathaceous or campanulate with 5 short teeth or lobes 4
4. Calyx campanulate, with 5 short teeth or lobes; fertile stamens 10; plants often conspicuously glandular .... **1. Adenolobus**
– Calyx spathaceous, splitting at the base down one side only, with the lobes fused or sometimes partly separated; fertile stamens 1–10, sometimes accompanied by staminodes; plants not glandular .... **2. Bauhinia**
5. Leaves with one pair of leaflets 6

* Key to fruiting material on p. 10.

– Leaves, or at least most of them, with 4–many pairs of leaflets 9
6. Bracteoles small or lacking, the fertile parts covered by the 4 imbricate sepals; leaflets with pellucid gland dots or resinous vesicles in the areoles visible at least in young leaves 7
– Bracteoles enveloping the bud, valvate, the sepals reduced; leaflets not pellucid punctate (beware black dots caused by fungal infections) but with a few surface glands near the base 46
7. Petals 5, all large or 3 large and 2 minute; stamens 10; leaflets with a single prominent main nerve and a twisted petiolule evident below the proximal base of the leaflet (can be sessile in cultivated species) . . . . **8. Hymenaea**
– Petals 0; stamens as next couplet; leaflets sessile at least on the proximal side, the distal margin sometimes slightly higher on a petiolule-like structure, often (not always) with more than one prominent nerve from the base, the margins at the base in only slightly different planes 8
8. Stamens 20–25; flowers in a slender raceme or panicle; leaflets with 7–12 similarly prominent nerves from the base, the bases of the two sessile leaflets contiguous and with a small appendage at the petiole-apex underneath . . . . **6. Colophospermum**
– Stamens (8)10(12); flowers in axillary or terminal panicles; leaflets with a somewhat excentric main nerve and sometimes 1–3 less prominent secondary lateral nerves from the base on the proximal side, the base of the leaflets asymmetric with the proximal margin of the leaflet arising from the apex of a petiolule-like structure, and without an appendage to the petiole . . . . **9. Guibourtia**
9. Leaves reduced, the pinna-rachilla green, ± terete or winged with numerous small to minute leaflets; plants covered with stout spines . . . . . **36. Parkinsonia**
– Leaves not so reduced, the pinna rachides bearing relatively well-developed leaflets; plants unarmed or with recurved prickles, rarely with spines 10
10. Leaves variously pinnate and bipinnate on the same shoot; leaflets crenulate; sepals and petals not clearly differentiated, small, the tepals greenish; cultivated tree or shrub, often with thorns . . . . **Gleditsia**
– Leaves all either pinnate or bipinnate; leaflets entire; sepals and petals markedly different, but one or other whorl sometimes lacking 11
11. Leaves bipinnate 12
– Leaves pinnate 28
12. Stamens 5, red, exserted; petals greenish, all similar, contrasting with red exposed inside of the hypanthium; style scarcely differentiated, shortly narrowed to a minute stigmatic pad, with the sutural groove extending to the tip; cultivated tree with flat winged pods . . . . **Acrocarpus**
– Stamens 10; petals and style more differentiated 13
13. Plants armed with prickles or spines 14
– Plants unarmed 20
14. Ovary and pod covered with stiff spines in the FZ area; seeds subglobose, hard, with horizontal fracture lines, 1–2 per pod; flowers unisexual, male flowers with reduced ovary and female flowers with sterile anthers . . . . **33. Guilandina**
– Ovary and pod not coarsely spiny; seeds compressed or if subglobose (*Mezoneuron*) then without fracture lines, generally several per pod; flowers bisexual 15
15. Trees or shrubs, not climbing, the prickles sparse and/or straight 16
– Climbers, generally with strong recurved prickles especially on the leaf rachis 17
16. Lower sepal with a plane margin; pods 2-valved, dehiscent; mostly shrubs in the FZ area . . . . **30. Caesalpinia**

– Lower sepal pectinate; pods oblong, thick, indehiscent; small cultivated tree .... **Tara**
17. Fruits samaroid, with a basal 1-seeded chamber and distal wing (ovary with marked asymmetry from an early stage); petals c. 3 mm long .... **29. Pterolobium**
– Fruits pod-like, sausage-shaped or winged along one edge; petals larger, the upper usually somewhat modified 18
18. Flowers subspicate; pods subtorulose with both sutures thickened ... **Moullava**
– Flowers distinctly pedicellate, the pedicels much longer than the calyx; fruit not as above 19
19. Fruits pod-like, not winged, ± dehiscent; calyx hairy in the FZ area .... **30. Caesalpinia**
– Fruits flattened, winged or expanded along the upper margin (evident from an early stage), indehiscent; calyx glabrous in the FZ area ...... **28. Mezoneuron**
20. Leaflets alternate; flowers small (2–5 mm long, white to cream or pale green); sepals open from an early stage 21
– Leaflets opposite; flowers usually larger; sepals valvate or imbricate 22
21. Flowers pedicellate; stamen filaments hairy; fruits dehiscent, (1)2–11-seeded; seeds without areoles; hairs on vegetative buds and young shoots grey-brown to yellowish .... **38. Erythrophleum**
– Flowers sessile; stamen filaments glabrous; fruits indehiscent, 1-seeded; seeds with an areole; hairs on vegetative buds and young shoots red-brown .... **39. Burkea**
22. Lowermost sepal like the uppermost sepal, both outside in bud, or sepals all valvate; stamens ± spreading; leaf axis adaxially grooved, sometimes with glands or bridges at leaflet insertions, scattered glands or glandular hairs usually not very evident (but do occur in *Peltophorum*) 23
– Lowermost sepal modified, ± boat-shaped, quite often forming a hood in bud; stamens more crowded around the gynoecium at least at the base; leaf axis adaxially ridged at least distally to each insertion of leaflets, mostly with scattered pustulate glands and glandular hairs 26
23. Sepals valvate; stamens several cm long in the FZ area, shorter or longer than petals; pods strap-shaped with seeds transverse in cavities .... **37. Delonix**
– Sepals imbricate; pods not as above; stamens less than 1.5 cm long, not or not much exceeding the petals 24
24. Stigma small; upper stamen embraced by upper petal and with a shorter anther; ovary stipe adnate to lower part of the hypanthium; fruit oblong-cuneate, opening at apex, with one seed in a papery endocarpic envelope; cultivated tree with very large bipinnate leaves .... **Schizolobium**
– Stigma peltate; anthers all similar; ovary stipe cenral in the hypanthium; fruit not as above 25
25. Fruits with two woody valves that recurve elastically on opening, the outside of each with a conspicuous longitudinal groove down the centre; pedicel not jointed in upper part; leaves usually mostly opposite; leaflets in the FZ area 3–15 pairs .... **34. Bussea**
– Fruits flattened, indehiscent, winged along both margins (ovary soon modified likewise); pedicel articulated to pedicel-like basal part of the hypanthium; leaves alternate; leaflets mostly 10–32 pairs per leaf .... **35. Peltophorum**
26. Trees; flowers in short dense racemes or panicles; calyx not glandular; fruits indehiscent, woody, curved or twisted in the FZ area .... **Libidibia**
– Herbs, shrubs or small trees; flowers in extended racemes; calyx conspicuously glandular at least towards tip of lower lobe; fruit dehiscent, often glandular 27

27. Stamens and petals spread around the ovary; glands in submarginal rows on leaflets or not evident; petals yellow (stamens may be red) **32. Erythrostemon***
– Stamens held in the gutter formed by the lower sepal, the lateral petals divergent in pairs on either side; glands scattered all over the leaflets; petals reddish .... **31. Pomaria**
28. Bracteoles not valvate, often caducous or absent, usually not enclosing the flower bud but if so then one or both margins of one bracteole overlapping the other, at least at the base 29
– Bracteoles paired, valvate throughout, well developed, completely enclosing the flower bud, usually persistent 42
29. Hypanthium expanded into a broad disk, with the stamens spread around the outside; petals absent 30
– Hypanthium obconic to tubular or negligible, with the stamens directed forwards; petals variably present or absent 31
30. Leaves imparipinnate; anthers basifixed; sepals well developed ... **24. Dialium**
– Leaves paripinnate; anthers dorsifixed; sepals negligible; cultivated tree .... **Ceratonia**
31. Anthers opening by terminal or basal pores or short slits, usually basifixed, but dorsifixed in *Cassia*; petals 5, all well developed; glands sometimes present on petiole or leaf rachis 32
– Anthers opening by longitudinal slits, dorsifixed; petals 1–5, often one larger or markedly modified; specialised glands lacking from leaf axes 34
32. Filaments of 3 abaxial filaments with a double hairpin or S-bend above the base, many times longer than the anthers; anthers opening by basal pores; bracteoles at or near the base of the pedicels; pods elongate, cylindrical to ± flattened, indehiscent .... **27. Cassia**
– Filaments straight, not more than twice as long as the anthers; anthers dehiscent by apical pores; bracteoles and pods as next couplet 33
33. Bracteoles 0; stamens all similar or progressively shorter abaxially; pods indehiscent or opening with valves that do not twist elastically, usually flattened .... **26. Senna**
– Bracteoles 2; stamens often in two whorls of different lengths but not progressively shorter abaxially; pods dehiscent with elastically coiled valves .... **25. Chamaecrista**
34. Petal 1, pubescent and green outside, dark red inside, with a long claw widened into a deeply bilobed lamina; fertile stamens 9, free, with 2 staminodes; leaflets with a small dot-like gland at the proximal side of the leaflet base either on the lower surface in the angle between the margin and midrib or on the margin itself .... **16. Afzelia**
– Petals 1–5, variously shaped but not as above; stamens 8–10; leaflets without marginal glands, but sometimes with 1–3 vesicles near base (*Micklethwaitia*) . 35
35. Flowers arranged spirally on the inflorescence axis 36
– Flowers distichous on the inflorescence axis 38
36. Flowers red with a conspicuous cup-shaped hypanthium and stamens extended like a brush .... **5. Schotia**
– Flowers white or yellow, with a green or inconspicuous hypanthium and stamens spreading, not brush-like 37
37. Leaflets alternate, ± symmetrical at base, with c. 20 pairs of closely spaced lateral nerves; flowers yellow; fruit samaroid, with a basal seed-chamber and distal wing; cultivated tree .... **Pterogyne**

* If petals red see *Caesalpinia pulcherrima*.

– Leaflets alternate to opposite, markedly asymmetrical at base, with 6–10 pairs of lateral nerves; flowers white; fruit a dehiscent pod 38
38. Leaflets with well-developed untwisted petiolules; petals 5, 3 large (1.5–2 cm long) and 2 minute; stamens about twice as long as the large petals . . . . **7. Daniellia**
– Leaflets sessile, subsessile or with very short twisted petiolules; petals 4–5, subequal, small (up to 1 cm long); stamens not much exceeding the petals . 39
39. Leaflets opposite, with short twisted petiolules, and with a few glands beneath near the base; hypanthium shallowly cupular . . . . **14. Micklethwaitia**
– Leaflets in middle part of the leaf alternate, sessile, without glands; hypanthium funnel-shaped, 3–4 mm long . . . . **13. Scorodophloeus**
40. Petals 0; leaflets with numerous pellucid gland dots, alternate . . **12. Copaifera**
– Petals 5; leaflets eglandular, opposite to alternate 41
41. Leaflets opposite (alternate elsewhere) and symmetrical at base, with lateral nerves extending to join the marginal nerve; petals unequal, the upper one narrower and differently coloured . . . . **10. Baikiaea**
– Leaflets alternate, asymmetrical at base, with lateral nerves looped to form a rather prominent submarginal nerve; petals subequal . . . . **11. Tessmannia**
42. Bracteoles petaloid, white or pink; fertile stamens 3 or rarely up to 6(8), sometimes with staminodes, all free from each other, with the 1(3) petals on the edge of a small but obvious, almost naked hypanthium; sepals minute or absent . . . . **17. Cryptosepalum**
– Bracteoles not petaloid, often thick and coriaceous; stamens 6–18(20), usually all fertile but, if not then at least some united at the base and/or sepals well developed 43
43. Bracteoles enclosing the young flower buds but soon caducous, exposing the bud enclosed by the calyx; filaments of stamens and staminodes all connate about halfway in a band; fertile stamens 3, alternating with 5 sterile teeth or short filaments; larger petals 3, gold with red lines . . . . **15. Tamarindus**
– Bracteoles enclosing the flower bud and persistent below the open flower; stamens and petals not as above 44
44. Leaflets all distinctly petiolulate, the petiolules not obviously twisted; base of leaflet often symmetric or nearly so 45
– Leaflets (at least the distal pair) sessile or rarely very shortly petiolulate and the petiolules obviously twisted as viewed from above; base of leaflet usually asymmetric with the proximal side decurrent almost or quite to the upper side of the leaf rachis and often twisted over the pulvinulus 46
45. Petals very unequal, one very large with a long narrow basal part and broad limb above; stamens with 9 shortly connate and 1 free, clustered around the gynoecium . . . . **18. Berlinia**
– Petals subequal in length, but 1–3 broader than the others; stamens 10, subequal, free, spreading . . . . **19. Isoberlinia**
46. Perianth 0 or if 1–7(11) parts, usually 4–7 all sepaloid and of similar form, grading inwards from broader to narrower, or 1–3(4) very small; if 5–11 and two distinct forms present then the inner narrower than the outer, even when (rarely) longer; stipular expansions or wings often present on the upper side of the leaf rachis . . . . **21. Brachystegia**
– Perianth clearly differentiated with 1–7 obvious petals, usually longer and broader than the 4–7 sepals, the latter very rarely reduced or absent; leaf rachis without expansions or wings 47
47. Sepals 5, well developed, subequal at least in length; petals variable, usually 5, well developed, subequal or one much larger; leaflets normally 4 or more . . . . **20. Julbernardia**

– Sepals 0 or 1–5 small and unequal; petals 1, large, alone, sometimes with 1–4 others rudimentary, or petals absent to one rudimentary; leaflets (1)2–numerous 48
48. Main nerve of leaflet marginal or almost so, the distal half of the leaflet thus completely reduced; one petal well developed; stamens 9–10, the 9 united at the base (the adaxial stamen free or absent) .... **23. Aphanocalyx**
– Main nerve of leaflet only slightly excentric; petals absent or with rudiment only; stamens 6–8, free .... **22. Icuria**

Key to genera based mainly on vegetative and pod characters

1. Leaves simple, bilobed, palmately nerved 2
– Leaves with 2–many leaflets (if a single pair sometimes bridged across their basal junction and with a small appendage at the fork beneath and exceptionally in *Guibourtia* fused into a single leaflet but then not bilobed) 5
2. Stems trailing or climbing, herbaceous, from a large tuber, generally with tendrils; pod woody, dehiscent or indehiscent, (1)2-seeded ...... **3. Tylosema**
– Stems forming shrubs or small trees, seldom climbing but then woody and without tendrils; pods more than 2-seeded 3
3. Pods oblong-reniform, with thin valves, dehiscent, often glandular; seeds flattened, wrinkled around the margins, with a small circular hilum wthout a U-shaped extension .... **1. Adenolobus**
– Pods longer, thicker walled, eglandular; seeds more turgid, smooth, with a hairpin- or U-shaped scar on the face developed from the funicle 4
4. Pods thinly woody, dehiscent; calyx remnant, if present, spathaceous .... **2. Bauhinia**
– Pods thickly woody, indehiscent; calyx shortly 5-lobed, falling before fruit matures .... **4. Piliostigma**
5. Leaves all with 1 pair of leaflets 6
– Leaves, or at least most of them, with more than 1 pair of leaflets 9
6. Pods without a raised lateral nerve; pellucid gland dots or resinous vesicles in the areoles generally visible at least in younger leaves 7
– Pods compressed, dehiscent, each valve with a strong lateral nerve running from base to upper part of pod on the side some way from the upper margin; leaflets not pellucid punctate (beware black dots caused by fungal infection), but with a few surface glands near base 45
7. Fruits turgid, thick-walled, in the native species irregularly shaped and resinous warted, 1–3-seeded; leaflets with a single prominent main nerve and a twisted petiolule evident below the proximal base of the leaflet (can be sessile in cultivated species) .... **8. Hymenaea**
– Fruits compressed, elliptic-oblong to obovate or ± reniform in outline, not particularly thick-walled, 1-seeded; leaflets sessile at least on the proximal side, the distal margin sometimes slightly higher on a petiolule-like structure, often (not always) with more than one prominent nerve from the base, the margins at the base in only slightly different planes 8
8. Pods pale yellowish brown, reniform to obliquely semi-circular, indehiscent; seed reniform, ± flattened, corrugated and with numerous small, sticky, reddish resin-glands, exarillate; leaflets with 7–12 similarly prominent nerves from the base, the bases of the two sessile leaflets contiguous and with a small appendage at the petiole-apex underneath .... **6. Colophospermum**
– Pods brown or reddish brown, obliquely semi-circular, ovate or ovate-oblong, dehiscent or not; seed not as above, with or without an aril; leaflets with a somewhat

excentric main nerve and sometimes 1–3 less prominent secondary lateral nerves from the base on the proximal side, the base of the leaflets asymmetric with the distal margin of the leaflet arising from the apex of a petiolule-like structure, and without an appendage to the petiole .... **9. Guibourtia**

9. Leaves reduced, the pinna-rachilla green, ± terete or winged with numerous small to minute leaflets; plants covered with stout spines ..... **36. Parkinsonia**
– Leaves not so reduced, the pinna rachides bearing relatively well-developed leaflets; plants unarmed or with recurved prickles, sometimes with spines ... 10

10. Leaves variously pinnate and bipinnate on the same shoot; leaflets crenulate; pods strap-shaped, often twisted, indehiscent; cultivated tree or shrub, often with straight or branched spines .... **Gleditsia**
– Leaves all either pinnate or bipinnate; leaflets entire; pods various 11

11. Leaves bipinnate 12
– Leaves simply pinnate 27

12. Plants armed with prickles or spines 13
– Plants unarmed 18

13. Fruits samaroid, with a basal 1-seeded chamber and a dorsal wing .... **29. Pterolobium**
– Fruits not samaroid 14

14. Seeds subglobose, hard, with horizontal fracture lines, 1–2 per pod; pod covered with stiff spines .... **33. Guilandina**
– Seeds compressed or if subglobose (*Mezoneuron*) then without fracture lines, generally several per pod; pods not conspicuously spiny 15

15. Pods dehiscent (rather tardily in *Caesalpinia decapetala*, but then shaped like a typical pea pod) .... **30. Caesalpinia**
– Pods indehiscent 16

16. Cultivated trees or shrubs; prickles principally at insertion of pinnae and leaflets; fruits oblong, thick, pinkish brown or crimson .... **Tara**
– Lianes or climbing shrubs, with scattered prickles; fruits not as above 17

17. Fruits compressed, winged or expanded along the upper margin, on well-developed pedicels .... **28. Mezoneuron**
– Fruits subtorulose with both sutures thickened, subspicate (pedicels less than 2 mm long) .... **Moullava**

18. Leaflets alternate 19
– Leaflets opposite 20

19. Pods dehiscent, woody or thinly woody, oblong or elliptic-oblong, mostly 2–11-seeded; seed without areoles; hairs when present on vegetative buds grey-brown to yellowish .... **38. Erythrophleum**
– Pods indehiscent, coriaceous, elliptic, 1-seeded; seed with an areole on each face; hairs on vegetative buds and young branchlets conspicuously rusty red ... . **39. Burkea**

20. Pods flat, winged along both margins, 1–2-seeded .... **35. Peltophorum**
– Pods 2-valved or thickened, not winged along both edges 21

21. Pods basally long-cuneate, with 1–few seeds arranged longitudinally at the apex, the valves flat or recurving 22
– Pods oblong to elliptic, several-seeded along their length, the valves flat, twisting or not opening 23

22. Seeds oblong-ellipsoid, not enclosed in a papery envelope on the recurved valves .... **34. Bussea**
– Seeds single at apex of pod, in a papery envelope resembling a wing, the valves spreading but not recurved; tall cultivated tree .... **Schizolobium**

23. Pods glandular or with plumose hairs; leaves generally ending in a single pinna, the leaflets often glandular 24
– Pods eglandular, without plumose hairs; leaves generally ending in a pair of pinnae, the leaflets without surface glands (sometimes gland-dotted) 25
24. Shrubs or small trees; glands stipitate . . . . . **32. Erythrostemon**
– Herbs or low shrubs up 0.5 m tall; glands sessile . . . . . **31. Pomaria**
25. Pods coiled and twisted, indehiscent, black . . . . . **Libidibia**
– Pods 2-valved, dehiscent 26
26. Pods not winged; leaflets numerous, small, oblong . . . . . **37. Delonix**
– Pods narrowly winged along upper margin; leaflets ovate to ovate-oblong, 2–7 cm wide . . . . . **Acrocarpus**
27. Leaves ending in a leaflet, the lateral leaflets, or at least the upper ones usually alternate, or if ending in a pair of leaflets the lateral leaflets definitely alternate; pods short, usually 1–2-seeded 28
– Leaves usually ending in a pair of leaflets, the lateral leaflets opposite or subopposite; pods various, generally elongate, usually several-seeded (except *Daniellia* with a single distal seed attached to each separated valve) 31
28. Pedicels (or their scars) spirally arranged on infructescence axis; leaflets not pellucid punctate 29
– Pedicels in 2 ranks (distichous); leaflets with gland dots (not always obvious in mature leaves) 30
29. Pods ellipsoid to spherical, indehiscent, not winged, with a brittle exocarp and a pulpy mesocarp*; leaflets petiolulate, almost symmetrical at base, usually with a rather dense vein reticulum . . . . . **24. Dialium**
– Pods oblong to oblong-obovate in outline, dehiscent, with a wing along the upper suture, woody; leaflets sessile, very asymmetrical at base, with a lax and irregular vein network . . . . . **13. Scorodophloeus**
30. Tree . . . . . **11. Tessmannia**
– Shrublet with a woody rootstock . . . . . **12. Copaifera**
31. Pods cylindrical to oblong-ellipsoid, turgid to rounded in section, if somewhat compressed then pulpy inside, usually indehiscent 32
– Pods flattened, generally dehiscent even if tardily so, not pulpy inside (sometimes with white pith) 35
32. Pods turgid, somewhat torulose (sausage-shaped), densely covered in a brown scurf, with a dry outer shell and pulpy inner layer; seeds with an areole; leaflets numerous, almost sessile, asymmetric at base, with 1–2 secondary nerves from the base on the proximal side of the main nerve . . . . . **15. Tamarindus**
– Pods and leaflets not as above; areole present or not; leaflets not as above . . 33
33. Seeds generally with an areole; glands often evident on petiole or leaf rachis; herbs, shrubs or trees . . . . . **26. Senna**
– Seeds without an areole; leaves eglandular; generally trees 34
34. Pods cylindrical . . . . . **27. Cassia**
– Pods somewhat compressed laterally . . . . . **Ceratonia**
35. Annual or perennial herbs; leaves with one or more glands on the petiole or with glandular hairs; seeds somewhat flattened, shiny, usually with lines or small pits in the testa; pods with rather thin elastically coiled valves . . . **25. Chamaecrista**
– Herbs, shrubs or trees; if leaves with glands (*Senna*) then valves of pods not elastically coiled 36
36. Seeds generally with an areole; glands often evident on petiole or leaf rachis; herbs, shrubs or trees . . . . . **26. Senna**

---

* If a cultivated tree with samaroid fruits, see *Pterogyne*.

– Seeds without an areole; glands absent from leaf axes; woody shrubs or trees, occasionally with annual shoots from a woody rootstock 37
37. Pods dehiscent with two valves, the coriaceous endocarp of which curls up and separates from the stiffly papery to thinly woody exocarp, to the distal end of which the solitary seed remains attached during dispersal . . . . **7. Daniellia**
– Pods with valves whose exocarp and endocarp remained attached to each other 38
38. Aril present as base of seed; valves of pod remaining flat after separation . . . 39
– Aril lacking; valves spirally twisted after separation 40
39. Seeds pale brown with a conspicuous yellow aril; pod with a hard margin or wing along the upper suture, often with the seeds attached after the eventual dehiscence of the valves . . . . **5. Schotia**
– Seeds black with a large orange, red or vermillion cupular aril; pods with flat, thick, woody valves, with seeds embedded in a white pith . . . . **16. Afzelia**
40. Upper suture of pod without any or with only a very slight wing-like ridge and without a strong lateral nerve 41
– Upper suture of pod flattened or concave, with an obvious projecting ridge or wing along each side, or if rarely without then with a strong lateral nerve running from base to upper part of pod on the side some way from the upper margin; ridges or wings spreading, suberect or revolute 42
41. Pedicels (or their scars) in 2 ranks on the infructescence axis; leaflets often with domatia-like swellings almost at the base of the proximal side and a few mm above the base on the distal side . . . . **10. Baikiaea**
– Pedicels arranged spirally; leaflets without such swellings . . . . . . **19. Isoberlinia**
42. Petiolules distinct, 3–7(9) mm long, not twisted . . . . **18. Berlinia**
– Petiolules 0–3 mm long, usually twisted 43
43. Pods ± densely brown tomentose when ripe; leaflets usually fringed with whitish pubescence on margins . . . . **20. Julbernardia**
– Pods glabrous or nearly so when ripe; leaflets without such a characteristic fringe of whitish pubescence 44
44. Pods with a strong lateral nerve running from base to upper part of pod on the side some way from the upper margin 45
– Pods without such a lateral nerve 46
45. Main nerve of leaflets marginal or almost so, the distal part of the leaflet thus completely reduced . . . . **23. Aphanocalyx**
– Main nerve of leaflets only slightly excentric . . . . **22. Icuria**
46. Pods broadest near the apex, almost obtriangular in outline, the sutural wings only slightly developed . . . . **14. Micklethwaitia**
– Pods broadest some way from the apex, more elliptic to obovate in outline, the sutural wings usually well developed . . **21. Brachystegia** and **17. Cryptosepalum**

1. ADENOLOBUS (Benth. & Hook. f.) Torre & Hillc.

Adenolobus (Benth. & Hook. f.) Torre & Hillc. in Bol. Soc. Brot., Sér. 2, **29**: 37 (1955). —Brummitt & Ross in Kew Bull. **31**: 399–406 (1976).
Bauhinia sect. *Adenolobus* Benth. & Hook. f., Gen. Pl. **1**: 576 (1865).

Erect to prostrate shrubs or small trees, sometimes with stalked glands. Tendrils absent. Leaves alternate or crowded on spur shoots, composed of 2 united leaflets, emarginate to shallowly bilobed, palmately nerved, petiolate, stipulate. Flowers in elongate racemes or 1–several on spur shoots, with bracts and paired bracteoles. Hypanthium obconic to cupular. Calyx with 5 relatively short lobes. Petals 5, yellow, ± marked red, subsimilar, long clawed, the adaxial one slightly modified. Stamens 10, all fertile, in 2 whorls of 5, of two lengths; anthers dorsifixed, dehiscing by longitudinal slits. Ovary long stipitate; style elongate; stigma small. Pods stipitate, oblong-reniform, dehiscent, the valves thinly textured and not twisting, 1–10-seeded. Seeds compressed, ovate in outline, slightly produced to the small subapical hilum.

Two species in southern Africa, fomerly included in *Bauhinia*, but easily recognised by the subequally lobed calyx.

The painting of *A. pechuelii* in Germishuizen (in Fl. Pl. Afr. **50**, 1988) and colour slides of *A. garipensis* (E. Mey.) Torre & Hillc. show that the two species have flowers that look very different. In *A. pechuelii* the petals are yellow and spread apart as in most African Bauhinias, the lower petals a little shorter than the lateral petals, the adaxial petal more obovate and flecked with red, turning orange and fading red. The style tip protrudes from mature buds, suggesting that the flowers are protogynous. In *A. garipensis*, which overlaps the distribution of *A. pechuelii*, ranging from S Angola to the Northern Cape Province of South Africa, the petals are strongly veined red and held in a tube around the fertile parts. Drawings of the two species in Flora of Southern Africa are misleading.

The seeds are very different from other genera of African Cercideae, relatively small, flattened and wrinkled around the edges, those of *A. pechuelii* with a white blister above the hilum. They would repay investigation in terms of their development, structure and dormancy characteristics.

Adenolobus pechuelii (Kuntze) Torre & Hillc. in Bol. Soc. Brot., Sér. 2, **29**: 38 (1955). —Schreiber in Merxmüller, Prodr. Fl. SW Afrika, fam. 59: 4 (1967). —Schmitz in Bull. Jard. Bot. Belg. **43**: 399 (1973). —Brummitt & Ross in Kew Bull. **31**: 401 (1976). —Coetzer & Ross in F.S.A. **16**(2): 56, fig.12/4–6 (1977). Neotype, selected by Brummitt & Ross (1976): Namibia, Swakompund Dist., S of Kuiseb, *Strey* 2592 (PRE neotype, K). FIGURE 3.2.**1**.

Bauhinia pechuelii Kuntze in Jahrb. Königl. Bot. Gart. Berlin **4**: 263 (1886). —Schinz in Mém. Herb. Boiss. **1**: 120 (1900). —Baker, Legum. Trop. Afr.: 658 (1930).

Bauhinia marlothii Engl. in Bot. Jahrb. Syst. **10**: 26 (1888). —Schinz in Mém. Herb. Boiss. **1**: 120 (1900). Type: Namibia, Karibib Dist., Usakos, *Marloth* 1184 (PRE isotype).

Small shrub up to 1.5 m, with short erect branches or prostrate and spreading. Branches glabrous, with scattered, shortly stalked, ± swollen-headed glands. Leaves: petiole 4–9 mm long; lamina glaucous, 1–3 × 1.2–4 cm, slightly emarginate to shallowly lobed at the apex, cordate at the base, the lobes semicircular, glabrous; stipules 2–4 mm long, subulate to lanceolate-acuminate. Racemes (5)10–20 cm long, 20–40-flowered, with scattered glands on the axes; pedicels 4–6(8 in fruit) mm long; bracts 2–4 × 1–2 mm, lanceolate to elliptic-ovate, acuminate; bracteoles on the pedicel, a little smaller. Hypanthium 3–4 mm long, obconic to cupular. Calyx 4–8 mm long, glandular; lobes 2–3 mm long, obtuse to rounded and mucronate. Petals spreading, yellow, with red spots on the adaxial petal, which turns orange and then red as it fades, 1.2–2 × 0.4–0.8 cm, elliptic above the well-developed claw, the 2 lower petals a little shorter than the lateral petals, the adaxial petal more obovate. Stamens with filaments 1–2.2 cm long. Ovary 4–8 mm long above a long stipe, glandular, up to c. 10-ovulate; style 8–12 mm long, slender. Pod pale yellow-brown to pinkish red, stipitate, 1.8–3 × 0.8–1.5 cm, oblong-reniform, slightly convex on the upper side, strongly rounded on the lower side, with 1–few seeds maturing. Seeds light brown, often mottled, 5–7 × 3–5 mm, slightly and irregularly tuberculate around the edges, with a white blister above the hilum.

Subsp. **pechuelii** Brummitt & Ross in Kew Bull. **31**: 402 (1976).

Botswana. N: Kuke (Kuki), fl. & imm.fr. ?1968, *Bate* 7 (K, SRGH).

Also in Namibia. In Botswana known only from the specimen cited without

Fig. 3.2.**1**. ADENOLOBUS PECHUELII subsp. PECHUELII. 1, flowering branch (× $^{2}/_{3}$), from *Kubirske* 55; 2, flower bud (× 4); 3, glands from calyx (× 20); 4, flower, longitudinal section (× 2); 5, upper petal (× 2); 6, anthers (× 6), 2–6 from *Oliver, Muller & Steenkamp* 6631; 7, pod (× 1) from *Kubirske* 60; 8, 9 seed, hilar and side views (× 4), from *M.K. Seely* s.n. Drawn by Juliet Williamson.

indication of habitat, but in Namibia mostly recorded from dry river beds.

Subsp. *mossamedensis* (Torre & Hillc.) Brummitt & Ross, from S Angola and N Namibia, differs from subsp. *pechuelii* by having no or very few glands.

2. BAUHINIA L.

Bauhinia L., Sp. Pl.: 374 (1753); Gen. Pl., ed. 5: 177 (1754). —De Wit in Reinwardtia **3**: 386, 390 (1956).

Trees or shrubs, seldom scandent or climbing, without tendrils, but branch tips sometimes coiling. Leaves bilobed or occasionally divided to the base, palmately nerved; stipules deciduous. Flowers in racemes or solitary, usually large and showy, bisexual, zygomorphic. Calyx spathaceous (the sepals ± cohering after the calyx has opened) above a variably developed hypanthium. Petals 5(6), free. Stamens 1–10, sometimes accompanied by staminodes, free; anthers opening by longitudinal slits. Ovary usually stipitate; style elongate; stigma capitate or small, sometimes unilateral. Pods linear-oblong to strap-shaped or broadened upwards, ± woody, dehiscent or elsewhere occasionally indehiscent, few- to many-seeded. Seeds ± compressed; hilum circular or crescent-shaped, with a U-shaped to hairpin-shaped scar of 2 aril lobes developed from the funicle.

About 160 species, pantropical, most numerous in the neotropics, generally in woodland and wooded grassland. Sometimes treated in a broad sense to include *Piliostigma* and *Tylosema*, but recent considerations of the systematics and phytogeography suggest that the circumscription generally used in African floristics, following the concepts of De Wit, Reinwardtia **3**: 386 (1954), should be retained (see Lewis in Lewis et al., Legumes of the World, 2005).

A number of species are grown as ornamentals, including *B. galpinii* and *B. tomentosa* among species native to the Flora area. Some medicinal uses are briefly outlined by B.-E. van Wyk & Gericke, People's Plants: 182 (2000).

Bauhinia acuminata L., Sp. Pl.: 375 (1753). —Steedman, Trees, Shrubs & Lianes S. Rhod.: 18 (1933). —Brenan in F.T.E.A., Legum.-Caesalp.: 207 (1967). Type from India.

Bauhinia linnaei Ali in Portugalia Acta Biol., Sér. B **8**: 244 (1966). Type from India.
Pauletia acuminata (L.) A. Schmitz in Bull. Jard. Bot. Belg. **43**: 387 (1973).

Small tree. Leaves rather large, (30)50–150 cm long, the lobes usually acutely pointed. Flower bud ± abruptly narrowed into a long beak up to 2 cm with the filiform sepal tips free at the end. Petals white, usually 3–4 cm long. Stamens 10.

Native of tropical Asia and widely cultivated. No specimens from the Flora area seen but recorded from Zimbabwe by Steedman (Trees Shrubs & Lianes S. Rhod.,1933).

Bauhinia bowkeri Harv. in F.C. **2**: 596 (1862). —Coetzer & Ross in F.S.A. **16**(2): 48 (1977). —Pooley, Trees Natal: 152 (1993). —M. Coates Palgrave, Trees Sthn. Africa: 334 (2002). Type: South Africa, Eastern Cape Province, along Bashee R. near Fort Bowker, *Bowker* 378 (TCD holotype).

Pauletia bowkeri (Harv.) A. Schmitz in Bull. Jard. Bot. Belg. **43**: 394 (1973).

Shrub or small tree. Leaves small, 0.5–4 cm long, lobed to more than half their length, glabrous. Petals white, c. 2.5–4.8 cm long. Stamens 10.

Zimbabwe. W: Bulawayo, Hillside, fl. & fr. xi.1958, *Guy* 59/58 (K, SRGH).
Native of the Eastern Cape Province of South Africa.

Bauhinia monandra Kurz in J. Asiatic Soc. Bengal, Nat. Hist. **42**(2): 73 (1873). —White, F.F.N.R.: 100 (1962). —Brenan in F.T.E.A., Legum.-Caesalp.: 207 (1967). —Schmitz in Bull. Jard. Bot. Belg. **43**: 373 (1973). Type a cultivated plant from Asia.

Tree up to 10 m. Leaves up to 13 cm long, lobed to less than half way. Hypanthium 2–2.5 cm long. Petals white to pink with purple speckles inside, the upper one deeper coloured, 3.5–5 cm long. Stamens 1. Pod 16–21 × 2–2.5 cm, linear-oblong, inflated, 10–18-seeded.

Zimbabwe. C: Harare, Enterprise Road, fl. 3.xi.1975, *Biegel* 5173 (K, SRGH). **Mozambique.** M: Maputo city, Zimbabwe Road, fl. 29.xi.1986, *Groenendijk & Almeida Sitoe* 2078 (K, LMU).

Widely cultivated in the tropics, formerly of unknown origin, but Du Puy, Leg. Madagascar (2002) is convinced it is a native of Madagascar. Also recorded as planted in Zambia by White in F.F.N.R. (1962).

Bauhinia purpurea L., Sp. Pl.: 375 (1753). —Steedman, Trees, Shrubs & Lianes S. Rhod.: 18 (1933). —Brenan in F.T.E.A., Legum.-Caesalp.: 208 (1967). —Coetzer & Ross in F.S.A. **16**(2): 54 (1977). Type from SE Asia.

Perlebia purpurea (L.) A. Schmitz in Bull. Jard. Bot. Belg. **43**: 382 (1973).

Tree. Leaves up to 13 cm long, the lobes usually rather narrow. Calyx folded into ridges at distal end in bud. Petals purple or pinkish-purple (or perhaps rarely white), c. 4.5–5.5 cm long. Fertile stamens 3 or 4.

Zambia. C: Chilanga, Mt. Makulu Research Station, fl. i.1979, *Critchett* 9/79 (K). **Zimbabwe.** C: Harare, Forest Nurseries, fl. 3.iv.1950, *Fisher* in GHS 27816 (K, SRGH). **Malawi.** C: Lilongwe, Capital Hotel, fl. 13.ii.1984, *Poynton* in Forestry Herb. 18096 (K). S: Blantyre Dist., Maone Estate, 2 km N of Limbe, fl. & fr. 12.iii.1970, *Brummitt* 9046 (K, MAL). **Mozambique.** M: Maputo, Jardim Tunduru (Vasco da Gama), fl. 23.v.1971, *Balsinhas* 1893 (K).

Native of SE Asia.

De Wit (Reinwardtia **3**: 409, 1956), recognises three varieties distinguished largely by different flower colours. It seems that material seen from the Flora area is referable to var. *purpurea*.

The petals in the Flora material are narrower and longer clawed than those of *B. variegata*, but elsewhere there is more variation, perhaps due to hybridisation.

Bauhinia variegata L., Sp. Pl.: 375 (1753). —Torre in Mendonça, Contr. Conhec. Fl. Moçamb. **2**: 79 (1954). —White, F.F.N.R.: 100 (1962). —Brenan in F.T.E.A., Legum.-Caesalp.: 208 (1967). —Richards & Morony, Check List Fl. Mbala Dist.: 67 (1969). —Coetzer & Ross in F.S.A. **16**(2): 54 (1977). —M. Coates Palgrave, Trees Sthn. Africa: 334 (2002). Type from SE Asia.

Perlebia variegata (L.) A. Schmitz in Bull. Jard. Bot. Belg. **43**: 386 (1973).

Tree. Leaves up to 10 cm long in the Flora area, lobed to about one third. Buds sometimes somewhat striate but not folded into ridges. Petals purple to pale red, sometimes striped or flecked, or white or with one petal yellow at the base, 3–6 cm long. Fertile stamens 5.

Zambia. N: Mbala (Abercorn), old Locust Office, fl. 21.v.1952, *Siame* 210 (K). W: Ndola, gardens and verges, fl. iv.1961, *Wilberforce* A/56 (K). C: Lusaka Forest Nursery, fl. 5.iii.1952, *White* 2197 (K). **Zimbabwe.** C: Harare (Salisbury), street tree, Third Street, fl. 18.iv.1950, *Pardy* P5/50 (K, SRGH). E: Mutare (Umtali), side street, fl. 14.ix.1961, *Methuen* 202 A & B (K). **Malawi.** S: Blantyre Dist., Maone Estate, 2 km N of Limbe, fl. 19.iii.1970, *Brummitt* 9223 (K). **Mozambique.** MS: Cidade de Chimoio

(Vila Pery), Hospital, fr. 16.ix.1944, *Mendonça* 253 (K, LISC). M: Maputo city, Rua D. João III, fl. & imm.fr. 15.v.1985, *Groenendijk* 1703 (K, LMU).

Widely cultivated, perhaps having originated in China; now probably the most commonly cultivated *Bauhinia* in the Flora area and frequently seen as a street tree.

The form with white flowers has been named var. *candida* Voigt, Hort. Suburb. Calcutt.: 253 (1845) (var. *alboflava* De Wit in Reinwardtia **3**: 412, 1956), but there seem to be several colour variants in cultivation and it is doubtful whether formal taxonomic rank is desirable for any of them. Such white-flowered plants are recorded from several parts of the Flora area: Zambia N: Mbala, old Locust Office, *Siame* 211 (K); Zimbabwe C: Gweru, *Jeffrey* 10/56 (K, SRGH); Malawi S: Maone Estate, *Brummitt* 9224 (K, MAL); Mozambique M: Maputo, Jardim Vasco da Gama, *Balsinhas* 1916 (K). It is also depicted in McColaugh, Gaborone in Bloom: 26 (1997), and in Schmidt, Lötter & McCleland, Trees Shrubs Mpumalanga & Kruger Nat. Park: 182 (2004).

1. Lateral nerves looped within the margin of the leaf 2
– Lateral nerves extending to the margin (all introduced) 7
2. Petals yellow, the upper one often with a purple blotch, broad, overlapping, forming a broad cup; hypanthium 2–6 mm long, obconic; pods pale, ± straw coloured to pale brown, shortly stipitate, parallel sided, finely tomentose to ± glabrescent . . . . **1.** *tomentosa*
– Petals red, pink or white, elliptic to spathulate, spreading to reflexed; hypanthium longer, tubular (often resembling a thickened pedicel); pods various but generally brown, with a stipe over 1 cm long and sometimes broadened upwards 3
3. Petals spathulate with a long claw equalling or exceeding the rounded blade in length, red; fertile stamens 3, with 7 staminodes; pods with the upper margin expanded into a flange up to 1 cm across; leaf lamina generally lobed no more than one third of its length . . . . **5.** *galpinii*
– Petals shortly clawed, elliptic to obovate, white to pink; stamens 5–10; pods not conspicuously flanged; leaf lamina lobed one third to two thirds 4
4. Stamens 10; stigma small, c. 1 mm across; leaves longer than broad, divided c. two thirds of their length, the lobes broadest above the base of the sinus between them (cultivated) . . . . *bowkeri*
– Stamens 5; stigma abruptly expanded from tip of the style, c. 3 mm across; leaves mostly broader than long, divided one third to two thirds of their length, the lobes broadest at the level of the base of the sinus or below 5
5. Hypanthium 0.5–1 cm long; calyx 0.8–1.3 cm long; petals pink; pods 1–2 cm broad; secondary venation of leaves somewhat sclariform, the cross nerves running ± straight and rather parallel between the main nerves . . **4.** *urbaniana*
– Hypanthium (1.5)2.5–5.5(6.5) cm long; calyx 1.7–4.5 cm long; petals white, occasionally with pink spots; pods (1.3)1.7–4.7 cm wide; secondary venation of leaves rather obscure to prominent and more reticulate 6
6. Staminodes lacking; branchlets rather finely brown or grey-brown, spreading or appressed-pubescent; secondary and tertiary veins of leaves fine and often rather obscure beneath . . . . **2.** *petersiana*
– Staminodes 5, united for most of their length into a yellow, linear-oblong or spathulate strap; branchlets tomentellous with brown, spreading hairs; secondary and tertiary veins of the leaves prominent beneath . . . . **3.** *mendoncae*
7. Stamens 10; hypanthium turbinate, c. 5 mm long; petals pure white; pods less than 2 cm wide, with raised sutures, 3–11-seeded; lobes of leaf obtuse to slightly acuminate . . . . *acuminata*
– Stamens 1–5; hypanthium tubular; petals speckled or flushed pink to purple or sometimes white; pods mostly 2–2.5 cm broad, without raised sutures, 10–20-

seeded; lobes of leaf mostly blunt or rounded 8
8. Stamen 1; petals ± spathulate, the long claw half to one third as long as the blade; hypanthium 2–3 cm long; buds smooth . . . . *monandra*
– Stamens 3–5; petals narrowly to broadly elliptic, more shortly clawed; hypanthium 1–2(2.5) cm long; buds as next couplet 9
9. Stamens 3; buds ridged distally (apparent also on mature calyx) . . . . . *purpurea*
– Stamens 5; buds not prominently ridged . . . . *variegata*

1. **Bauhinia tomentosa** L., Sp. Pl.: 375 (1753). —Harvey in F.C. **2**: 275 (1862). —Oliver in F.T.A. **2**: 290 (1871). —Schinz, Pl. Menyharth.: 51 (1905). —Baker, Legum. Trop. Afr.: 654 (1930). —Gomes e Sousa, Pl. Menyharth.: 70 (1936).—Wilczek in F.C.B. **3**: 271 (1952). —Torre in Mendonça, Contrib. Conhec. Fl. Moçamb. **2**: 75 (1954). —De Wit in Reinwartia **2**: 409 (1956). —Torre & Hillcoat in C.F.A. **2**: 192 (1956). —Roti-Michelozzi in Webbia **13**: 153, figs.3 & 4 (1957). —White, F.F.N.R.: 99 (1962). —Drummond & Corby, List. Legum. S. Rhod.: 6 (1964). —Gomes e Sousa, Dendrol. Moçamb. Estudo Geral **1**: 247, fig.50 (1966). —Brenan in F.T.E.A., Legum.-Caesalp.: 209 (1967). —Richards & Morony, Check List Fl. Mbala Dist.: 67 (1969). —Drummond in Kirkia **8**: 212 (1972). —Corby in Kirkia **9**: 313 (1974). —Coetzer & Ross in F.S.A. **16**(2): 49 (1977). —Ross in Fl. Pl. Afr. **46**: fig.1817 (1980). —Polhill & Thulin in Fl. Eth. **3**: 68 (1989). —F. & J.A. Venter, Indig. Trees: 284, photos (1996). —M. Coates Palgrave, Trees Sthn. Africa: 336 (2002). —Schmidt, Lötter & McCleland, Trees Shrubs Mpumalanga & Kruger Nat. Park: 182 (2004). Type: Sri Lanka, *Burmann*, Thesaurus Zeylanicus, fig.18 (1737).

Alvesia bauhinioides Welw. in Ann. Consel. Ultramar, parte não official, sér. 1 **1858**: 587 (1859). Type from Angola.

Bauhinia mucora Bolle in Peters, Naturw. Reise Mossamb. **6**(1): 22 (1862). Type: Mozambique, near Tete, *Peters* (B, not seen).

Bauhinia tomentosa var. *glabrata* Hook. f. in Curtis, Bot. Mag.: fig.5560 (1866) (also given as var. *glabra* on same page) —Torre in Mendonça, Contrib. Conhec. Fl. Moçamb. **2**: 76 (1954). Type from Angola, cultivated at Kew.

Bauhinia volkensii Taub. in Engler, Pflanzenw. Ost-Afrikas **C**: 200 (1895). —Baker, Legum. Trop. Afr.: 654 (1930). —Brenan, Check-list For. Trees Shrubs Tang. Terr.: 88 (1949). Type: Kenya, Taita Dist., NE of L. Chala, *Volkens* 1765a (B† holotype).

Bauhinia wituensis Harms in Bot. Jahrb. Syst. **26**: 275 (1899). —Baker, Legum. Trop. Afr.: 654 (1930). —Brenan, Check-list For. Trees Shrubs Tang. Terr.: 89 (1949). Type: Kenya, Lamu Dist., Witu, *F. Thomas* 132 (B† holotype, BM, K).

Shrub or small tree 1–5(8) m high. Young branches glabrous to pubescent. Leaves: petiole 0.5–2.5 cm; lamina 1–6(8) cm from base of midrib to tip of lobes, lobed from one third to one half its length with the angle between the lobes acute, somewhat cordate to truncate or rarely slightly cuneate at the base, the lobes rounded to slightly obtusely pointed, glabrous to minutely to fairly densely pubescent on both surfaces. Inflorescences 1–2(7)-flowered at the ends of short branches or these aggregating into panicles of up to 25(40) flowers; axes and pedicels brown-pubescent to -tomentellous; pedicels 0.3–1.2 cm or up to 2.7 cm in fruit. Hypanthium 0.2–0.6 cm, funnel-shaped, smooth. Calyx 1–2.2(2.8) cm, smooth, in bud varying from rounded at the apex to abruptly narrowed into a short apiculum or narrow beak up to 6 mm long. Petals 1.7–5.5 cm long, suborbicular to obovate or elliptic, not or scarcely clawed, entirely yellow (forma *concolor*) or yellow with a deep red to purplish or brownish blotch at the base of the upper petal or occasionally on the three uppermost petals (forma *tomentosa*). Fertile stamens 10. Ovary densely pubescent to tomentose; stigma obliquely peltate or the margins on the lower side decurrent down the upper style. Pods (6.5)8–13 × 1.2–1.7 cm, linear-oblong, usually markedly parallel-sided, thinly woody, ± straw-coloured to pale brown, glabrescent to occasionally tomentose, usually 8–12-seeded. Seeds blackish-brown, 7–10 mm long, 5–7 mm broad, strongly compressed, ellipsoid.

Zambia. N: Mbala Dist., Sisya (Sizi) Forest, Kalambo road, fl. 28.v.1961, *Richards* 15147 (K). C: Luangwa Dist., Luangwa (Feira), fl. & fr. 5.ii.1963, *Grout* 284 (K). S: Choma Dist., Mapanza, fr. 4.v.1959, *E.A. Robinson* 2859 (K). **Zimbabwe.** N: Binga Dist., Lubu R., fl. 12.i.1958, *Lovemore* 544 (K, SRGH). W: Hwange (Wankie) Nat. Park, 0.6 m from Sinamatella Camp along Mandavu road, fl. 28.i.1969, *Rushworth* 1468 (K, LISC, SRGH). C: Harare Dist., grounds of University College, fl. 14.x.1967, *Corby* 1948 (K, SRGH). E: Mutare Dist., just S of White Waters, fl. & fr. 8.iii.1966, *Chase* 8405 (K, LISC, SRGH). S: Chiredzi Dist., Chidumo Store, between Chiredzi and Mana Pools, fl. 15.i.1971, *Kelly* 396 (K, SRGH). **Malawi.** S: Chikwawa Dist., Lengwe Nat. Park (Game Reserve), NE corner, fr. 5.iii.1970, *Brummitt* 8898 (K, LISC, MAL, PRE, SRGH). **Mozambique.** N: Pemba (Porto Amélia), 5 km S of town, fl. 14.iii.1960, *Gomes e Sousa* 4535 (K). T: Cabora Bassa, fr. 27.iv.1972, *Pereira & Correia* 2250 (K, LISC). MS: Umhlohlo, Lower Búzi, fl. 2.i.1911, *Swynnerton* 1254 (K). GI: Xai-Xai Dist., between Lumane and Xai-Xai (Vila João Belo), fr. 7.ii.1948, *Mendonça* 7293 (K). M: Maputo, Parque José Cabral, fl. & fr. 18.xii.1973, *Balsinhas* 2595 (K).

From Ethiopia and Somalia southwards to Angola and South Africa (Mpumalanga and KwaZulu-Natal), and in tropical Asia. Usually at lower altitudes, in riverine forest and thicket, coastal forest, etc.; recorded up to 1500 m in Mbala Dist. but otherwise below 1000 m.

A variable species, particularly in pubescence, leaf size, aggregation of inflorescence, development of a beak on the end of the bud, flower size, and blotching of the petals. Brenan in F.T.E.A (1967) commented that the inflorescences tend to aggregate into panicles particularly in Zimbabwe and Mozambique. Occasional herbarium specimens (e.g. *Plowes* 2298, Zimbabwe) show consistently small leaves up to only 2 cm long. Observations are required on the variability of these characters in populations and in local districts, and in individual plants from year to year.

Sometimes cultivated, e.g. Malawi, Zomba Botanic Garden, fl. & fr. 21.x.1987, *Tawakali & Thera* 1153 (K, MAL).

2. **Bauhinia petersiana** Bolle in Peters, Naturw. Reise Mossamb. **6**, 1: 24 (1861). —Gomes e Sousa, Dendrol. Moçamb. Estudo Geral **1**: 246, fig.49 (1966). —Tölken in Fl. Pl. Afr. **39**: fig.1532 (1969). —Coetzer & Ross in F.S.A. **16**(2): 51, fig.11 (1977). —M. Coates Palgrave, Trees Sthn. Africa: 335, t.84 (2002). Type: Mozambique, in Rios de Senna, *Peters* s.n. (K isotype).

Perlebia petersiana (Bolle) A. Schmitz in Bull. Jard. Bot. Belg. **43**: 385 (1973).

Shrub or small tree 1–7(10) m high, or sometimes scrambling up to 10 m by means of branch tendrils, or occasionally a suffrutex up to only 0.4 m high. Young branches rather finely brown- or grey-brown, spreading- or appressed-pubescent. Leaves: petiole 0.4–2.2 cm; lamina 1–7(8) cm from base of midrib to tip of lobes, lobed from one third to two thirds of its length with the angle between the lobes acute to obtuse, cordate to truncate at the base, the lobes rounded to slightly obtusely pointed, glabrous to minutely pubescent above, minutely appressed-pubescent or fairly densely spreading- or ascending-pubescent beneath. Inflorescences 1–10-flowered and these sometimes aggregating into panicles; axes usually brown-pubescent to -tomentellous; pedicels up to c. 1 cm, merging into the hypanthium. Hypanthium (1.5)2.5–5.5(6.5) cm, ± linear but broadening somewhat above, densely pubescent to tomentellous, longitudinally ribbed. Calyx (1.7)2.2–4.5 cm, in bud usually narrowed into a terminal beak with free tips of sepals showing. Petals white (rarely with pink blotches), 2.2–8.4 cm long, 0.6–2.2(4.2) cm broad, elliptic to elliptic-oblanceolate with strongly undulate-crisped margins, with a claw 0.2–1.5 cm long. Fertile stamens 5; filaments white to deep red or purplish. Ovary pubescent to tomentose; stigma capitate-peltate. Pods dark reddish-brown, woody, 10–24 cm long, (1.3)1.7–4.7 cm broad, ± linear-oblong or sometimes narrowed towards the base, not markedly parallel-sided, glabrescent, 4–10-seeded. Seeds dark purplish brown, 1.3–3 cm long, 0.8–1.8 cm broad, strongly compressed, oblong to circular.

From Tanzania and SE Congo to Namibia and South Africa (Limpopo and North-West Province, Gauteng).

Undersurface of leaf with minute closely appressed hairs; hypanthium and calyx closely appressed-tomentellous; pods (2.3)3–4.7 cm broad ... subsp. *petersiana*
Undersurface of leaf rather densely pubescent with irregularly spreading or ascending hairs; hypanthium and calyx with spreading or ascending hairs; pods (1.3)1.7–3 cm broad .... subsp. *macrantha*

Subsp. **petersiana** Coetzer & Ross in F.S.A. **16**(2): 53, fig.11 (1977). —Brummitt & Ross in Kew Bull. **30**: 594 (1976). —Lock, Leg. Afr. Check-list: 42 (1989). —M. Coates Palgrave, Trees Sthn. Africa: 335 (2002).

Bauhinia petersiana sensu Oliver in F.T.A. **2**: 288 (1871). —R.E. Fries, Wiss. Ergebn. Schwed. Rhod.-Kongo-Exped. **1**: 70 (1914). —Eyles in Trans. Roy. Soc. S. Afr. **5**: 366 (1916). —Baker, Legum. Trop. Afr.: 656 (1930). —Burtt Davy & Hoyle, Check-list For. Trees Shrubs, Nyasaland: 35 (1936). —Gomes e Sousa, Dendrol. Moçamb. **4**: 48 (1949). —Wilczek in F.C.B. **3**: 274 (1952). —Torre in Mendonça, Conhec. Fl. Moçamb. **2**: 76 (1954). —O. Coates Palgrave, Trees Central Africa: 70 (1957). —Topham in Burtt Davy & Hoyle, Check List For. Trees Shrubs Nyasaland Prot.: 34 (1958). —White, F.F.N.R.: 99 (1962).—Boughey in J. S. Afr. Bot. **30**: 159 (1964). —Brenan in F.T.E.A. Legum.-Caesalp.: 211, fig.47 (1967). —Gomes e Sousa, Dendrol. Moçamb. Estudo Geral **1**: 246, fig.49 (1967). —Richards & Morony, Check List Fl. Mbala Dist.: 67 (1969). —Drummond in Kirkia **8**: 212 (1972). —Corby in Kirkia **9**: 313 (1974).

Perlebia petersiana (Bolle) A. Schmitz in Bull. Jard. Bot. Belg. **43**: 385 (1973).

Usually an erect shrub or small tree 2–7(10) m high, sometimes scrambling. Leaves (2)3–7(8) cm long, lobed from one third to one half their length, the angle between the lobes (30)45–100°. Inflorescences 2–10-flowered and tending to aggregate together into panicles. Hypanthium and calyx closely appressed-tomentellous. Petal width 0.6–1.5(2.7) cm. Pod width (2.3)3–4.7 cm. Seeds 1.5–3 × 1–1.8 cm.

Caprivi Strip. 2 km E of Katima Mulilo on road to Ngoma, fr. 31.i.1975, *Vahrmeijer* 2507 (PRE). **Zambia.** B: Zambezi Dist., Chavuma, fl. 13.x.1952, *Holmes* 952 (K). N: Mbala Dist., Kalambo Falls, fl. 28.x.1956, *Richards* 6805 (K). W: Solwezi Dist., few km E of Kabompo R., fr. 31.vii.1930, *Milne Redhead* 804 (K). C: Lusaka Dist., below road to Rufunsa, c. 6 km from Luangwa Bridge, fl. 6.ix.1947, *Brenan & Greenway* 7819 (K). E: Chipata (Fort Jameson), fl. 1.i.1936, *Winterbottom* 20 (K). S: Choma Dist., Choma Nat. Forest (Siamambo Forest Res.), fr. 30.vii.1952, *Angus* 89 (K). **Zimbabwe.** N: Hurungwe Dist., half way down Zambezi escarpment on Makuti–Chirundu road, fl. 29.ix.1965, *Corby* 1400 (K, SRGH). W: Hwange Dist., Chamabondo, Victoria Falls Nat. Park, fl. & fr. 23.i.1975, *Gonde* 131/75 (K, SRGH). C: Chegutu Dist., c. 16 km S of Chegutu (Hartley), fl. 2.i.1967, *Biegel* 1621 (K, SRGH). E: Chipinge Dist., Sabi Valley, Honde Dip, fr. 30.v.1951, *Whellan* 503 (K, SRGH). **Malawi.** N: near Emanyeleni, 19 km S of Rumphi on road to Ekwendeni, fr. 21.v.1970, *Brummitt* 10986 (K, LISC, MAL, SRGH). C: Dedza Dist., Ndelema Village, Mua-Livulezi Forest Res., fr. 1.iv.1967, *Jeke* 88 (K, SRGH). S: Zomba, fl. xii.1900, *Purves* 49 (K). **Mozambique.** N: near Ribáuè, fl. xii.1941, *Gomes e Sousa* 2284 (K, PRE). Z: near Mocuba, fl. 11.xii.1942, *Torre* 4790 (K, LISC). T: between Fíngoè and Chicoa, 11.9 km from Fíngoè, fr. 29.vi.1949, *Barbosa & Carvalho* 3384 (K). MS: Manica, fl. ii.1922, *Honey* 703 (K). GI: between junction from Vilanculos–Nova Mambone–Mabote and that of Mabote, 8.7 km from junction, fr. 28.iii.1962, *Barbosa & Balsinhas* 5042 (K, LISC, LMA).

Also in Tanzania and SE Congo. In various types of woodland (*Brachystegia, Acacia, Sclerocarya*, etc.), usually on escarpment slopes; recorded 30–1500 m.

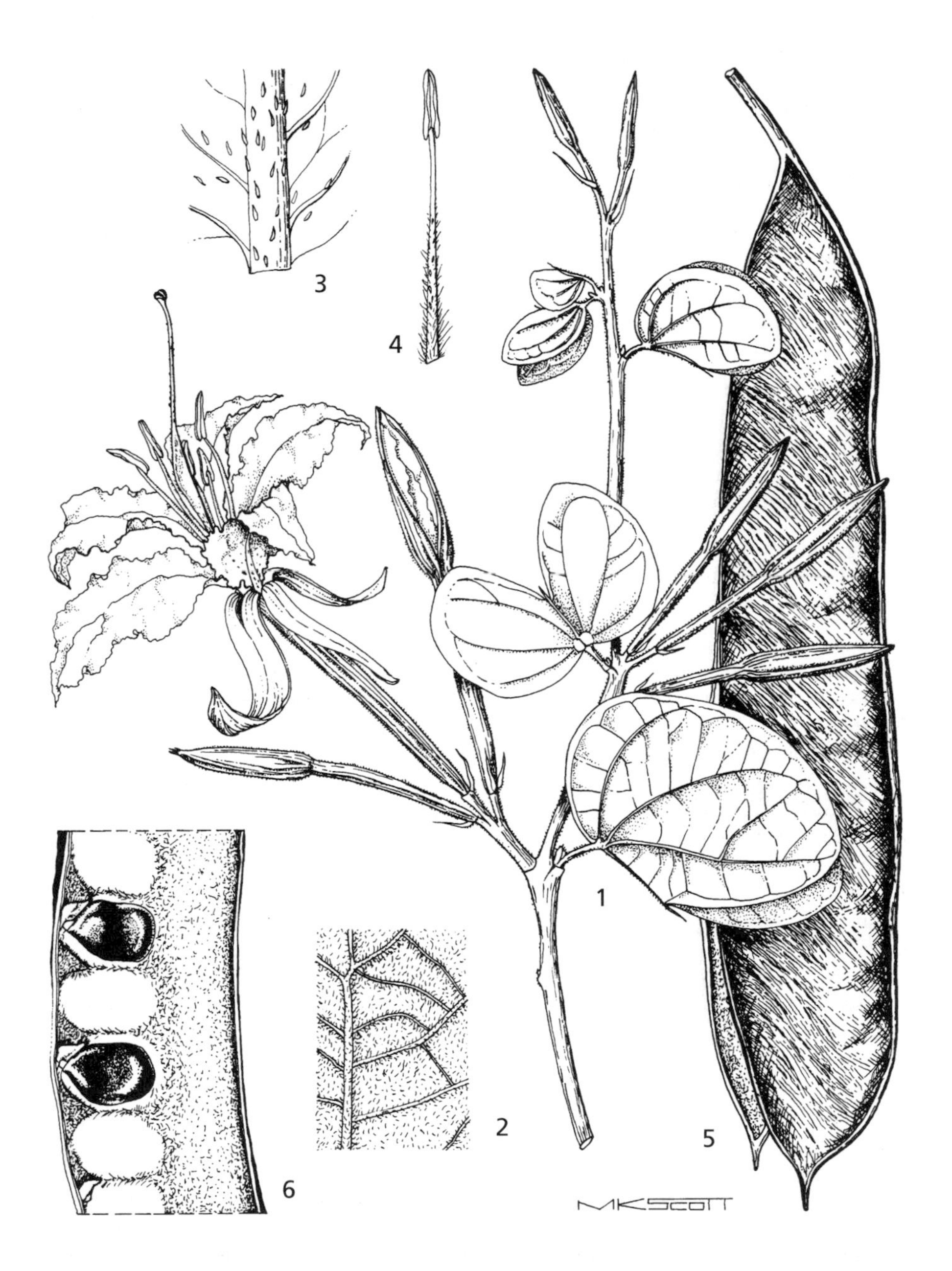

Fig. 3.2.**2**. BAUHINIA PETERSIANA subsp. MACRANTHA. 1, flowering twig (× 1); 2, part of lower surface of leaf showing hairs (× 4); 3, part of petal showing glands (× 4); 4, stamen (× 2), 1–4 from *Killick & Leistner* 3199; 5, pod (× 1); 6, part of dehisced pod showing seeds (× 1), 5, 6 from *Hodgson* s.n. Drawn by M.K. Scott. Reproduced with permission from Flora of Southern Africa.

Subsp. **macrantha** (Oliv.) Brummitt & J.H. Ross in Kew Bull. **37**: 236 (1982). —F. & J.A. Venter, Indig. Trees: 248 (1996). —M. Coates Palgrave, Trees Sthn. Africa: 335 (2002). Type: Botswana, Lake Ngami, 1856, *McCabe* in *Atherstone* 7 (K holotype). FIGURE 3.2.**2**.

Bauhinia macrantha Oliv. in F.T.A. **2**: 289 (1871). —Harms in Warburg, Kunene-Samb.-Exped. Baum: 251 (1903). —Eyles in Trans. Roy. Soc. S. Afr. **5**: 366 (1916). —Baker, Legum. Trop. Afr. 656 (1930). —Miller, Check-list For. Trees Shrubs Bech. Prot.: 23 (1948); in J. S. Afr. Bot. **18**: 28 (1952). —Torre & Hillcoat in C.F.A. **2**: 194 (1956). —O. Coates Palgrave, Trees Central Africa: 73 (1957). —White, F.F.N.R.: 99 (1962). —Drummond & Corby, List. Legum. S. Rhod.: 6 (1964). —Schreiber in Merxmüller, Prodr. Fl. SW Afrika, fam. 59: 6 (1967). —Tölken in Fl. Pl. Africa **39**: fig.1531 (1969). —Drummond in Kirkia **8**: (1972). —Corby in Kirkia **9**: 313 (1974).

Bauhinia serpae Ficalho & Hiern in Trans. Linn. Soc. London, Bot. **2**: 20 (1881). —Baker in J. Bot. **66**: 138 (1928) in part; Legum. Trop. Afr.: 656 (1930). Type: Angola, Ninda, *Serpa Pinto* 9 (LISU holotype).

Perlebia macrantha (Oliv.) A. Schmitz in Bull. Jard. Bot. Belg. **43**: 384 (1973).

Perlebia macrantha subsp. *serpae* (Ficalho & Hiern) A. Schmitz in Bull. Jard. Bot. Belg. **43**: 384 (1973).

Bauhinia petersiana subsp. *serpae* (Ficalho & Hiern) Brummitt & J.H. Ross in Kew Bull. **30**: 594 (1976). —Coetzer & Ross in F.S.A. **16**(2): 53, fig.11 (1977).

Usually a bushy or erect shrub 1–3 m high, sometimes a suffrutex only 0.4 m high, sometimes a small tree up to 4 m high, or scrambling up to 10 m. Leaves 1–4.5 cm long, lobed from half to two thirds of their length, the angle between the lobes 25–45°. Inflorescences 1–3(4)-flowered, each inflorescence fairly distinct. Hypanthium and calyx fairly densely pubescent to tomentellous with spreading or ascending hairs. Petal width 0.8–2.2(4.2) cm. Pod width (1.3)1.7–3 cm. Seeds 1.3–1.5 × 0.7–1 cm.

Caprivi Strip. c. 97 km from Katima Mulilo on road (Finaughty's) to Singalamwe, fl. 30.xii.1958, *Killick & Leistner* 3199 (K, PRE). **Botswana.** N: Ngamiland Dist., Khwebe, fl. i.1897, *E.J. Lugard* 144 (K). SW: c. 80 km N of Kang, fr. 18.ii.1960, *Wild* 5069 (K, SRGH). SE: Kgatleng Dist., c. 8 km S of Mosomane (Artesia), fl. 15.i.1960, *Leach & Noel* 76 (K, SRGH). **Zambia.** B: Senanga Dist., Siloana, fr. 28.xii.1960, *Mubita* C4 (K, SRGH). W: Mwinilunga Dist., 60 km S of Mwinilunga on road to Kabompo, fl. & fr. 25.i.1975, *Brummitt, Chisumpa & Polhill* 14108 (K, NDO). S: Kalomo Dist., Katombora (Katambora), fl. & fr. 2.xii.1955, *Gilges* 488 (K, SRGH). **Zimbabwe.** N: Binga Dist., near Lungwalala Flats, fl. 4.xi.1956, *Lovemore* 469 (K, SRGH). W: Nyamandhlovu Dist., road to Bonisa Farm, fl. 12.xii.1971, *Best* 942 (K, SRGH). C: Gweru Dist., Lower Gwelo T.T.L., edge of old Hunters Road, fr. 21.vii.1966, *Biegel* 1280 (K, SRGH).

Also in S Angola, Namibia and South Africa (Limpopo Province, North-West Province and Gauteng). Largely confined to Kalahari sands, often associated with *Baikiaea, Copaifera, Baphia massaiensis* subsp. *obovata, Colophospermum, Terminalia sericea,* etc.; 850–1250 m.

The two subspecies are geographically distinct except for a slight overlap in the area surrounding the Victoria Falls (see distribution map in Kew Bull. **30**: 594, 1976). However, within this small area they are probably kept apart by their ecological preferences, subsp. *macrantha* being characteristic of sandy flats and subsp. *petersiana* of escarpment woodlands. Although subsp. *petersiana* is recorded from all of the provinces of Zambia, in this Flora it is probably absent for most of the SW part of the country from Kabompo District almost to Livingstone, where it is replaced by subsp. *macrantha.*

3. **Bauhinia mendoncae** Torre & Hillc. in Bol. Soc. Brot., Sér. 2, **29**: 36 (1955); in C.F.A. **2**: 193 (1956). —White, F.F.N.R.: 100 (1962). Type from Angola.

Perlebia mendoncae (Torre & Hillc.) A. Schmitz in Bull. Jard. Bot. Belg. **43**: 384 (1973).

Suffrutex 40 cm high to small tree 3.3 m high. Young branches tomentellous with spreading brown hairs. Leaves: petiole 0.6–1.5 cm; lamina (2)3–7(9.2) cm from base of midrib to tip of lobes, lobed for half to two thirds of its length with the angle between the lobes 20–45°, cordate to sometimes truncate at the base, the lobes rather narrow and often rather distinctly obtusely pointed, glabrescent above, fairly densely covered with irregularly spreading brown hairs beneath, especially on the rather prominent veins. Inflorescence usually 1–5-flowered, these sometimes aggregating into small lax panicles; axes brown-tomentose; pedicels c. 4 mm. Hypanthium 2.5–5.2 cm, ± linear but broadening somewhat above, brown-tomentose, longitudinally ribbed. Calyx 1.8–3.2 cm, the bud obtusely pointed but often with filiform sepal tips extending beyond, tomentose like the hypanthium. Petals 3.2–6 cm long, 1.4–2.8 cm broad, elliptic to obovate with undulate-crisped margins, with a claw 0.5–1.8 cm, white. Fertile stamens 5; 5 staminodes united for nearly all their length into a yellow linear-oblong or spathulate strap, 1.5–3 × 1–2.5 cm. Ovary brown villous; stigma capitate-peltate. Pods 7–16.5 × 2–2.5 cm, irregularly oblong to linear-oblong, rusty brown-tomentose when young but somewhat glabrescent when older and revealing a dark reddish-brown surface. Seeds reddish brown, 1.2–1.6 × 0.9–1.2 cm,.

Zambia. B: Kaoma (Mankoya) to Mongu, mile 31, fl. & fr. 22.ii.1952, *White* 2119 (FHO, K). W: Mwinilunga Dist., near source of R. Isongailu, fl. 31.xii.1937, *Milne-Redhead* 3893 (K).

Also in Angola. On Kalahari sands, often in *Cryptosepalum* woodland (mavunda), also with *Baikiaea, Brachystegia,* etc.; 900–1250 m.

Similar in appearance to *B. petersiana* subsp. *macrantha,* which seems to occur further south on the Kalahari sands in Zambia, and from which it is fairly readily distinguished by its denser rusty-brown tomentum, more prominent secondary and tertiary venation on the undersurface of the leaf and the yellow staminodial strap in the flowers. The inclusion of *B. mendoncae* in a list of trees from Zimbabwe by Boughey (J. S. Afr. Bot. **30**, 1964) seems to be an error.

4. **Bauhinia urbaniana** Schinz in Verh. Bot. Vereins Prov. Brandenburg **30**: 169 (1888). —Baker in J. Bot. **66**: 138 (1928); Legum. Trop. Afr.: 656 (1930). —Miller, Check-list For. Trees Shrubs Bech. Prot.: 23 (1948); in J. S. Afr. Bot. **18**: 29 (1952). —Torre & Hillcoat in C.F.A. **2**: 193 (1956). —White, F.F.N.R.: 100 (1962). —Schreiber in Merxmüller, Prodr. Fl. SW Afrika, fam. 59: 6 (1967). —Coetzer & Ross in F.S.A. **16**(2): 51 (1977). —M. Coates Palgrave, Trees Sthn. Africa: 336 (2002). Type: Namibia, NW Kalahari between Karakobis and Levisfontein, *Schinz* 275 (Z holotype).

Perlebia urbaniana (Schinz) A. Schmitz in Bull. Jard. Bot. Belg. **43**: 386 (1973).

Shrub to small tree 1–3 m or scrambling up to 5 m. Young branches grey- or brown-pubescent to tomentellous. Leaves: petiole 0.3–1.1 cm; lamina 1–4.5 cm from base of midrib to tip of lobes, lobed for a half to two thirds of its length with the notch angle between the lobes 20–50°, weakly to strongly cordate at the base, the lobes ± rounded at the apex, glabrous above, irregularly spreading pubescent beneath, the veins rather prominent beneath. Inflorescences 3–15-flowered, the flowers densely crowded onto short lateral shoots 1–3 cm long, these sometimes tending to aggregate into narrow leafy panicles; axes brown- or greyish-tomentellous; pedicels up to 4 mm. Hypanthium 0.5–1 cm, linear-oblong to obconical, brown- or grey-tomentose to villous, rather obscurely longitudinally ridged. Calyx 0.8–1.3 cm, ± villous like the hypanthium. Petals pink or less often white, 1.2–2.5(3.2) cm long, 0.4–1(1.4) cm broad, elliptic to obovate with somewhat undulate margins and a short claw. Fertile stamens 5; staminodes united into a strap up to 2 mm broad and almost equalling the stamens. Ovary villous; stigma capitate-peltate. Pods dark reddish-brown, 8–14 × 1.7–2.2 cm, linear-oblong, pubescent when young but glabrescent. Seeds not known.

Botswana. N: Chobe Nat. Park, near Kasane, fl. 18.iv.1986, *R.M. & D. Polhill* 5245 (K). **Zambia.** B: Mongu, fl. & imm.fr. 15.vii.1962, *E.A. Robinson* 5420 (K). S: Sesheke

Dist., Mulobezi, fl. 30.i.1970, *Anton-Smith* in *GHS* 208046 (K, SRGH).

Also in Angola and Namibia. On Kalahari sands, in woodland of *Baikiaea, Guibourtia, Brachystegia, Pterocarpus*; 900–1050 m.

Readily distinguished from *B. mendoncae*, which also occurs on the Kalahari sands of this part of Zambia, by its much smaller flowers crowded into dense inflorescences and usually with pink instead of white petals. Fruiting material of *B. urbaniana* is poorly represented in the herbaria seen and should be supplemented.

5. **Bauhinia galpinii** N.E. Br. in Gard. Chron., Sér. 3, **9**: 728 (1891). —Oliver in Hooker's Icon. Pl. **20**: fig.1994 (1891). —Eyles in Trans. Roy. Soc. S. Afr. **5**: 366 (1916). —Pole Evans, Fl. Pl. S. Afr. **2**: fig.79 (1922). —Steedman, Trees, Shrubs & Lianes S. Rhod.: 18 (1933). —O. Palgrave, Trees Central Africa: 74 (1957). —Drummond & Corby, List Legum. S. Rhod.: 6 (1964). —Gomes e Sousa, Dendrol. Moçamb. Estudo Geral **1**: 248 (1966). —Brenan in F.T.E.A., Legum.-Caesalp.: 207 (1967). —Drummond in Kirkia **8**: 212 (1972). —Palmer & Pitman, Trees Sthn. Africa: 867 (1973). —Corby in Kirkia **9**: 313 (1974). —Coetzer & Ross in F.S.A. **16**(2): 50 (1977). —Pooley, Trees Natal: 150 (1993). —F. & J.A. Venter, Indig. Trees: 170 (1996). —M. Coates Palgrave, Trees Sthn. Africa: 334 (2002). —Schmidt, Lötter & McCleland, Trees Shrubs Mpumalanga & Kruger Nat. Park: 182 (2004). Types: South Africa, Mpumalanga, Doorn Spruit near Spelonken, *Nelson* 409 (K syntype); near Barberton, *Saunders* under *Wood* 3885 (K syntype); and Barberton, *Galpin* 421 (K syntype, PRE).

Bauhinia punctata Bolle in Peters, Naturw. Reise Mossambique **6**, 1: 23 (1861) non Jacq. (1780). —Oliver in F.T.A. **2**: 292 (1871). —Baker, Legum. Trop. Afr.: 659 (1930). —Burtt Davy, Fl. Pl. Ferns Transvaal: 323 (1932). —Martineau, Rhod. Wild Fl.: 33 (1954). —Torre in Mendonça, Contr. Conhec. Fl. Moçamb. **2**: 77 (1954). —Pardy in Rhod. Agric. J. **52**: 35 (1955). —White, F.F.N.R.: 100 (1962). —Breitenbach, Indig. Trees Sthn. Africa **3**: 340 (1965). Type: Mozambique, "in Mossimboa und Boror", *Peters* s.n. (B† holotype, K).

Perlebia galpinii (N.E. Br.) A. Schmitz in Bull. Jard. Bot. Belg. **43**: 382 (1973).

Shrub 1–4 m high or scrambling up to 5 m or more. Young branches puberulous. Leaves: petiole 0.5–1.5(2.3) cm; lamina 2–4.5(6.5) cm from base of midrib to tip of lobes, lobed for less than one third of its length and sometimes appearing only emarginate at the apex, the angle between the lobes commonly 90° or more but occasionally acute, cordate to truncate at the base, the lobes rounded or broadly obtusely pointed, glabrous above, the lower surface regularly covered with minute, usually appressed, greyish hairs and also with occasional scattered, much shorter, orange medifixed trichomes. Inflorescences 2–10-flowered; axes densely puberulous to pubescent with appressed brown hairs; pedicels 2–5 mm, merging into the hypanthium. Hypanthium 1.7–3.2 cm, brown-puberulous to pubescent, obscurely to conspicuously longitudinally ribbed. Calyx 1.9–2.8 cm, brown puberulous to pubescent. Petals bright red (variously described as scarlet, crimson, brick red, nasturtium, geranium red, salmon, or bluish red), 2.6–4.6 cm long, the limb suborbicular, cuneate to cordate at the base, and equalled or exceeded in length by the conspicuous claw, the undersurface furnished with orange trichomes like those on the leaf undersurface. Stamens 3, staminodes 7, filiform, up to 6 mm long. Ovary densely appressed-pubescent; stigma clavate. Pods dark reddish brown, 8–13 × 1.8–2.8 cm, ± linear-oblong, expanded along the upper margin into a flange up to 1 cm across, puberulent to pubescent when young but glabrescent later. Seeds dark reddish brown, c. 11 × 8.5 mm.

Zambia. S: tentatively recorded by F. White (F.F.N.R., 1962) from the Zambezi Valley, fide Fanshawe, and recorded for this province by Fanshawe (Check-list Woody Pl. Zambia, 1973), but no specimens seen. **Zimbabwe.** N: Mutoko Dist., c. 32 km W of Mutoko, fl. 30.xi.1950, *Whellan* 479 (K, SRGH). C: Gweru Dist., near Willoughby's, fl. & fr. 25.ii.1931, *Norlindh & Weimarck* 5130 (K). E: Nyanga Dist., lower Nyamarora

R., fl. xii.1958, *Davies* 2538 (K, SRGH). S: Chiredzi Dist., Gonarezhou Nat. Park, between Nyala and Tswiza, fr. 4.vi.1971, *Grosvenor* 617 (K, SRGH). **Mozambique.** N: between Nampula and Nametil, 2.1 km from Nampula, fl. 15.x.1952, *Barbosa & Balsinhas* 5154 (K, LISC). Z: Pebane Dist., between Pebane and Mualama, 8.3 km from Pebane, fr. 5.x.1949, *Barbosa & Carvalho* 4299 (K). MS: between Chimoio (Vila Pery) and Macequece, fr. 25.vii.1941, *Torre* 3177 (K, LISC). M: Namaacha, fl. ii.1931, *Gomes e Sousa* 427 (K).

Also in Swaziland and South Africa (Limpopo Province, Mpumalanga and KwaZulu-Natal). On river banks, in woodland, on termite mounds, sandveld, etc; up to 1370 m.

B. galpinii has also been widely cultivated as an ornamental in the tropics, and is known from cultivation in other parts of the Flora area. Zambia. B: Zambezi (Balovale), fl. vii.1952, *Gilges* 86 (K). W: Ndola Park, fl. 24.xii.1951, *White* 1804 (FHO, K). Zimbabwe. W: Matopos Park, fl. iii.1918, *Eyles* 1255 (K). C: Harare (Salisbury) Forest Nursery, fl. i.1956, *Cant* 1/66 (K, SRGH). Also depicted in McColaugh, Gaborone in Bloom: 60 (1997).

The species is usually said to be a shrub or climber up to about 5 m. One collection from Zimbabwe E, c. 6 km along Tanganda road from Chipinge road turn-off, high open forest, fl. 7.x.1961, *Methuen* 284 (K), is described as "climber reaching the top of the highest trees, about 60' (21 m), but limited to this locality".

3. TYLOSEMA (Schweinf.) Torre & Hillc.

Tylosema (Schweinf.) Torre & Hillc. in Bol. Soc. Brot., Sér. 2, **29**: 38 (1955). — Castro et al. in Bot. J. Linn. Soc. **147**: 99 (2005).
Bauhinia sect. *Tylosema* Schweinf., Reliq. Kotschy.: 17 (1868).

Stems trailing or climbing, herbaceous, becoming woody below, arising from a large woody underground tuber. Tendrils usually present, axillary, forked. Leaves alternate, comprised of two united leaflets, shortly to deeply lobed, palmately nerved, petiolate; stipules subpersistent. Flowers in racemes, bisexual, heterostylous, irregular, with bracts and bracteoles. Hypanthium short. Sepals 5, the upper pair often partially to completely fused. Petals 5, the adaxial smaller than the rest and bicallose basally. Stamens: 2 fertile, the other 7–8 staminodial, unequal, some ± flattened; anthers dehiscing by longitudinal slits. Ovary long-stipitate, few-ovulate; style elongate; stigma small. Pods short, woody, dehiscent with slightly inrolled valves, 1–2(6)-seeded. Seeds compressed to globose, with a small subapical hilum, from which extends a short U-shapedÏ line.

5 species from north-eastern to southern Africa.

Leaf lobes 6–12(20) cm long, separated up to a third of the way from apex to attachment of the petiole; petioles 3–8 cm long; ovary hairy; seeds compressed .... **1.** *fassoglense*

Leaf lobes 3–7.5 cm long, separated more than halfway from apex to attachment of the petiole; petioles 1.5–3.5 cm long; ovary glabrous or nearly so; seeds ovoid or globular .... **2.** *esculentum*

1. **Tylosema fassoglense** (Schweinf.) Torre & Hillc. in Bol. Soc. Brot., Sér. 2, **29**: 38 (1955); in C.F.A. **2**: 198 (1956). —Drummond & Corby, List. Legum. S. Rhod.: 27 (1964). —Brenan in F.T.E.A., Legum.-Caesalp.: 213, fig.48 (1967). —Schreiber in Merxmüller, Prodr. Fl. SW Afrika, fam. 59: 20 (1967). —Binns, Herb. Fl. Malawi: 24 (1968). —Drummond in Kirkia **8**: 213 (1972). —Ross, Fl. Natal: 195 (1973). —Schmitz in Bull. Jard. Bot. Belg. **43**: 376 (1973). —Corby in

Kirkia **9**: 314 (1974): all as *fassoglensis*. —Coetzer & Ross in F.S.A. **16**(2): 61, fig.14 (1977). —Polhill & Thulin in Fl. Ethiopia **3**: 70, fig.95.6.1 (1989). —Germishuizen, Wild Flowers Nthn. S Africa: 172 (1997). —Pooley, Wild Flowers KwaZulu-Natal & E Region: 256 (1998). —Castro et al. in Bot. J. Linn.Soc. **147**: 99 (2005). Types: Sudan, Fazoghli, *Boriani* 131 (W), and Metemma, Gallabat, *Schweinfurth* 2250 (BM), 2252 (BM, K), 2253 (B†). FIGURE 3.2.**3**.

Bauhinia fassoglensis Schweinf., Reliq. Kotschy.: 14, figs.12, 13 (1868). —Oliver in F.T.A. **2**: 286 (1871). —Taubert in Engler, Pflanzenw. Ost-Afrikas **C**: 199 (1895). —Harms in Warburg, Kunene-Samb.-Exped. Baum: 251 (1904). —R.E.Fries, Wiss. Ergebn. Schwed. Rhod.-Kongo-Exped. **1**: 70 (1914). —Eyles in Trans. Roy. Soc. S. Afr. **5**: 366 (1916). —Baker, Legum. Trop. Afr.: 659 (1930). —Burtt Davy & Hoyle, Check-list For. Trees Shrubs, Nyasaland: 35 (1936). —Wilcek in F.C.B. **3**: 272 (1952). —Miller in J. S. Afr. Bot. **18**: 28 (1952). —Torre in Mendonça, Contr. Conhec. Fl. Moçamb. 2: 78 (1954). —Martineau, Rhod. Wild Fl.: 33 (1954). —Roti-Michelozzi in Webbia **13**: 163 (1957). —Topham in Check List For. Trees Shrubs Nyasaland Prot.: 34 (1958). —White, F.F.N.R.: 99 (1962).

Bauhinia cissoides Oliv. in F.T.A. **2**: 287 (1871). —Taubert in Engler & Prantl, Nat. Pflanzenfam. III. **3**: 151 (1892). —Hiern, Cat. Afr. Pl. Welw. **1**: 295 (1896). —Baker, Legum. Trop. Afr.: 659 (1930). Type: Angola, Ambaca, *Welwitsch* 552 (LISU holotype, BM, K).

Bauhinia welwitschii Oliv. in F.T.A. **2**: 287 (1871). —Taubert in Engler & Prantl, Nat. Pflanzenfam. III. **3**: 151 (1892). —Hiern, Cat. Afr. Pl. Welw. **1**: 296 (1896). —Baker, Legum. Trop. Afr.: 659 (1930). Type: Angola, Pungo Andongo, Tunda Quilombo, *Welwitsch* 554 (LISU holotype, BM, K).

Bauhinia kirkii Oliv. in F.T.A. **2**: 288 (1871). —Baker, Legum. Trop. Afr.: 660 (1930). —Burtt Davy, Fl. Pl. Ferns Transvaal: 322 (1932). Type: Zambia, Batoka Highlands, *Kirk* s.n. (K holotype).

Bauhinia fassoglensis forma *cissoides* (Oliv.) Baker f. in J. Bot. **66**: 139 (1928).

Stems prostrate and trailing or climbing, up to 6 m or more, herbaceous or woody below. Branches reddish pubescent to tomentose when young, becoming greyish or glabrescent. Tendrils forked, 4–8 cm long. Leaves: petiole 3–8 cm long; lamina 6–12(20) × 5–12(18) cm, bilobed one-tenth to $^1/_3$ at the apex, deeply cordate at the base, the lobes half-ovate to half-elliptic or half-obovate, usually reddish pubescent beneath especially on the nerves but varying from subglabrous to tomentose; stipules 2–6 mm long, lanceolate-caudate to broadly ovate. Racemes (2)5–25(40) cm long, the numerous flowers fairly closely arranged on the upper part; bracts 5–8 mm long, lanceolate-caudate; bracteoles on the pedicel, a little smaller; pedicels (1.5)2–6 cm long. Sepals 1–1.5(2) cm long, oblong-lanceolate, tapering in the upper part, the upper pair fused, hairy. Petals yellow, fading pink, the larger ones 2–4 × 1–3 cm, obovate to obovate-circular, crinkled, tapering into a claw. Stamens: filaments of fertile stamens 8–15 mm long, those of staminodes 3–6 mm long. Ovary 5–6 mm long, hairy. Pod shortly stipitate, 5–11 × 3–7 cm, oblong-ovate to asymmetrically elliptic to obovate in outline, glabrescent, the valves slightly inrolled. Seeds chestnut-brown to blackish, 1.5–2.8 × 1–2 cm, subcircular to elliptic in outline, somewhat compressed.

Botswana. N: near Shaile airstrip, st. 20.i.1979, *P.A. Smith* 26226 (K, SRGH). **Zambia.** B: Zambezi (Balovale), fl. 1.xi.1952, *Gilges* 231 (K). N: 10.5 km (6.5 miles) NE of Chiengi, fl. 13.x.1949, *Bullock* 1256 (K). W: Ndola, fl. 21.ix.1953, *Fanshawe* 367 (K, NDO). C: Lusaka, Chipongwe Cave, fl. 11.i.1995, *Bingham & Truluck* 10268 (K). E: Petauke, Old Boma Road, fl. 3.xii.1958, *Robson* 814 (K, LISC). S: Mazabuka, Central Research Station, fl. 8.xii.1931, *Trapnell* in CRS 572 (K). **Zimbabwe.** N: near Darwendale, imm.fr. 20.iv.1948, *Rodin* 4311 (K). W: 48 km W of Hwange (Wankie) on Bulawayo–Victoria Falls Road, fl. xi.1974, *Raymond* 285 (K, SRGH). C: Marondera, Grasslands Res. Station, fl. 4.xii.1960, *Corby* 953 (K, SRGH). E: Chimanimani Dist., Nyahodi (Nyhodi) Valley, fl. 8.xi.1963, *Plowes* 2373 (K, LISC, SRGH). S: Mashava (Mashaba), fl. 28.x.1953, *Stayner* in SRGH 44207 (K, SRGH). **Malawi.** N: c. 21 km SW of Chisenga, path to Musitu Forest above Wenya road, fl. 11.vii.1970, *Brummitt* 12005

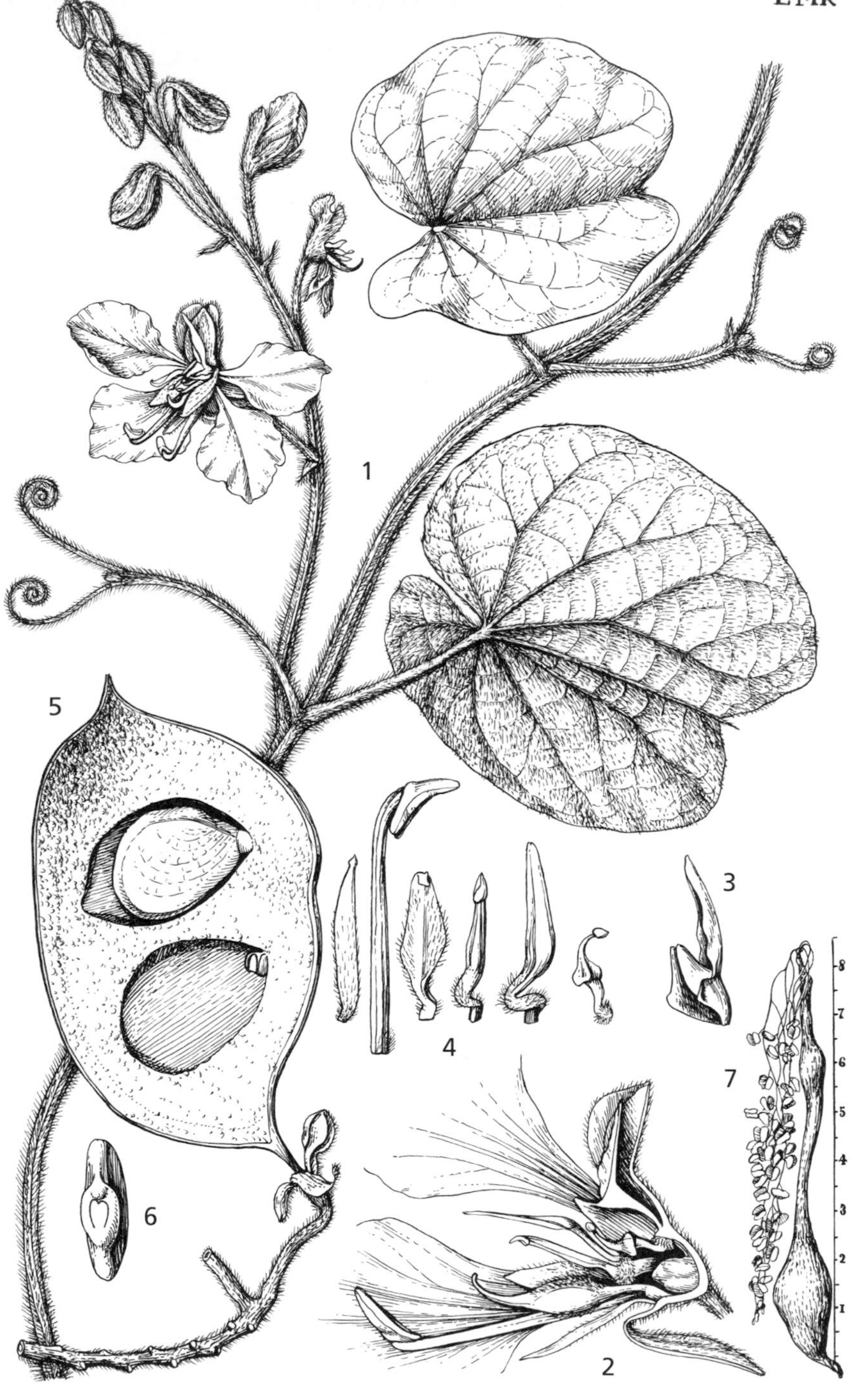

Fig. 3.2.3. TYLOSEMA FASSOGLENSE. 1, part of flowering stem (× ²⁄₃); 2, longitudinal section of flower (× 2); 3, upper petal, side view (× 2); 4, one fertile stamen and five staminodes (× 2), 1–4 from *Harley* 9410; 5, valve of pod viewed from inner side (× ²⁄₃), from *Rodin* 4341; 6, seed, hilar view (× 1), from *Chandler* 1126; 7, tuberous root with scale in feet, from a photograph by J.H. Hopkins at Kew. Drawn by Lura Ripley. From F.T.E.A.

(K). C: Ntcheu Dist., near Bilira turnoff, fl. 23.xi.1967, *Salubeni* 903 (K, SRGH). S: Shire Valley, Liwonde, fl. 1.ii.1959, *G. Jackson* 1201 (K). **Mozambique.** N: Erati, Namapa, at base of Malala Range, fr. 30.iii.1961, *Balsinhas & Marrime* 336 (K, LISC, LMA). T: between Mualadze (Vila Gamito) and Furancungo, 37.9 km from Mualadze, fr. 11.vii.1949, *Barbosa & Carvalho* 3574 (K, LMA). MS: Chimoio, fr. 9.iii.1948, *Barbosa* 1150 (K, LISC). M: Goba, fl. 23.xi.1944, *Mendonça* 3084 (K, LISC).

Sudan, Ethiopia and East Africa south to Angola, South Africa (North-West and Limpopo Province, Gauteng, Mpumalanga, KwaZulu-Natal) and Swaziland. Edges of forest, miombo woodland, mopane and mixed woodland, wooded grassland, riverine communities and dambos; up to 1700 m.

A number of minor uses are recorded but the species is not valued nearly so highly as *T. esculentum.*

2. **Tylosema esculentum** (Burch.) Schreiber in Mitt. Bot. Staatssamml. München **3**: 611 (1960); in Merxmüller, Prodr. Fl. SW Afrika, fam. 59: 20 (1967). —Coetzer & Ross in F.S.A. **16**(2): 62 (1977). —van Wyk & Malan, Wild Fl. Highveld, ed. 2: 124 (1997). —Castro et al. in Bot. J. Linn. Soc. **147**: 99 (2005). Type: South Africa, North-West Province (N Cape), Kalahari, Litakun near a branch of Moshewa R., *Burchell* 2414 (K holotype, PRE photo).

Bauhinia esculenta Burch., Trav. **2**: 589 (1824). —Schinz in Mém. Herb. Boiss. **1**: 121 (1900). —Baker, Legum. Trop. Afr.: 659 (1930). —Burtt Davy, Fl. Pl. Ferns Transvaal: 322 (1932). —Verdoorn in Fl. Pl. S. Afr. **33**: fig.1311 (1959). —Letty, Wild Fl. Transvaal: 157 (1962).

Bauhinia burkeana (Benth.) Harv. in F.C. **2**: 275 (1862). —Taubert in Engler & Prantl, Nat. Pflanzenfam. III. **3**: 151 (1892). Type: South Africa, Mooi R., *Burke & Zeyher* s.n. (K holotype).

Bauhinia bainesii Schinz in Mém. Herb. Boiss. **1**: 121 (1900). —Brummitt & Ross in Kew Bull. **31**: 219 (1976). Lectotype, chosen by Brummitt & Ross (1976): Namibia, E of Gobabis at Oas, *Schinz* 2061 (Z).

Stems prostrate and trailing up to several metres long, herbaceous or woody below. Branches reddish to greyish pubescent at first, glabrescent. Tendrils forked, 1–4 cm long. Leaves: petiole 1.5–3.5 cm long; lamina 3–7.5 × 4–10 cm, bilobed halfway or more, cordate at the base, the lobes reniform; stipules 3–5 mm long, elliptic-oblong to obovate. Racemes 4–12 cm long, c. 8–20-flowered, rather shortly pedunculate; bracts 3–4 mm long, linear-lanceolate; bracteoles on the pedicel, a little smaller; pedicels 2–4.5 cm long. Sepals 8–12 mm mm long, oblong-lanceolate, tapered distally, pubescent, the upper pair fused. Petals yellow, the 4 larger ones 1.5–2.5 cm long, obovate above a long claw, slighly crinkled. Stamens: filaments of fertile stamens 10–12 mm long, those of stamonodes 3–6 mm long. Ovary 5–6 mm long, glabrous or nearly so. Pod red turning brown, shortly stipitate, 5–9 × 4.5–6.5 cm, ovate-oblong, glabrous. Seeds brown to brownish-black, 1.3–2 × 1.2–1.8 cm, ovoid to subspherical.

Botswana. N/SE: Kwebe to Palapye, fl. i.1899, *E. Lugard* 242 (K). SW: 24 km (15 miles) N of Ghanzi, st. 21.iv.1963, *Ballance* 623 (K, SRGH). SE: Mahalapye to Lephepe road, fl. 13.i.1958, *de Beer* 556 (K, SRGH). **Zimbabwe.** W: Victoria Falls, 1.xii.1913, *Peter* L.7858 (K).

Also in Namibia and South Africa (Limpopo and North-West Provinces, Gauteng). Habitat not recorded in Flora area, but generally in dry sandy places, though also associated with dolomite in grassland or wooded grassland in southern Africa.

Widely used as an edible plant, the roasted seeds rather like cashew nuts. The tubers can also be roasted and eaten.

Verdoorn (Fl. Pl. S. Afr. **33**, 1959) says the seeds are at first pale yellow, covered with a clear viscid substance, becoming dark brown when ripe with a brittle shell.

4. PILIOSTIGMA Hochst.

Piliostigma Hochst. in Flora **29**: 598 (1846). —De Wit in Reinwardtia **3**: 530 (1956).
Bauhinia sect. *Piliostigma* (Hochst.) Benth. in Bentham & Hooker, Gen. Pl. **1**: 576 (1865).

Deciduous trees or shrubs. Leaves composed of 2 united leaflets, bilobed, petiolate, with deciduous stipules. Flowers in terminal, leaf-opposed or axillary racemes or panicles, medium to small, unisexual with the plants dioecious; bracts small; bracteoles similar at top of the pedicel at an articulation. Hypanthium infilled. Calyx deeply cupular, with 4–5 short lobes. Petals 5, usually white to pink. Stamens 10(11), all fertile in male flowers, reduced to staminodes in female flowers; filaments villous below; anthers dehiscing by longitudinal slits. Stigma thick, capitate, flattened-globose, sessile on the ovary. Pods leathery to woody, oblong, indehiscent, many-seeded. Seeds embedded in pulp, irregularly arranged, with a U-shaped line below the small subapical hilum.

3 species in tropical Africa, Asia and Australia.

Piliostigma thonningii (Schumach.) Milne-Redh. in Hooker's Icon. Pl. **35**: fig.3460, p.2 (1947). —Codd, Trees & Shrubs Kruger Nat. Park: 66 (1951). —Miller in J. S. Afr. Bot. **18**: 35 (1952). —Pardy in Rhod. Agric. J. **49**: 219 (1952). —Torre & Hillcoat in C.F.A. **2**: 199 (1956). —Roti-Michelozzi in Webbia **13**: 174 (1957). —O. Coates Palgrave, Trees Central Africa: 115 (1957). —Keay in F.W.T.A. ed.2, **1**: 444 (1958). —White, F.F.N.R.: 126, fig.20 (1962). —Boughey in J. S. Afr. Bot. **30**: 159 (1964). —Drummond & Corby, List Legum. S. Rhod.: 22 (1964). —Breitenbach, Indig. Trees Sthn. Africa **3**: 342 (1965). —Brenan in F.T.E.A., Legum.-Caesalp.: 206, fig.46 (1967). —Gomes e Sousa, Dendrol. Moçamb. Estudo Geral **1**: 249 (1966). —Schreiber in Merxmüller, Prodr. Fl. SW Afrika, fam. 59: 18 (1967). —Richards & Morony, Check List Fl. Mbala Dist.: 74 (1969). —Van Wyk, Trees Kruger Nat. Park: 194 (1972). —Drummond in Kirkia **8**: 213 (1972). —Palmer & Pitman, Trees Sthn. Africa **2**: 871 (1973). —Corby in Kirkia **9**: 314 (1974). —Coetzer & Ross in F.S.A. **16**(2): 59, fig.13 (1977). —Polhill & Thulin in Fl. Ethiopia **3**: 68, fig.95.5.2 (1989). —F. & J.A. Venter, Indig. Trees: 46 (1996). —M. Coates Palgrave, Trees Sthn. Africa: 337 (2002). Type: Ghana, Aquapim, *Thonning* s.n. (C holotype, PRE photo). FIGURE 3.2.**4**.

Bauhinia thonningii Schumacher & Thonn. in Schumacher, Beskr. Guin. Pl.: 203 (1827). —Baker, Legum. Trop. Afr.: 657 (1930). —Burtt Davy, Fl. Pl. Ferns Transvaal: 322 (1932). —Steedman, Trees, Shrubs & Lianes S. Rhod.: 18 (1933). —Gomes e Sousa in Bol. Soc. Estud. Colón. Moçamb. **26**: 45 (1935). —Bremekamp & Obermeyer in Ann. Transvaal Mus. **16**: 417 (1935). —Burtt Davy & Hoyle, Check-list For. Trees Shrubs, Nyasaland: 35 (1936). —Miller, Check-list For. Trees Shrubs Bech. Prot.: 23 (1948). —Gomes e Sousa, Dendrol. Moçamb. **4**: 54 (1949). —Wilczek in F.C.B. **3**: 275 (1952). —Suessenguth & Merxmüller in Mitt. Bot. Staatssamml. München **1**: 155 (1952). —Topham in Burtt Davy & Hoyle, Check List For. Trees Shrubs Nyasaland Prot.: 34 (1958). —K. Coates Palgrave, Trees Sthn. Africa, ed. 2: 284 (1988). —Van Wyk, Trees Kruger Nat. Park, ed.3: 87 (1994).

Locellaria bauhinioides Welw. in Ann. Consel. Ultram., parte Não Official, Sér. 1 [Apontamentos Phytogeographicos] **1**: 588 (1858). Type: Angola, Calumguembo, *Welwitsch* 486 (BM holotype).

Bauhinia articulata sensu auctt., sphalm. pro *reticulata* DC., sed non DC. sensu stricto; Oliver in F.T.A. **2**: 290 (1871). —Eyles in Trans. Roy. Soc. S. Afr. **5**: 366 (1916). —Steedman in Proc. Rhod. Sci. Assoc. **24**: 6 (1925).

Bauhinia reticulata sensu auctt. non DC. sensu stricto.; Hiern, Cat. Afr. Pl. Welw. **1**: 296 (1896). —Harms in Warberg, Kunene-Sambezi-Exped. Baum: 251 (1903). —Schinz, Pl. Menyharth.: 51 (1905). —Sim, For. Fl. Port. E. Afr.: 48 (1909). —R.E. Fries, Wiss. Ergebn. Schwed. Rhod.-Kongo-Exped. **1**: 71 (1914). —Gomes e Sousa, Pl. Menyharth.: 70 (1936).

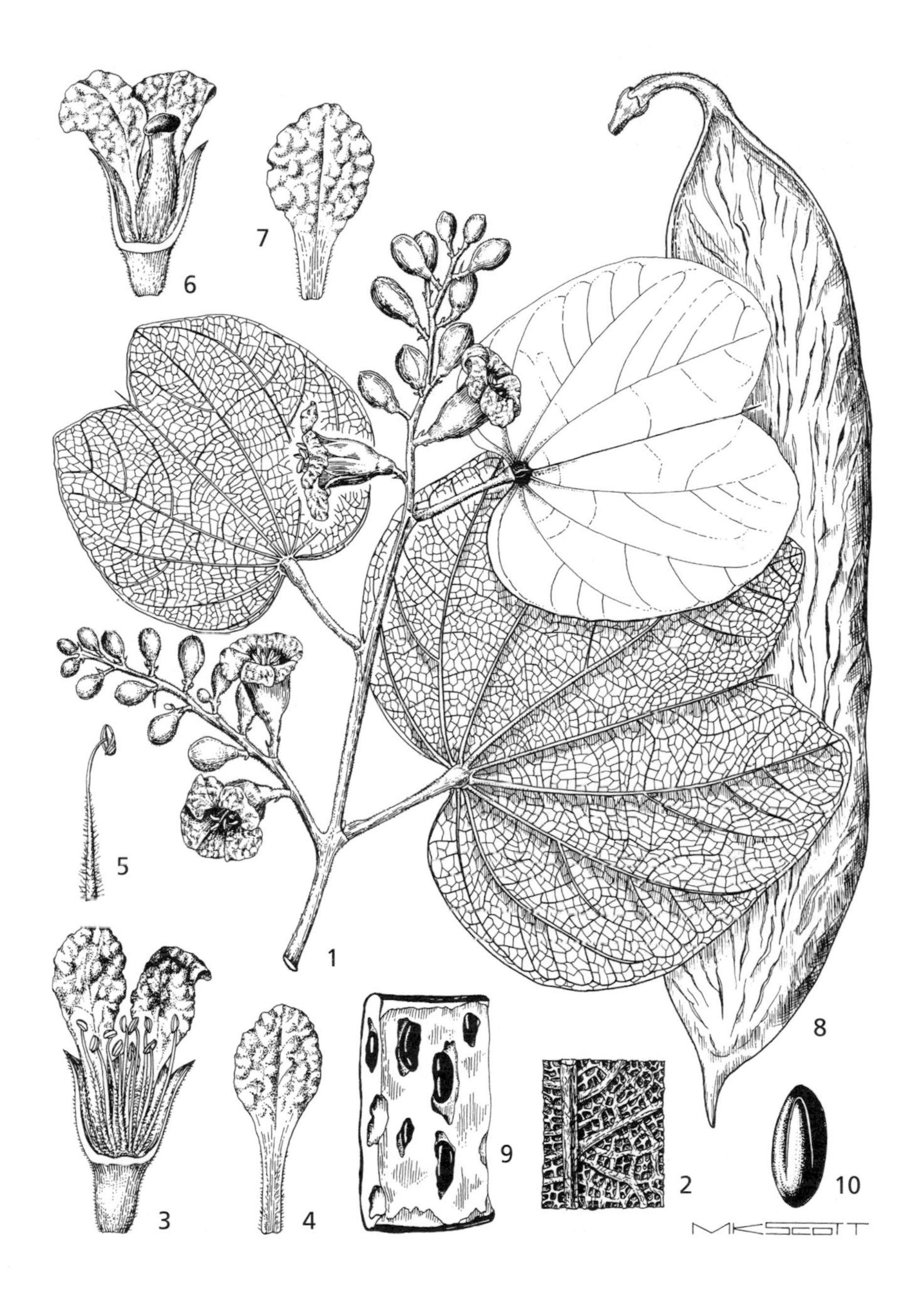

Fig. 3.2.4. PILIOSTIGMA THONNINGII. 1, male flowering branchlet (× $^{2}/_{3}$); 2, lower surface of leaf (× 6), both from *Scheepers* 1195; 3, male flower, with part of calyx and corolla removed (× $1^{1}/_{3}$); 4, petal of male flower (× $1^{1}/_{3}$); 5, stamen (× 2), 3–5 from *Story* 5384; 6, female flower, with part of calyx and corolla removed (× $1^{1}/_{3}$); 7, petal of female flower (× $1^{1}/_{3}$), 6, 7 from *Rodin* 4123; 8, pod (× $^{2}/_{3}$); 9, part of pod cut open to show seeds (× $^{2}/_{3}$); 10, seed, hilar view (× 2), 8–10 from *Codd* 7077. Drawn by M.K. Scott. Reproduced with permission from Flora of Southern Africa.

Tree up to 10 m with a spreading crown, or occasionally flowering as a shrub. Young branches with rusty-brown tomentum, somewhat glabrescent later. Leaves: petiole 2–5(7) cm; lamina up to 16 cm from base of midrib to tip of lobes, lobed for $^1/_8$ – $^1/_3$ of its length, the angle between the lobes slightly less than to slightly more than 90°, usually strongly cordate at the base, the lobes rounded to acutely pointed, glabrous above, with rusty-brown crisped hairs beneath, the lower surface with a conspicuously prominent finely reticulate venation. Dioecious, inflorescences of male trees forming very narrowly pyramidal panicles up to 20 × 5.5 cm; those of female trees forming short few-flowered panicles up to 7 cm long; axes and pedicels brown-tomentellous. Calyx 0.8–1.5 cm long, brown-tomentellous. Petals white to lilac or pink, 1.1–2 cm long, obovate, the margins strongly undulate. Stamens 10(11), scarcely reaching the throat of the flower. Ovary brown-tomentose; style very short or absent; stigma capitate. Pods woody, 12–30(37) × 3–6.5 cm, varying from 2 to 10 times as long as broad, oblong to linear oblong, brown-pubescent when young but later glabrescent. Seeds dark brown to blackish, 7–9 × 5–7 × 3–4 mm, obovoid to ellipsoid, compressed.

Caprivi Strip. c. 13 km NW of Ngoma Ferry, fr. 16.vii.1952, *Codd* 7077 (K, PRE). **Botswana.** N: Ngamiland Dist., Mboma, Okavango Swamps, fr. vi.1971, *P.A. Smith* 102 (K, LISC, SRGH). **Zambia.** B: Sesheke Dist., fl. & fr. c. 1911, *Gairdner* 146 (K). N: Mbala Dist., Mpulungu, fl. 14.iii.1950, *Bullock* 2639 (K). W: Mwinilunga Dist., Kalenda Plain, fl. 24.xii.1937, *Milne-Redhead* 3808 (K). C: Lusaka, fl. 1953, *Puffett* s.n. (K, PRE). S: Mazabuka Dist., Magoye Forest Res., fr. 14.iv.1952, *White* 2657 (FHO, K). **Zimbabwe.** N: Hurungwe Dist., Mwami (Miami), K34 Experimental Farm, fr. 15.vi.1946, *Wild* 1144 (K, SRGH). W: Bulawayo, fl. 26.xii.1911, *Rogers* 5494 (K). C: Gweru Dist., Senga, fl. 6.i.1963, *Loveridge* 539 (K, LISC, SRGH). E: Mutare Dist., Odzani R. Valley, fl. & fr. xii.1915, *Teague* 532 (K). S: Masvingo Dist., Great Zimbabwe Ruins (Zimbabwe), fr. 19.x.1930, *Fries, Norlindh & Weimarck* 2094 (K). **Malawi.** N: Nkhata Bay, fl. 9.v.1971, *Pawek* 4798 (K, MAL). C: Lilongwe Dist., Bunda Agric. College farm, 23 km S of Lilongwe, fl. 19.ii.1970, *Brummitt* 8658 (K, MAL, SRGH). S: Zomba, fl. xii.1915, *Purves* 227 (K). **Mozambique.** N: Ribáuè, fl. iii.1943, *Gomes e Sousa* 2303 (K). Z: Mocuba Dist., between Mocuba and Mugeba, 7 km from Mocuba, fl. 16.v.1949, *Barbosa & Carvalho* 2652 (K). T: near acampamento of G.P.Z., fl. 24.i.1972, *Macêdo* 4659 (K, LISC). MS: Manica, Macequece, fl. & fr. vii.1923, *Honey* 800 (K). GI: Massinga, fl. iv.1936, *Gomes e Sousa* 1722 (K, LISC).

From Senegal to Ethiopia and south to Angola, Namibia and South Africa (Limpopo Province and Mpumalanga). A common tree of deciduous woodland or grassland; up to 1500 m in the Flora area.

5. SCHOTIA Jacq.

Schotia Jacq., Collect. **1**: 931 (1787). —Codd in Bothalia **6**: 515–533 (1956) nom. conserv.

Trees or shrubs, evergreen or deciduous, unarmed. Leaves usually paripinnate; leaflets mostly in 3–18 pairs, petiolulate, without gland-dots; stipules small, extending in front of the petiole, falling very early. Inflorescence a short lateral or terminal panicle or raceme, often congested and sometimes produced before the leaves; flowers spirally arranged on the ultimate axes; bracts small, caducous; bracteoles 2, small, caducous, not enclosing the bud. Flowers usually red. Hypanthium prominent, cup-shaped. Sepals 4, unequal, imbricate. Petals 5 (reduced or absent in one species), red. Stamens usually 10, free or connate below; anthers dorsifixed, dehiscing by longitudinal slits. Ovary stipitate, the stalk fused to the side of the hypanthium; ovules several; style elongate, filiform; stigma terminal, capitate. Pods oblong or elliptic, flattened, the central portion tending to fall away leaving the sutures, sometimes with the seeds attached. Seeds several, without areole, with a fleshy yellow aril partially enveloping the seed.

A southern African genus of 5 species. The red flowers produce copious nectar and are visited by birds. The three Guineo-Congolian species formerly included in *Schotia* have now been transferred to the genera *Leonardoxa* Aubrév. and *Normandiodendron* J. Léonard (see Aubréville in Adansonia **8**: 177–179 (1968) and Léonard in Bull. Jard. Bot. Belg. **62**: 433–451 (1993)).

Tree; leaflets all rounded or emarginate at the apex; petals reduced to linear filaments .... **1.** *brachypetala*
Shrub, sometimes scrambling; at least some leaflets acuminate or mucronate at the apex; petals not reduced, oblanceolate .... **2.** *capitata*

1. **Schotia brachypetala** Sond. in Linnaea **23**: 39 (1850). —Sim, For. Fl. Port. E. Afr.: 51 (1909). —Harms in Engler, Pflanzenw. Afrikas **3**(1): 452 (1915). —Eyles in Trans. Roy. Soc. S. Afr. **5**: 365 (1916). —Baker, Legum. Trop. Afr.: 709 (1930). —Gomes e Sousa, Dendrol. Moçamb. **2**: 95 (1949); Dendrol. Moçamb., Estudo Geral **1**: 262 (1966). —Pardy in Rhod. Agric. J. **49**: 173 (1952). —Torre in Mendonça, Contrib. Conhec. Fl. Moçamb. 2: 83 (1954). —Codd in Bothalia **6**: 526–529 (1956). —Schreiber in Merxmüller, Prodr. Fl. SW Afrika, fam. 59: 18 (1967). —Palmer & Pitman, Trees Sthn. Africa **2**: 855 (1972). —Drummond in Kirkia **8**: 219 (1972); in **10**: 246 (1977). —Ross in F.S.A. **16**(2): 28, fig.5 (1977). —Pooley, Trees Natal: 146 (1993). —Van Wyk, Trees Kruger Nat. Park: 85 (1994). —M. Coates Palgrave, Trees Sthn. Africa: 327 (2002). Type: South Africa, KwaZulu-Natal, Durban (Port Natal), *Gueinzius* 33 (S holotype, K photo, SAM). FIGURE 3.2.**5**.

Schotia brachypetala var. *pubescens* Burtt Davy, Fl. Pl. Ferns Transvaal: 326 (1932). Type: South Africa, Mpumulanga, Letaba Dist., Shiluvane, *Junod* 635 (K holotype, PRE).

Schotia rogersii Burtt Davy, Fl. Pl. Ferns Transvaal **2**: 326 (1932). Type: South Africa, Limpopo Prov., Soutspansberg Dist., Waterpoort, *Rogers* 21246 (K holotype, GRA, PRE, SAM).

Schotia semireducta Merxm. in Mitt. Bot. Staatssamml. München **6**: 199 (1953). Type: Zimbabwe, Rusape, *Dehn* 42 in *GHS* 40239 (SRGH).

Tree, (5)10–15 m tall; bark dark grey-brown, furrowed. Branchlets dark brown, smooth, puberulous, glabrescent. Leaves paripinnate (occasionally imparipinnate); stipules narrowly triangular, 5 × 1.5 mm, early caducous; petiole 1.2–2 cm long, sparsely pubescent at first, glabrescent; rachis 3.5–17.5 cm long, sparsely pubescent at first, glabrescent, terete, furrowed, at most very narrowly winged; leaflets opposite to subopposite, in (3)4–6(7) pairs, 2.5–6.5 × 1.0–3.2 cm, elliptic to rhombic-elliptic or rhombic-ovate, often asymmetric, apex rounded or retuse (mucronate in very young developing leaves), base cuneate, pubescent to glabrous, glabrescent; venation slightly prominent on both surfaces. Inflorescences often cauliflorous on the older wood, paniculate, congested; axis puberulous; flowers deep red or scarlet; bracts c. 2.5 × 2 mm, broadly ovate-elliptic, concave, puberulous; bracteoles paired, c. 3 × 2 mm, broadly elliptic, puberulous; bracts and bracteoles early caducous leaving prominent raised scars; pedicel 3–5 mm long, glabrous or puberulous. Hypanthium 3–5 mm long, glabrous or pubescent especially on the ridges. Calyx lobes 4, ovate, 8–11 × 4–6 mm, gland-dotted and glabrous to sparsely pubescent outside, mainly towards the base, glabrous inside. Petals absent or reduced to one or more strap-shaped rudiments 2–3 mm long, or 1–4 normal, narrowly obovate, clawed, up to 1.8 × 0.6 cm. Stamens 10; filaments c. 17–19 mm long, expanded and fused in the basal 2–3 mm, glabrous; anthers dorsifixed. Ovary c. 10 × 3 mm, narrowly elliptic, flattened, glabrous but verrucose near the margins; stipe c. 3 mm long; style c. 9 mm long, glabrous, stigma capitate, papillose. Pod 10–17 × 3.5–4.5 cm, oblong, flattened, glabrous, with numerous small resin-filled vesicles in the pod wall, several-seeded, tardily dehiscent leaving the dorsal and ventral sutures attached to the pedicel. Seeds pale brown, 15–20 × 10–15 mm, broadly elliptic to rhomboid, flattened, smooth; aril 0.9–1.2 cm long, cup-shaped, basal, yellow.

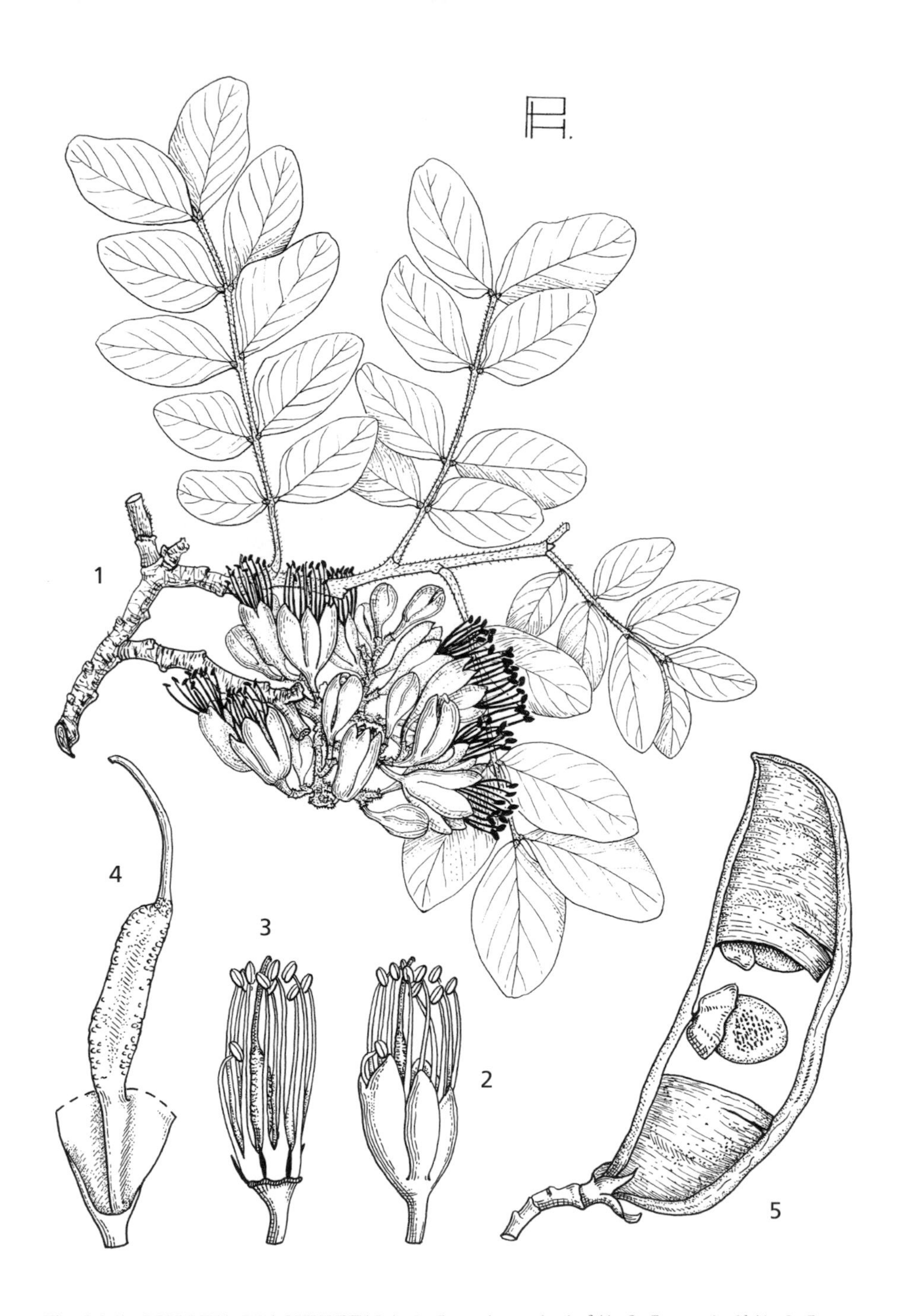

Fig. 3.2.**5**. SCHOTIA BRACHYPETALA. 1, flowering twig (× 3/5); 2, flower (× 1 1/2); 3, flower with sepals removed (× 1 1/2); 4, gynoecium (× 2), 1–4 from *Codd* 6695; 5, pod, one valve partly cut way to show arillate seed (× 2/3), from *Howes* 20. Drawn by Pat Halliday. Reproduced with permission from Flora of Southern Africa.

Botswana. SE: c. 190 km from Palapye, c. 20 km E of Baines' Drift, N bank of Limpopo R., fr. 17.xii.1976, *Mott* 1061 (K, UCBG). **Zimbabwe.** C: Harare (Salisbury), fl. & imm.fr. ix.1917, *Eyles* 844 (BM, K). E: Mutare (Umtali), commonage, fl. 15.ix.1952, *Chase* 4595 (BM, LISC, SRGH). S: Masvingo (Fort Victoria), fl. 10.vii.1955, *Plowes* 1865 (K, LISC, SRGH). **Mozambique.** M: Maputo, Quinta da Pedra, fl. 29.ix.1948, *Gomes e Sousa* 3854 (COI, K, LISC, PRE).

Also in South Africa. Zambezian woodland and thicket, often on termite mounds; near sea-level up to 1350 m.

Conservation notes: A widespread taxon; Lower Risk, Least Concern.

The leaflets are sometimes attacked by a gall-forming organism which causes the formation of patches of dense persistent pubescence along the midrib. The flowers produce copious nectar and are visited by sunbirds. Ross (1977) states that the timber is of good quality.

2. **Schotia capitata** Bolle in Peters, Naturw. Reise Mossambique **6**(1): 18 (1861). —Oliver in F.T.A. **2**: 310 (1871). —Harms in Engler, Pflanzenw. Afrikas **3**(1): 453 (1915). —Sim, For. Fl. Port. E. Afr.: 51 (1909). —Baker, Legum. Trop. Afr.: 710 (1930). —Codd in Bothalia **6**: 521–523 (1956). —Palmer & Pitman, Trees Sthn. Africa **2**: 853 (1972). —Ross in F.S.A. **16**(2): 26, fig.5/6 (1977). —Pooley, Trees Natal: 146 (1993). —M. Coates Palgrave, Trees Sthn. Africa: 327 (2002). Type: Mozambique, Inhambane, *Peters* s.n (B†, BM sketch). Neotype (selected by Ross): Mozambique, Maputo, Goba, near R. Maivavo, fl. 5.xi.1960, *Balsinhas* 204 (P holotype, K).

Schotia tamarindifolia Sims var. *forbesiana* Baill. in Adansonia **6**: 197 (1866). Type: Mozambique, Maputo (Delagoa) Bay, *Forbes* 4323 (K isotype).

Schotia transvaalensis Rolfe in Bull. Misc. Inform., Kew **1906**: 248 (1906). —Burtt Davy, Fl. Pl. Ferns Transvaal: 356 (1932). —Miller, Check-list For. Trees Shrubs Bech. Prot.: 24 (1948); in J. S. Afr. Bot. **18**: 36 (1952). —Torre in Mendonça, Contrib. Conhec. Fl. Moçamb. **2**: 83 (1954). —Codd, Trees & Shrubs Kruger Nat. Park: 68, fig.63 (1951). Type: South Africa, Mpumulanga, Barberton Dist., *Orange* s.n. (K holotype).

Theodora capitata (Bolle) Taub. in Engler & Prantl, Nat. Pflanzenfam. III. **3**: 138 (1892); in Engler, Pflanzenw. Ost-Afrikas **C**: 198 (1895).

Shrub or small tree, 1.5–7 m tall, sometimes scrambling or scandent. Branchlets brown, soon becoming grey or dark brown, smooth, glabrous or sparsely puberulous, glabrescent. Leaves paripinnate; stipules up to 8 × 4.5 mm, broadly and asymmetrically triangular, acuminate at apex, rounded and sometimes tooothed at base, caducous; petiole 0.2–0.7 cm long, glabrous or sparsely puberulous; rachis 3.5–6.5 cm long, glabrous or sparsely puberulous, narrowly winged, the wing broadening distally between each pair of leaflets; leaflets opposite, (3)4–6-jugate, 2.7–4.5 × 1.2–3.2 cm, broadly elliptic to broadly obovate, almost symmetrical, glabrous, apex mucronate or apiculate, base cuneate, slightly asymmetrical; venation slightly prominent on both surfaces. Inflorescence a dense subcapitate panicle; axis puberulous; bracts and bracteoles c. 2 × 1 mm, broadly oblong-elliptic, early caducous, sparsely puberulous, margins shortly and sparsely ciliate. Flowers bright red, almost sessile. Hypanthium 3–4 mm long, glabrous. Calyx lobes 4, broadly ovate, the outer two c. 9 × 7 mm, the inner two narrower, glabrous. Petals 5, bright red, c. 15 × 4–5 mm, narrowly obovate, tapering gradually to the base. Stamens 10, alternately long and short, basally connate into a distinct sheath; filaments c. 20 mm long at anthesis, glabrous; anthers 2–2.5 mm long, elliptic, dorsifixed. Ovary oblong, c. 3 × 1.5 mm, glabrous; stipe short; style c. 15 mm long, glabrous, stigma capitate, papillose. Pod 7.5–10.5 × 2.5–4 cm, oblong, flattened, glabrous, brown, with a lignified upper suture which may persist, sometimes with the seeds and/or arils attached after the eventual dehiscence of the pod. Seeds c. 10 × 10 × 5 mm, broadly ovoid, with a cup-shaped yellow aril extending about halfway up the seed.

Mozambique. GI: Inhambane, Vilankulo (Vilanculos) area, fl. & fr. 24.xi.1942, *Mendonça* 1488 (K, LISC). M: Maputo, Moamba, Machava–Moamba road, fl. & fr.

23.x.1980, *Zunguze & Boane* 290 (K, LMU).

Also in South Africa (Mpumalanga and KwaZulu-Natal) and Swaziland. Dry woodland and scrub forest; 0–300 m.

Conservation notes: Of restricted distribution within the Flora area, but extends widely outside it. Lower Risk, Least Concern.

Drummond (Kirkia **8**: 219 (1972); **10**: 246 (1977)) records *S. capitata* from Zimbabwe. The material at Kew of the specimen he cites — *Drummond* 7717, from the Zimbabwe/Mozambique border near Chicualacuala (Vila Eduardo Mondlane) — is sterile and is not typical *S. capitata*. It may be of hybrid origin; Codd (1956) claims that hybrids are not infrequent.

6. COLOPHOSPERMUM J. Léonard

Colophospermum J. Léonard in Bull. Jard. Bot. État. **19**: 390 (1949); in Bull. Jard. Bot. Nat. Belg. **67**: 21–43 (1999). —P.P. Smith, Timberlake & A.E. Van Wyk in Taxon **47**: 751–752 (1998).

Hardwickia sensu Breteler in Adansonia **19**: 289 (1997) pro parte, non Roxb.

Medium to large tree or sometimes a shrub, deciduous, unarmed. Leaves petiolate, with a single pair of leaflets, the basal pulvinus of each being contiguous with the opposite one across the top of the petiole, with a small appendage at the petiole apex; leaflets with c. 8–12 main veins and no midrib, with numerous pellucid glands; stipules small, caducous. Inflorescence of axillary racemes, these sometimes aggregating into a panicle at branch ends; bracteoles absent. Hypanthium absent. Sepals 4, 2 outer overlapping, 2 inner. Petals absent. Stamens c. 20–25, free; anthers dorsifixed, dehiscing by longitudinal slits. Ovary compressed, with style attached on upper margin. Fruit compressed, asymmetric, rounded at apex, with attachment of the style evident on upper margin, glandular, 1-seeded, indehiscent. Seeds compressed, narrowly winged, deeply corrugated and glandular on surface.

A monotypic genus confined to southern tropical Africa. The generic name refers to the resinous seeds.

Colophospermum was considered as congeneric with the genus *Hardwickia* Roxb. from the drier parts of peninsular India by Breteler (1997), and since the latter is the earlier name he took this up in preference to *Colophospermum*. A proposal to conserve the latter was therefore made by Smith, Timberlake and van Wyk (1998). However, Léonard (1999) provided ample reasons for keeping the two genera separate. Furthermore, the conservation proposal has been accepted (Taxon **54**: 1096, 2005) so that if future authors do wish to combine the genera, the name *Colophospermum* will be used. See also the molecular data in Fougere-Danezan, Maumont & Bruneau in Adv. Legum. Syst. **10**: 161–180 (2003).

Krüger, Tiedt & Wessels in Bot. J. Linn. Soc. **131**: 223–233 (1999) have looked at the floral development. They suggest that the two outer 'sepals' are in fact bracteoles with no elongation of the pedicel above them. The 20–25 stamens are initiated in a single whorl. There is no trace of petals. The anthers are basifixed early in development and then become dorsifixed. Comparable data are needed from *Hardwickia* to check the supposed relationship.

Colophospermum mopane (Benth.) J. Léonard in Bull. Jard. Bot. État **19**: 390 (1949); in Publ. I.N.E.A.C., Sér. Scient. **45**: 88, fig.9 (1950). —Miller in J. S. Afr. Bot. **18**: 30 (1952). —Pardy in Rhod. Agric. J. **50**: 152 (1953). —Torre & Hillcoat in C.F.A. **2**: 239 (1956). —O. Coates Palgrave, Trees Central Africa: 101 (1957). —Topham in Burtt Davy & Hoyle, Check List For. Trees Shrubs Nyasaland Prot.: 37 (1958). —Palmer & Pitman, Trees Sthn. Africa **2**: 843 (1977). —White, F.F.N.R.: 121

(1962). —Drummond & Corby, Leg. Indig. Nat. S. Rhod.: 7 (1964). —Breitenbach, Indig. Trees South Africa **3**: 324 (1965). —Gomes e Sousa, Dendrol. Moçamb. Estudo Geral **1**: 250, fig.53 (1966). —Schreiber in Merxmüller, Prodr. Fl. SW Afrika, fam. 59: 13 (1967). —Drummond in Kirkia **8**: 213 (1972). —Ross, F.S.A. **16**(2): 17 (1977). —Timberlake in Zimbabwe Bull. For. Res. **11** (1995). —F. & J.A. Venter, Indig. Trees: 142 (1996). —M. Coates Palgrave, Trees Sthn. Africa: 317 (2002). Types: Mozambique, Shiramba, 25.i.1860, *Kirk* s.n. (K syntype); Lupata, 28.i.1860, *Kirk* s.n. (K syntype). FIGURE 3.2.**6**.

Copaifera mopane Benth. in Trans. Linn. Soc. London **25**: 317 (1865). —Oliver in F.T.A. **2**: 315 (1871). —Harms in Warburg, Kunene-Samb.-Exped. Baum: 248 (1903). —Sim, For. Fl. Port. E. Afr.: 51 (1909). —Harms in Engler, Pflanzenw. Afrikas **3**(1): 443, fig.244 (1915). —Eyles in Trans. Roy. Soc. S. Afr. **5**: 364 (1916). —Baker, Legum. Trop. Afr.: 750 (1930). —Hutchinson in Bull. Misc. Inform., Kew **1931**: 226–229 (1931). —Burtt Davy, Fl. Pl. Ferns Transvaal: 326 (1932). —Gomes e Sousa, Pl. Menyharth.: 70 (1936). —Burtt Davy & Hoyle, Check-list For. Trees Shrubs, Nyasaland: 36 (1936). —Pole Evans in Bot. Surv. S. Afr. Mem. **21**: 24, 30 (1948). —Miller, Check-list For. Trees Shrubs Bech. Prot.: 24 (1948).

Copaiba mopane (Benth.) Kuntze, Rev. Gen. Pl. **1**: 172 (1891). —Gilg in Engler, Pflanzenw. Ost-Afrikas **B**: 305, 419 (1895). —Taubert in Engler, Pflanzenw. Ost-Afrikas **C**: 197 (1895). —Hiern., Cat. Afr. Pl. Welw. **1**: 303 (1896). —Schinz in Mém. Herb. Boissier **1**: 119 (1900).

Hardwickia mopane (Benth.) Breteler in Adansonia **19**: 289 (1997).

Small to medium tree 4–15 m tall, or sometimes larger up to 22 m, with 1–several main trunks from the base and rather few main branches diverging at a narrow angle or sometimes remaining stunted as a shrub; bark deeply vertically fissured, often in an elongate-reticulate pattern; deciduous. Young branches glabrous, smooth, grey. Leaves consisting of a petiole, one pair of leaflets, and a small flat appendage 2–4(5) × 1–1.5 mm at the petiole apex on the ventral side; petiole (1)1.5–4(4.8) cm long, glabrous; leaflets (2)4–10(13) × (1.5)2.5–5(6.5) cm, obliquely ovate or lanceolate to falcate-triangular, somewhat asymmetric at the base and with a basal pulvinus broader than long and (7)8–12(14) prominent nerves radiating from it (a distinct midrib lacking), acute to obtuse at the apex, glabrous, coriaceous, with a raised reticulum of secondary veins, with numerous pellucid gland dots; stipules up to 5 × 3.5 mm, ovate, falling very early. Inflorescences of usually 7–13-flowered axillary racemes up to 7 cm long (including peduncle), simple or sometimes with one or two branches at the base, or occasionally almost forming a panicle towards the ends of branches; bracts broadly ovate to suborbicular, caducous leaving a raised scar below the pedicel; pedicels 4–8 mm long. Flower buds spherical, up to 4(6) mm diameter. Sepals 4, 2 outer enclosing two inner in bud, suborbicular, green or yellowish, scarious towards margins, glabrous, all reflexed in flower. Stamens 20–25; filaments c. 6 mm long, whitish; anthers 1.3–2(2.5) mm long, yellowish. Ovary c. 3 × 1.5 mm, flat, obovate, rounded at distal end, with style attached on upper margin, glabrous; style c. 2 mm long, glabrous; stigma broad, asymmetrically peltate. Fruits (2.7)3–4.5(6) × 1.8–2.5(3.2) cm, asymmetrically obovate to laterally reniform (i.e. reniform with the pedicel attached at one side, not in the middle of the curve), rounded distally and with the attachment of the style half to two-thirds of the way along the upper side, compressed, glabrous, straw-coloured with minute sunken glandular flecks of darker brown, usually with raised reticulate venation, indehiscent. Seed filling most of the fruit, compressed, subreniform, with a flattened margin c. 1 mm wide, the surface deeply contorted into a pattern of ridges and valleys and pitted with numerous reddish glands, sticky.

Botswana. N: Makgadikgadi Pan, fl. & imm.fr. i.1957, *M. Paterson* 30 (K, SRGH). SE: c. 5 km N of Radisele Siding, fl. 20.i.1960, *Leach & Noel* 257 (K, SRGH). **Zambia.** N: Isoka Dist., Luangwa Valley, fl. & imm.fr. 15.i.1938, *Trapnell* 1823 (K). C: Kafue, st. 26.iii.1963, *van Rensburg* KBS 1798 (K). E: Chipata (Fort Jameson), Nsefu–Luangwa road, fr. 11.v.1963, *van Rensburg* KBS 2134 (K). S: Kalomo Dist., Katombora, st. 22.ii.1956, *Gilges* 604 (K, SRGH). **Zimbabwe.** N: Mount Darwin Dist., c. 3 km towards escarpment from Mukumbura (Mkumburu) R., fl. 23.i.1960, *Phipps* 2414 (K, SRGH). W: Hwange Dist., c. 1 km from Sinamatella Camp along road to Mandavu, fl. 28.i.1969, *Rushworth* 1459 (K, LISC, SRGH). C: Gweru Dist., Mlezu Agriculture School Farm, c. 29 km SSE of Kwekwe (Que Que), fl. 11.i.1966, *Biegel* 791 (K, SRGH).

Fig. 3.2.**6**. COLOPHOSPERMUM MOPANE. 1, flowering branch (× 2/3); 2, leaf (× 1); 3, flower (× 3); 3, gynoecium (× 6), 1–4 from *Codd* 4827; 5, fruiting twig (× 2/3); 6, seed (× 1); 5, 6 from *Schlieben & Hartman* 2229. Drawn by Pat Halliday. Reproduced with permission from Flora of Southern Africa.

E: Chipinge Dist., Sabi Valley, fl. 23.i.1964, *Corby* 1071 (K, SRGH). S: turn off to Limpopo Ranch, c. 16 km N of Beitbridge, fl. 25.iii.1959, *Drummond* 6020 (K, LISC, SRGH). **Malawi.** N: Vwaza Marsh, c. 16 km from Katowo, past Khuramaji, fr. iv.1953, *Chapman* 78 (FHO). S: Machinga Dist., between Liwonde and Balaka, fr. 4.vii.1936, *B.D. Burtt* 5999 (K). **Mozambique.** T: Cahora Bassa Dist., 7.9 km from Nhancapirire (Inhacapirire) towards Chicoa, fr. 19.ii.1972, *Macêdo* 4859 (K, LISC). MS: near Nhamatanda (Vila Machado), fr. 23.iv.1948, *Mendonça* 4038 (K, LISC). GI: Guijá, Chicholo, fr. 10.vi.1947, *Pedrógão* 287 (K, LISC).

Also in Angola, Namibia and South Africa (Limpopo Province and northern tip of Mpumalanga). Characteristic of hot, dry river valleys with low rainfall, where it is often dominant and may form almost pure stands over large areas — the well-known "Mopane woodland"; recorded in the Flora area usually between 60 and 1000 m altitude, exceptionally up to 1300 m.

Conservation notes: A widespread species; not threatened.

Mopane is an important tree as an ecological dominant and fodder plant for many animals, particularly elephant. The pendulous leaves fold their single pair of leaflets together in hot sun and afford little shade, the large tracts of woodland being infamous for their monotony and shadelessness. Leaves and fruits are very glandular and smell strongly of turpentine. The wood is dark, hard and heavy and difficult to work, and is used mainly for fence posts, mine props or firewood, rarely for furniture. The distribution and survival of mopane seeds are discussed by Jarman & Thomas in Kirkia **7**: 103–107 (1969). A review and bibliography is given by Timberlake (1995).

The author citation "(Benth.) J. Léonard" is here preferred to the commonly given "(Kirk ex Benth.) Kirk ex J. Léonard" even though Kirk originally used the name *Colophospermum mopane* in manuscript notes before the species was described, since Bentham did not take up the full name used by Kirk, and Kirk did not propose a new combination based on Bentham's publication.

7. DANIELLIA Benn.

Daniellia Benn. in Pharm. Journ. **14**: 252 (1855). —Léonard in Publ. I.N.E.A.C., Sér. Scient. **45**: 90 (1950); in Mém. Acad. Roy. Sci. Belg. **30**(2): 110 (1957). —Brenan in F.T.E.A., Legum.-Caesalp.: 130 (1967).

Unarmed deciduous or evergreen trees. Leaves simply pinnate; petiolules not twisted; leaflets opposite or subopposite, with translucent gland dots; stipules intrapetiolar, connate into a narrowly ovate-ellipsoid structure enfolding the bud, early caducous. Inflorescence paniculate; flowers spirally arranged; pedicels jointed at base; bracteoles well developed, almost completely concealing the young flower buds, both caducous before anthesis. Hypanthium elongate. Sepals 4, imbricate (2 outer, 2 inner). Petals (in the Flora area) 3 large and 2 small. Stamens 10, 9 shortly connate at the base. Ovary stipitate, the stipe adnate to the hypanthium; style long; stigma small. Pods 1-seeded, dehiscing into 2 valves; endocarp coriaceous, curling and separating from the exocarp. Seeds solitary, large, with a smooth hard testa, affixed near the distal end of the pod, wind-dispersed attached to the exocarp of one valve of the pod.

About 9 species, all African, mainly in equatorial forest and seasonal tropical wetter woodlands; virtually absent from East Africa, except for NW Uganda, and only just reaching the Flora Zambesiaca area.

Daniellia alsteeniana Duvign. in Bull. Soc. Roy. Bot. Belgique **71**: 28 (1949); in Bull. Séances Inst. Roy. Colon. Belge **20**: 677 (1949). —Léonard in Publ. I.N.É.A.C., Sér. Scient. **45**: 112, figs. 13, 14 (1950); in F.C.B. **3**: 344, fig.25 (1952). Type from Congo (Katanga). FIGURE 3.2.**7**.

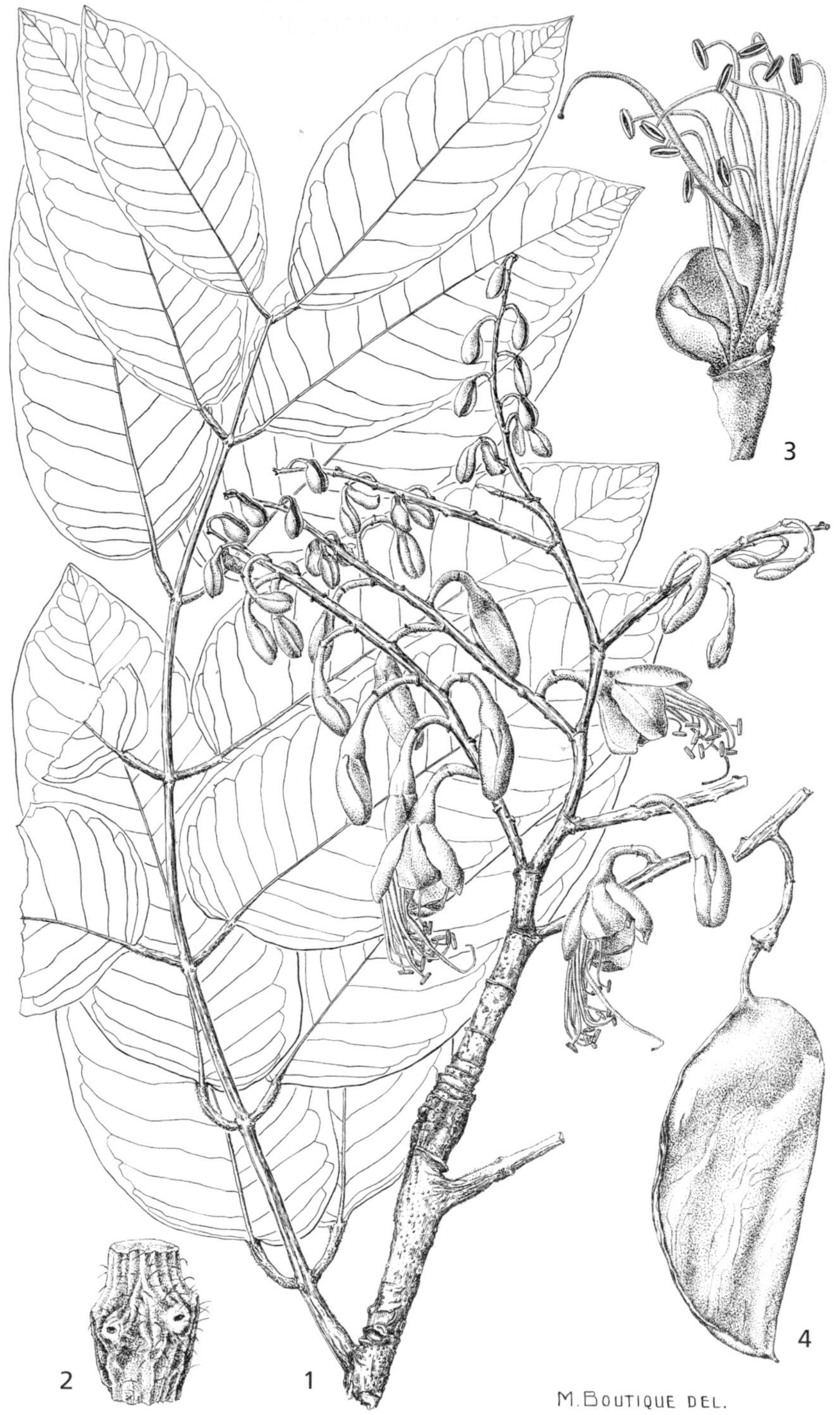

Fig. 3.2.7. DANIELLIA ALSTEENIANA. 1, leafy branch with inflorescence (× 1/2); 2, glandular orifices on leaf rachis (× 5), from *Bequaert* 19; 3, flower with sepals and large petal removed showing 1 large, 1 medium and 1 small petal as well as filaments joined at base (× 1); 1, 3 from *Duvigneaud* 950a; 4, pod (× 1/2), from *Duvigneaud* 1414. Drawn by M. Boutique. Reproduced from Flora du Congo Belge with permission of the National Botanic Garden of Belgium.

Tree to 20 m with a straight trunk to 1 m in diameter at the base; bark greyish, furrowed. Twigs pubescent. Leaves paripinnate; leaflets in 4–7 pairs, 5–18 × 3–9 cm, ovate-elliptic, elliptic or oblong, the proximal ones largest, acute, glabrous and shiny above, glaucous and sparsely pubescent beneath with scattered whitish spreading hairs, particularly on the midrib, lateral nerves in 8–12 pairs; stipules very unequal, the larger up to 6 × 1 cm, early caducous. Panicles spreading, held above the crown, to 30 cm long in total; axis and receptacle pubescent to glabrous; bracts obovate; pedicels 2–3 cm long including the hypanthium, to 4 cm in fruit. Hypanthium pubescent within, c. 5 mm in diameter. Sepals 19–23 × 12–23 mm, oblong to oblong-obovate, glabrous to glabrescent outside. Petals: two 16–18 × 12–18 mm, oblong to suborbicular; one 13–17 × 9–11 mm, oblong to obovate-oblong; two 3 × 2 mm, ovate. Stamens 10, united at the base, to 5 cm long, hairy below. Ovary 8–10 × 4–6 mm, 8–12-ovulate, glabrous or pubescent on the margins; stipe 1.5 cm long; style to 4 cm long. Pods 9–10 × 4–6 cm, asymmetrically ovate, 1(2)-seeded. Seeds brown, shiny, 2 × 1.5 cm, ovoid, flattened.

Zambia. N: Kasama, fr. 20.x.1957, *Lawton* 283 (FHO).

Also in Angola and Congo (Katanga). In chipya woodland; c. 1500 m.

Conservation notes: Best regarded as Vulnerable in the Flora area as it is restricted to forest patches and strips near the forest-savanna boundary. Data Deficient worldwide.

Only 3 specimens have been seen from the Flora area, and the species is not included in White's Forest Flora of Northern Rhodesia (1962). The descriptions of the flowering parts have been taken from the account in Flore du Congo Belge, and from Congo specimens at Kew.

The timber is said to be exploited in the Katanga, and the tree is described in F.C.B. as producing 'a magnificent clear copal' (gum).

The panicles of upward-facing creamy white flowers of the West African *D. oliveri* (Rolfe) Hutch. & Dalziel have been shown to be pollinated by bats; the flowers of the present species seem to be similarly presented.

8. HYMENAEA L.

Hymenaea L., Sp. Pl.: 1192 (1753); Gen. Pl., ed. 5: 499 (1754). —Lee & Langenheim in Univ. Calif. Publ. Bot. **69**: 1–109 (1975).

Trachylobium Hayne in Flora **10**: 743 (1827). —Léonard in Publ. I.N.E.A.C., Sér. Scient. **45**: 125 (1950); in Mém. Acad. Roy. Sci. Belg. **30**(2): 115 (1957).

Trees, evergreen, unarmed. Leaves with a single pair of leaflets, with pellucid gland dots; petiolules short, twisted; stipules small, early caducous. Flowers in panicles, spirally arranged on the ultimate axes; bracts concave, caducous; bracteoles 2, enclosing the young bud, imbricate, caducous. Hypanthium short. Sepals 4, unequal (2 + 2), imbricate. Petals 3 large and 2 minute; sometimes all 5 large. Stamens 10; anthers dorsifixed, dehiscing by longitudinal slits. Ovary stipitate; ovules usually 4; style elongated, filiform; stigma terminal, small. Pods ovoid or ellipsoidal, the epicarp hard with numerous prominent resin-filled vesicles, indehiscent; endocarp pithy-powdery. Seeds 1–3, without areole or aril.

A genus of about 15 species, almost all in the New World tropics.

Hymenaea coubaril L., native of South America, may be grown as an ornamental tree. **Zimbabwe**. N: Makonde Dist., Rainbow Nurseries, st. 18.xii.1974, *Biegel* 4729 (K, SRGH). The identity of this sterile specimen is not certain. The species differs from *H. verrucosa* by having 5 subequal petals, longer stamens (filaments 2–3 cm long) and a smooth woody fruit.

Hymenaea verrucosa Gaertn., Fruct. Sem. Pl. **2**: 306, fig.139/7 (1791). —Langenheim & Lee in Brittonia **26**: 3–21 (1974). —Beentje, Kenya Trees, Shrubs & Lianas: 243 (1994). —M. Coates Palgrave, Trees Sthn. Africa: 319 (2002). Type of uncertain origin. FIGURE 3.2.**8**.

Trachylobium verrucosum (Gaertn.) Oliv. in F.T.A. **2**: 311 (1871). —Harms in Engler, Pflanzenw. Afrikas **3**(1): 450 (1915). —Baker, Legum. Trop. Afr.: 737 (1930). —Brenan, Check-list For. Trees Shrubs Tang. Terr.: 107 (1949); in F.T.E.A., Legum.-Caesalp.: 134 (1967). —K. Coates Palgrave, Trees Sthn. Africa: 269 (1988).

Trachylobium mossambicense Klotzsch in Peters, Naturw. Reise Mossambique **6**(1): 21, fig.2 (1861). —Kirk in J. Linn. Soc. **11**: 1 (1869). —Sim, For. Fl. Port. E. Afr.: 51 (1909) in part, see note below. Type: Mozambique, mainland of Quirimba [Querimba] I., Quissanga, *Peters* s.n. (B† holotype, K).

Trachylobium hornemannianum Hayne in Flora **1827**: 744 (1827); Getreue Darstell. Gew. **11**: fig.18 (1830). —Oliver in F.T.A. **2**: 311 (1871). —Taubert in Engler, Pflanzenw. Ost-Afrikas **B**: 305, 414 (1895). —Gomes e Sousa, Dendrol. Moçamb. **2**: 98–101 (1949); Dendrol. Moçamb. Estudo Geral **1**: 254, fig.56 (1966). Type from Mauritius.

Large tree (or occasionally a shrub) to at least 20 m, with broad crown. Young branchlets pale brown, lenticellate, glabrous. Leaves unijugate; stipules tiny, triangular with ciliate margins, early caducous; petiole 1.2–3 cm long, glabrous; leaflets 5.3–10.5 × 2.8–5.5 cm, ovate-elliptic, asymmetric, the inner margin weakly curved, cuneate at the base, the outer strongly curved and cuneate to subcordate at the base, apex bluntly acuminate, glabrous; midrib distinct, venation prominulous on both surfaces, resin vesicles (dark spots) scattered in the areolae beneath. Inflorescences axillary and terminal, paniculate; axis terete, densely to sparsely appressed-pubescent, glabrescent; bracts and bracteoles c. 2.5 mm in diameter, subcircular, concave, very early caducous. Flowers c. 15 mm long; pedicels 5–7 mm long, terete. Calyx lobes 4, c. 10 × 5–7 mm, broadly elliptic to obovate, densely pubescent outside, densely sericeous inside. Petals 5; the three upper ones white, clawed, the limb c. 9 × 8 mm, suborbicular, the claw c. 10 mm long; the two lower ones minute, c. 1.5 mm long, oblong. Stamens 10; filaments 12–14 mm long, glabrous; anthers dorsifixed. Ovary 4-ovulate, oblong, c. 3 × 2 mm, pilose; stipe very short, densely spreading-hairy at the apex, glabrous below; style c. 6 mm long, glabrous, stigma capitate. Pod c. 5 × 3 cm, ovoid to ellipsoidal, glabrous except for a ring of hairs at the extreme base, with numerous superficial resin-filled vesicles, 1–3-seeded, indehiscent. Seed dark brown, shiny, c. 20 × 12 mm, ellipsoidal, enclosed in the cream-coloured mealy endocarp.

Mozambique. N: between Diaca and Mocímboa da Praia, 20 km from Mocímboa da Praia, fl. 25.iii.1961, *Gomes e Sousa* 4668 (COI, K). Z: Maganja da Costa, Floresta de Gobene, 42 km from Maganja da Costa town, coast road, fl. 14.ii.1966, *Torre & Correia* 14602 (LISC). MS: Muanza, Cheringoma coastal area, Chinizúia R., fr. v.1973, *Tinley* 2877 (K, LISC, SRGH). M: Ponta Vermelha, fl. 20.ii.1961, *Mogg* 29744 (K, LISC, SRGH).

Along the East African coast from Kenya southwards; also Madagascar, Mauritius and the Seychelles. Coastal forests, woodland and thicket, usually on sandy soils; mostly below 100 m.

Conservation notes: On a global scale, Lower Risk, Least Concern. Widespread in a specific and probably somewhat vulnerable habitat in the Flora area.

Mendonça 1471 (BM) from Manica e Sofala, has fruits which are almost smooth although of mature size. They may still be immature.

An important source of resin (East African gum copal, Zanzibar copal). It is possible that the range may have been extended by planting; many herbarium specimens (not cited here) are from planted trees. Probably under-represented in herbaria because of its distinctiveness.

Fig. 3.2.**8**. HYMENAEA VERRUCOSA. 1, flowering branchlet (× 2/3); 2, part of leaflet showing glands (× 4); 3, flower bud, showing bracteoles (× 3); 4, section through hypanthium, showing ovary and three of four sepals (× 2); 5, one of three large upper petals (× 2); 6, one of two minute lower petals (× 2); 7, stamen (× 2); 8, gynoecium, with ovary opened to show ovules (× 3); 9, stigma (× 8), 1–9 from *Trump* 100; 10, pods (× 2/3); 11, seed (× 2/3), 10 & 11 from *Wallace* 832. Drawn by Lura Ripley. From F.T.E.A.

3
7
4
5
6
1
2
10
9
8
11
LMR

Sim (For. Fl. Port. E. Afr.: 51, fig.56 (1909)) somewhat surprisingly confused this taxon with *Balanites maughamii* Sprague. His illustration depicts *Balanites* while the description is a mixture of elements from both *Balanites* and *Trachylobium.*

9. GUIBOURTIA Benn.

Guibourtia Benn. in J. Linn. Soc. **1**: 290 (1857). —Léonard in Bull. Jard. Bot. État **19**: 400–406 (1949), in Bull. Jard. Bot. État **20**: 269–284 (1950).

Trees or shrubs, evergreen, unarmed. Leaves unijugate, elsewhere sometimes unifoliolate and in *G. conjugata* leaflets occasionally fused; leaflets with a main nerve and sometimes several secondary lateral nerves from the base, often with pellucid gland dots; petiolule ± twisted (the distal margin of the leaflet arising higher and in a different plane); stipules small, falling very early. Inflorescence paniculate, of groups of spikes. Flowers spirally arranged on the inflorescence axis; bracts small, caducous or subpersistent; bracteoles 2, small, caducous or subpersistent, not enclosing the bud. Hypanthium absent. Calyx lobes 4, unequal (2 outer, 2 inner), imbricate. Petals absent. Stamens usually 10; filaments alternately long and short in the bud; anthers dorsifixed, dehiscing by longitudinal slits. Ovary stipitate or sessile; ovules usually 2; style elongate, filiform; stigma terminal, capitate. Pods suborbicular or obliquely elliptic, either coriaceous and tardily dehiscent, or chartaceous and indehiscent. Seeds 1(2) without an areole, with a fleshy aril enveloping the seed, or without.

A tropical genus with a few species in South America and about 14 in Africa. Generic limits between *Copaifera* and *Guibourtia* are under study and may change.

1. Leaves with a distinct midrib substantially larger than the other veins; fruits thick-walled, tardily dehiscent; seed ellipsoid, completely covered by an aril . . . . **1.** *coleosperma*
– Leaves with 3–6 subequal nerves from the base or close to it; fruits (where known) thin-walled, indehiscent; seed elliptic, flattened, without an aril 2
2. Flower-buds ellipsoidal, minutely puberulous, grey when dry, resin glands present on calyx but not very prominent; at least some brown inflorescence bracts and bracteoles present at the mature bud stage; fruits unknown . . . . **2.** *sousae*
– Flower buds spherical, glabrous, blackish when dry, resin glands prominent on outside of calyx lobes; inflorescence bracts and bracteoles falling very early and absent at the mature bud stage 3
3. Leaflets rounded at apex; ovary glabrous (stipe pilose) . . . . **3.** *conjugata*
– Leaflets acuminate at apex; ovary pilose at least on the lower suture . . . . **4.** *schliebenii*

1. **Guibourtia coleosperma** (Benth.) J. Léonard in Bull. Jard. Bot. État **19**: 402 (1949). —Miller in J. S. Afr. Bot. **18**: 32 (1952). —Pardy in Rhod. Agric. J. **49**: 172 (1952). —Léonard in F.C.B. **3**: 364 (1952). —Torre & Hillcoat in C.F.A. **2**: 242 (1956). —White, F.F.N.R.: 124, fig.21 (1962). —Schreiber in Merxmüller, Prodr. Fl. SW Afrika, part 59: 14 (1967). —Drummond in Kirkia **8**: 219 (1972); in **10**: 246 (1977). —Palmer & Pitman, Trees Sthn. Africa **2**: 841 (1972). —Ross in F.S.A. **16**(2): 22, fig.4 (1977). —K. Coates Palgrave, Trees Sthn. Africa, ed.2: 267 (1988). —M. Coates Palgrave, Trees Sthn. Africa: 318 (2002). Type: Zambia, Batoka Highlands, ix.1860, *Kirk* s.n. (K holotype). FIGURE 3.2.**9**/1–5.

Copaifera coleosperma Benth. in Trans. Linn. Soc. **25**: 316, fig.43A (1865). —Oliver in F.T.A. **2**: 314 (1871). —Burkill in Append. II to Johnson Brit. Centr. Afr.: 245 (1897). —Sim, For. Fl. Port. E. Afr.: 52 (1909). —Harms in Engler, Pflanzenw. Afrikas **3**(1): 441, fig.243 (1915). —Eyles in Trans. Roy. Soc. S. Afr. **5**: 364 (1916). —Baker, Legum. Trop. Afr.:

Fig. 3.2.**9**. GUIBOURTIA COLEOSPERMA. 1, flowering branchlet (× $^{2}/_{3}$); 2, flower (× 4); 3, gynoecium (× 4), 1–3 from *De Winter* 3809; 4, fruiting branchlet (× $^{2}/_{3}$); 5, dehisced pod showing attached seed (× 1), 4 & 5 from *Codd* 7069. GUIBOURTIA CONJUGATA. 6, leaf (× $^{2}/_{3}$), from *Van der Schijff* 3338. Drawn by Pat Halliday. Reproduced with permission from Flora of Southern Africa.

752 (1930). —Miller, Check-list For. Trees Shrubs Bech. Prot.: 24 (1948). —Gomes e Sousa, Dendrol. Moçamb. **1**: 132–136 (1949).

Tree to 30 m tall; bark grey, smooth, sometimes flaking. Branchlets brown, smooth, becoming grey with fine longitudinal cracks and red-brown lenticels, or entirely red-brown, glabrous. Leaves paripinnate, unijugate; stipules 15 × 2 mm, very narrowly ovate, very early caducous; petiole 1.5–3 cm long, glabrous; leaflets 3.5–8 × 2.3–4.5 cm, ovate to elliptic, asymmetric, apex acuminate, base cuneate, glabrous; a single main vein from the base, dividing the blade in a ratio of 2–3:1, venation slightly prominent on both surfaces, areolae with small resin vesicles. Inflorescences axillary and terminal, paniculate; axis angular, glabrous to very sparsely puberulous; bracts and bracteoles very early caducous, the bracts c. 2 mm long, broadly ovate, pubescent, the bracteoles c. 1.5 mm long, broadly ovate, pubescent. Flowers 7–8 mm long; pedicels 2–5 mm long, angled. Sepals 4(5), unequal, c. 6 × 2–4 mm, ovate, glabrous outside, densely puberulous inside. Petals absent. Stamens 10; filaments c. 7 mm long, glabrous; anthers dorsifixed. Ovary 2-ovulate, c. 1 mm diameter, subcircular, glabrous; stipe very short, densely pubescent; style c. 5 mm long, glabrous; stigma capitate, bilobed. Pod 2.5–3.5 × 1.6–2.2 cm, asymmetrically elliptic, flattened, glabrous, the wall wrinkled and with numerous resin-filled vesicles, 1-seeded, tardily dehiscent. Seed dark brown, shiny, c. 15 × 10 mm, broadly ellipsoidal, somewhat flattened; aril completely enclosing the seed, red when fresh, pale brown and brittle when dry. Germination epigeal; cotyledons suborbicular to very broadly ovate, slightly asymmetric, 3–4-nerved from the base, glabrous. First true leaves similar to foliage leaves.

Caprivi Strip. Katima Mulilo, fl. 28.ii.1982, *M. Müller* 1852 (K, WIND). **Botswana.** N: Ngamiland Dist., Shishikola Pan, fl. 1.i.1973, *P.A. Smith* 325 (K, SRGH). **Zambia.** B: Mongu Dist., 63 miles from Mongu on road to Kaoma, fl. & fr. 31.i.1975, *Brummitt, Chisumpa & Polhill* 14179 (K, LISC). W: Mwinilunga Dist., Kalene Hill Mission, fl. 28.ix.1952, *Holmes* 919 (FHO). C: Kabwe Dist., Luwonde Forest, st. viii.1934, *Trapnell* 1547 (K). S: Livingstone, fl. 25.xi.1949, *Wild* 3207 (K, LISC, SRGH). **Zimbabwe.** W: Nyamandhlovu Dist., Gwayi (Gwaai), Umphindo (Mpindo), Gusu forest, fl. ii.1949, *Davies* 239 (K, LISC, SRGH).

Also in Namibia, Angola and southern Congo. Forest and woodland on Kalahari sands, sometimes planted; 750–1400 m.

Conservation notes: Widely distributed with no apparent threats; Lower Risk, Least Concern.

The aril is edible and is used in times of famine ('The aril which is red is made into a drink said to be very nourishing and by this in times of hunger the people have been saved' — J. Kirk: note on type sheet). The seeds are also cooked, pounded and eaten.

2. **Guibourtia sousae** J. Léonard in Bull. Jard. Bot. État **20**: 270 (1950). Type: Mozambique, Mawayela (Maueele), *Gomes e Sousa* 1927 (K holotype, COI).

Tree 12–15 m tall. Branchlets grey-brown, later grey, smooth with fine longitudinal cracks and lenticels, glabrous. Leaves paripinnate; stipules not seen, probably early caducous; petiole 12–15 mm long, glabrous; leaflets 1-jugate, 3.8–4.5 × 2.5–3 cm, broadly elliptic to broadly obovate, asymmetric, rounded or narrowed to the retuse apex, base cuneate, glabrous; venation slightly prominent on both surfaces, midrib dividing the blade in a ratio of c. 2:1, areolae often with small raised dots (resin vesicles?). Inflorescences terminal, paniculate; axis puberulous; bracts apparently early caducous; bracteoles c. 2.5 × 2 mm, broadly ovate, cordate at base, puberulous, persisting at least to late bud stage. Flower buds only seen, 4–4.5 mm long, ellipsoidal, spirally arranged on the axis; pedicel c. 0.2 mm long. Calyx lobes 4, c. 4 × 2.5 mm, elliptic, tightly imbricate, minutely puberulous and with numerous resin vesicles outside, puberulous inside. Petals absent. Stamens 10; anthers dorsifixed. Ovary c. 2 mm diameter, subcircular, 2-ovulate; stipe very short; style c. 3 mm long, curled in bud; stigma capitate, bilobed. Pod and seeds not known.

Mozambique. GI: Panda Dist., Mawayela (Maueele), fl. buds, xii.1936, *Gomes e Sousa* 1927 (COI, K).

Sandy soil, with *Androstachys johnsonii* (Euphorbiaceae); 150 m.

Conservation notes: Endemic to the Flora area and known only from the type. A formal category is inappropriate, or Data Deficient.

Only known from the type, which has apparently mature buds. These are very different in shape from those of *G. conjugata* and larger, resembling those of the Angolan taxon *G. carrissoana* (M.A. Exell) J. Léonard var. *gossweileri* (M.A. Exell) J. Léonard. More material would be most welcome. In the absence of buds, *G. sousae* and *G. conjugata* are difficult to distinguish. The presence of numerous resin-vesicles in the areolae of the reticulation in *G. sousae* may be diagnostic, but it is difficult to generalise from a single specimen. None of the flowering specimens of *G. conjugata* at Kew shows this feature, and the sterile and fruiting material has therefore also been referred to *G. conjugata.* The collector recorded the flowers as white, and the wood as black or reddish and good for carpentry. *Gomes e Sousa* 1928 (COI, K, LISC) and 1929 (COI), collected on the same date and in the same site, are *G. conjugata.*

Gomes e Sousa, (Dendrol. Moçamb. Estudo Geral **1**: 251, 1966), mentions this species. He cites *Copaifera coleosperma* Benth. (*Guibourtia coleosperma*) as a synonym. His illustration appears to represent this taxon, but the description seems to apply mainly to *G. sousae.*

3. **Guibourtia conjugata** (Bolle) J. Léonard in Bull. Jard. Bot. État **19**: 402 (1949). —Pardy in Rhod. Agric. J. **51**: 111 (1954). —White, F.F.N.R.: 124 (1962). —Gomes e Sousa, Dendrol. Moçamb. Estudo Geral **1**: 253, fig.55 (1966). —Drummond in Kirkia **10**: 246 (1977). —Palmer & Pitman, Trees Sthn. Africa **2**: 839 (1972). —Ross in F.S.A. **16**(2): 20, fig.4/6 (1977). —Van Wyk, Trees Kruger Nat. Park: 84 (1994). —M. Coates Palgrave, Trees Sthn. Africa: 319 (2002). Type: Mozambique, near Sena and Tete, *Peters* (B holotype† (sketch of B material at BM), BR, K). FIGURE 3.2.**9**/6.

Gorskia conjugata Bolle in Peters, Naturw. Reise Mossambique **6**(1): 16, fig.3 (1861).

Copaifera gorskiana Benth. in Trans. Linn. Soc. **25**: 317 (1865). —Oliver in F.T.A. **2**: 315 (1871). —Harms in Engler, Pflanzenw. Afrikas **3**(1): 441 (1915). Sim, For. Fl. Port. E. Afr.: 51 (1909) as *Copiafera.* —Baker, Legum. Trop. Afr.: 751 (1930) nom. superfl. Type as above.

Copaiba conjugata (Bolle) Kuntze, Rev. Gen. Pl.: 172 (1891). —Taubert in Engler, Pflanzenw. Ost-Afrikas **C**: 197 (1895).

Copaifera conjugata (Bolle) Milne-Redh. in Bull. Misc. Inform., Kew **1934**: 400 (1934). —Gomes e Sousa, Dendrol. Moçamb. **1**: 136 (1948).

Tree 6–18 m tall, or sometimes multi-stemmed, to 4.5 m; bark grey, smooth, lightly longitudinally fissured and sometimes peeling in rectangular pieces. Branchlets grey, smooth, becoming grey with fine longitudinal cracks and red-brown lenticels, or entirely red-brown, sparsely pubescent, soon glabrescent. Leaves paripinnate, unijugate; stipules 3–4 × 2–2.5 mm, ovate to elliptic, early caducous; petiole 0.9–2.2 cm long, glabrous; leaflets 3.2–7(9) × 2–4.5(6.5) cm, ovate, asymmetric, apex rounded to acute, base rounded, glabrous; one major and 3–4 veins arising from the base, venation prominulous on both surfaces, resin vesicles in areolae few or none. Occasionally leaflets are fused to give a unifoliolate very broadly ovate leaf with a cordate base (e.g. *Barbosa & Balsinhas* 5101). Inflorescences axillary and terminal, paniculate; axis terete, pubescent with a mixture of longer and shorter hairs; bracts and bracteoles very early caducous (not seen). Flowers 6–7 mm long; pedicels 1.5–2 mm long, terete. Calyx lobes 4, c. 6 × 2–4 mm, elliptic, glabrous and with numerous resin vesicles outside, puberulous inside. Petals absent. Stamens 10; filaments c. 6 mm long, glabrous; anthers dorsifixed. Ovary 2-ovulate, c. 1 mm diameter, subcircular, glabrous; stipe very short, with a few spreading hairs; style c. 3 mm long, glabrous; stigma capitate, bilobed. Pod 3.5 × 2.5 cm, broadly elliptic-oblong, slightly asymmetric, flattened, glabrous, the wall chartaceous, with numerous small resin-dots, 1-seeded, indehiscent, with a small very narrowly ovate dorsal wing adjacent to the style remnant. Seed dark brown, shiny, c. 22 × 18 mm, broadly elliptic, flattened, exarillate. Germination epigeal; cotyledons suborbicular to depressed-elliptic, slightly asymmetric, 4–6-nerved from the base, hypocotyl and petiolules pubescent. First true leaves similar to foliage leaves.

Zambia. S: Gwembe Valley, fr. iii.1934, *Trapnell* 1472 (K). **Zimbabwe.** N: Binga, fl. 24.xi.1964, *Jarman* 315 (K, SRGH). E: Save/Runde (Sabi/Lundi) Junction, fr. 17.vii.1941, *Chorley* in *GHS* 8077 (K). S: Mwenezi Dist., Mateke Hills, north side, fl. 5.xii.1974, *Cleghorn* 3012 (K, SRGH). **Mozambique.** T: Zambezi Valley, near Tete, fr. 8.vi.1947, *Hornby* 2728 (K, SRGH). MS: Chibabava Dist., Madanda Forest, fr. ix.1911, *Dawe* 410 (K). GI: Aldeia da Barragem, road to Pafuri, near Missão Chirundzo (Xirrunso), fl. 16.xi.1957, *Barbosa & Lemos* 8152 (COI, LISC, K). M: Magude Dist., near Mapulanguene, fl. 1.xii.1944, *Mendonça* 3197 (BM, LISC, K).

Also in South Africa (Limpopo and Mpumalanga Provinces). Dry forest and woodland, usually on sandy soils; sea level to 1500 m.

Conservation notes: Widespread in the Flora area and also outside it; Lower Risk, Least Concern.

There appears to be some disjunction between the coastal populations and those of Zambia and Zimbabwe, and specimens from those countries tend to have leaflets with acute rather than rounded apices, but both forms can be found on the same specimen. Some have even been named as *G. schliebenii*, but that species differs in its ovary which is pilose at least on the lower suture.

4. **Guibourtia schliebenii** (Harms) J. Léonard in Bull. Jard. Bot. État **19**: 404 (1949). —Torre in Mendonça, Contrib. Conhec. Fl. Moçamb. **2**: 85 (1954). —Brenan in F.T.E.A., Legum.-Caesalp.: 136, fig.27 (1967). Type: Tanzania, Lindi Dist., L. Lutamba, *Schlieben* 6123 (B† holotype, BM, BR, EA, K).

Copaifera schliebenii Harms in Notizbl. Bot. Gart. Berlin-Dahlem **13**: 415 (1936). — Brenan, Check-list For. Trees Shrubs Tang. Terr.: 99 (1949).

Tree 6–8 m (to 20 m in Tanzania) tall. Branchlets pale brown, smooth, soon becoming grey, glabrous. Leaves paripinnate, unijugate; stipules not seen, apparently early caducous; petiole 0.3–0.9 cm long, sparsely puberulous; leaflets 2–3.8 × 1.2–2.1 cm, ovate, asymmetric, the inner margin almost straight, the outer curved and subcordate at the base, apex acuminate, glabrous; midrib distinct, with 2–3 lateral veins arising close to the base, venation slightly prominent on both surfaces, resin vesicles in areolae few or none. Inflorescences axillary and terminal, paniculate; axis terete, pubescent with a mixture of longer and shorter hairs; bracts and bracteoles very early caducous (not seen). Flowers 4–5 mm long; pedicels 0.4–0.8 mm long, terete. Calyx lobes 4, c. 3 × 1–2 mm, elliptic, glabrous and with numerous resin vesicles outside, puberulous inside. Petals absent. Stamens 10; filaments c. 4 mm long, glabrous; anthers dorsifixed. Ovary c. 1 mm diameter, subcircular, 2-ovulate, pilose mainly on the lower suture; stipe very short, with a few spreading hairs; style c. 4 mm long, glabrous; stigma capitate, bilobed. Pod 3 × 2.1 cm, broadly obovate, slightly asymmetric, flattened, glabrous, the wall chartaceous, reticulation prominent, 1-seeded, indehiscent, with a small semilunate dorsal wing adjacent to the style remnant. Seed dark brown, shiny, c. 15 × 12 mm, broadly elliptic to suborbicular, flattened, exarillate.

Mozambique. N: hills at mouth of Messalo (Msalu) R., fl. ii.1912, *Allen* 112 (K). GI: between road crossing of Vilanculos–Nova Mambone and Mabote, 102.3 km from crossing, fr. (picked from ground) 28.iii.1952, *Barbosa & Balsinhas* 5062 (K, LISC).

Also in Tanzania. Coastal forest and woodland; probably below 100 m.

Conservation notes: Apparently very restricted in its distribution, but inadequately known. Probably Endangered (B1), although further research and collection needed.

Brenan, in a note on the Kew sheet, pointed out that *Barbosa & Balsinhas* 5062 has longer petioles and usually more basal nerves than the material of *G. schliebenii* from Cabo Delgado and Tanzania, 1200 km to the north, and that the form of the junction of the leaflets with the petiole is somewhat different. I treat this as *G. schliebenii* here, but when complete material is available a rethink may be needed. There are fewer than 10 collections of *G. schliebenii*, which cannot be regarded as well known. Material from Tanzania, provisionally attributed to this species by Brenan (1967) has leaflets up to 5.8 cm long, but no similar material has been seen from the Flora area.

10. BAIKIAEA Benth.

Baikiaea Benth. in Bentham & J.D. Hooker, Gen. Pl. **1**: 581 (1865).

Trees, evergreen or deciduous. Leaves paripinnate or with alternate leaflets (not in our area); leaflets with well-developed petiolules, with a marginal nerve, usually with one or two marked swellings (?domatia) on the margin near the base. Flowers sometimes (not in the F.Z. area) very large (up to 25 cm long), in terminal or axillary racemes, the axes usually brown-tomentellous with flowers arranged distichously; bracts caducous, often leaving a prominent scar; bracteoles not enclosing flower bud, caducous. Sepals 4, very narrowly imbricate, brown-tomentellous outside and velutinous-subsericeous inside, inserted on the rim of an infilled ± campanulate hypanthium. Petals 5, four of them equal, the fifth narrower and differently coloured, obovate-spathulate, crinkled at the margins, imbricate, villous along and near the midrib. Stamens 10, one of them free and the other nine united towards the base into a short sheath, glabrous or villous below; anthers dorsifixed, dehiscing longitudinally. Ovary shortly stipitate, tomentose; ovules 1–many; style glabrous; stigma ± peltate. Fruit woody, compressed, dehiscing into two valves, oblong-oblanceolate to suborbicular. Seeds large, compressed, without an areole.

A genus of 5 or 6 species extending from Nigeria and Uganda southwards to the Flora area. The one species in the Flora area is exceptional in having relatively small flowers and regularly paripinnate leaves with small leaflets; the others have flowers 6–25 cm long and alternate (or occasionally the basal ones opposite) large coriaceous leaflets and are mostly evergreen species of rain forests.

Baikiaea plurijuga Harms in Warburg, Kunene-Sambesi Exped. Baum: 248 (1903). —Eyles in Trans. Roy. Soc. S. Afr. **5**: 365 (1916). —Baker, Legum. Trop. Afr.: 705 (1930). —Pardy in Rhod. Agric. J. **48**: 402 (1951). —Miller, Check-list For. Trees Shrubs Bech. Prot.: 23 (1948); in J. S. Afr. Bot. **18**: 28 (1952). —Torre & Hillcoat in C.F.A. **2**: 220, fig.45 (1956). —Léonard in Mém. Acad. Roy. Sci. Belg. **30**(2): 74 (1957). —O. Coates Palgrave, Trees Central Africa **1**: 66–69 (1957). —White, F.F.N.R.: 98 (1962). —Breitenbach, Indig. Trees Sthn. Africa **3**: 336 (1965). —Schreiber in Merxmüller, Prodr. Fl. SW Afrika, fam. 59: 5 (1967). —Drummond in Kirkia **10**: 243 (1975). —Palmer & Pitman, Trees Sthn. Africa **2**: 861 (1973). —Ross in F.S.A. **16**(2): 37, fig.7 (1977). —F. & J.A. Venter, Indig. Trees: 116 (1996). —M. Coates Palgrave, Trees Sthn. Africa: 329 (2002). Type from Angola. FIGURE 3.2.**10**.

Round-topped tree usually 10–17 m high, occasionally reaching 25 m; young bark smooth and grey but becoming rough and fissured when old; deciduous. Young branches pubescent to tomentose with appressed or ascending rusty-brown hairs, but quickly glabrescent to leave a smooth grey bark. Leaves paripinnate: stipules 5–9 × 1–3 mm, obliquely linear-triangular, pubescent to tomentose, caducous; petiole and rachis together (4)6–11(13) cm long, of which the petiole is (0.8)1.5–2.5(3) cm, pubescent to tomentose with rusty brown hairs; petiolules 1–3 mm long; leaflets in (3)4–5(very rarely 6) pairs, strictly opposite or rarely with up to 1 mm separating members of a pair, each (3)4–7(9) × (1)1.5–2.5(4) cm, variable in shape from lanceolate (or smaller basal leaflets even ovate) to elliptic or elliptic-oblong or occasionally (particularly uppermost leaflets) oblong-lanceolate, rounded at the base (sometimes slightly asymmetric) or the uppermost pair markedly asymmetric with rounded proximal side and cuneate distal side, usually rounded at the apex (sometimes broadly so) and either apiculate or distinctly emarginate, but occasionally acute, pubescent with brownish hairs on both surfaces when young but quickly glabrescent except on the midrib beneath, sometimes with conspicuous pellucid gland-dots or dashes between the tertiary veins, often with swellings (?domatia) on the leaflet margin almost at the base on the proximal side and a few mm above the base on the distal side; lateral nerves extending to the marginal nerve. Racemes 4–35 cm long, ± erect, brown-pubescent to -tomentose, with up to 40 flowers regularly alternately arranged on opposite sides, the rachis consequently somewhat zigzag, most flowers (except usually one or two which produce fruit) falling with pedicel to leave a prominent ledge (bract scar) at each node of the otherwise denuded rachis; bracts c. 4 × 5 mm, ovate to semicircular, caducous; pedicels c. 9–13 mm, with prominent bracteole scars c. 1–3 mm. from the base; bracteoles 4–5 × 3–5 mm, broadly ovate to broadly elliptic, usually rounded at apex, quickly caducous. Calyx 13–20 mm long, divided to the base into 4 lobes of which the upper (adaxial) is 7–10 mm broad, the lateral one 4–7 mm broad, and the lower (abaxial) one 3–5 mm broad and slightly the longest through curvature, all brown-pubescent to -tomentose outside except on the overlapped margins 0.5 mm broad, and pale brown-tomentose inside. Petals 5, 2.3–3.0(3.5) cm long, obovate-spathulate, pale mauve to bluish-purple, undulate at lower margins, villous towards midrib especially near base. Stamens 10, about equalling the petals, one free, nine lightly fused towards the base, the lower part villous and the free part glabrous; anthers (3.5)4–5 mm long, versatile. Ovary 12–15 × 4–5 mm, obliquely oblanceolate, brown-tomentose; style slightly exceeding petals and stamens; stigma capitate. Pods 8–14 × 3.5–4.7(5) cm, oblanceolate with oblique apex, woody, brown-tomentose, dehiscing into two spirally twisted valves. Seeds dark reddish brown, 2–2.5 × 1.5–1.8 × 3 mm, elliptic-orbicular, compressed.

Caprivi Strip. Katima Mulilo, *Killick & Leistner* 3050 (PRE). **Botswana.** N: Chobe Dist., E flank of southernmost Goha Hills, fr. 19.v.1977, *P.A. Smith* 2064 (K, SRGH). **Zambia.** B: Sesheke Dist., Masese, fl. 10.iii.1973, *Chisumpa* 9 (K, NDO). S: Livingstone, fr. i.1910, *Rogers* 7256 (K). **Zimbabwe.** N: Gokwe South Dist., upper Lutope R., fr. 16.viii.1951, *Whellan* 538 (K, LISC, SRGH). W: Lupane Dist., Gwayi (Gwaai) Forest Land, fl. 20.iv.1972, *Grosvenor* 727 (K, LISC, SRGH). C: Lower Gweru (Gwelo), Stephenson Block, fl. 17.ii.1951, *Harvie* 1/51 (K, LISC, SRGH).

Also in northern Namibia and southern Angola. Confined to Kalahari sands, where it is often abundant and dominant, often associated with *Guibourtia coleosperma* and *Schinziophyton rautanenii*; c. 900–1200 m.

Conservation notes: widespread and usually abundant where it occurs; not threatened.

B. plurijuga, known as Zambezi Redwood, Rhodesian Teak or African Teak, is an important timber tree producing a hard dark reddish wood extensively used for railway sleepers, mine supports, parquet floors and furniture.

11. TESSMANNIA Harms

Tessmannia Harms in Bot. Jahrb. Syst. **45**: 295, fig.2 (1910). —Léonard in Bull. Jard. Bot. État **19**: 384 (1949).

Trees, evergreen, unarmed. Leaves simply pinnate, ending in one leaflet or a pair; leaflets alternate, with pellucid gland dots (not always obvious in mature dried leaves); lateral nerves looped into a submarginal nerve; petiolules short, twisted; stipules intrapetiolar, often early

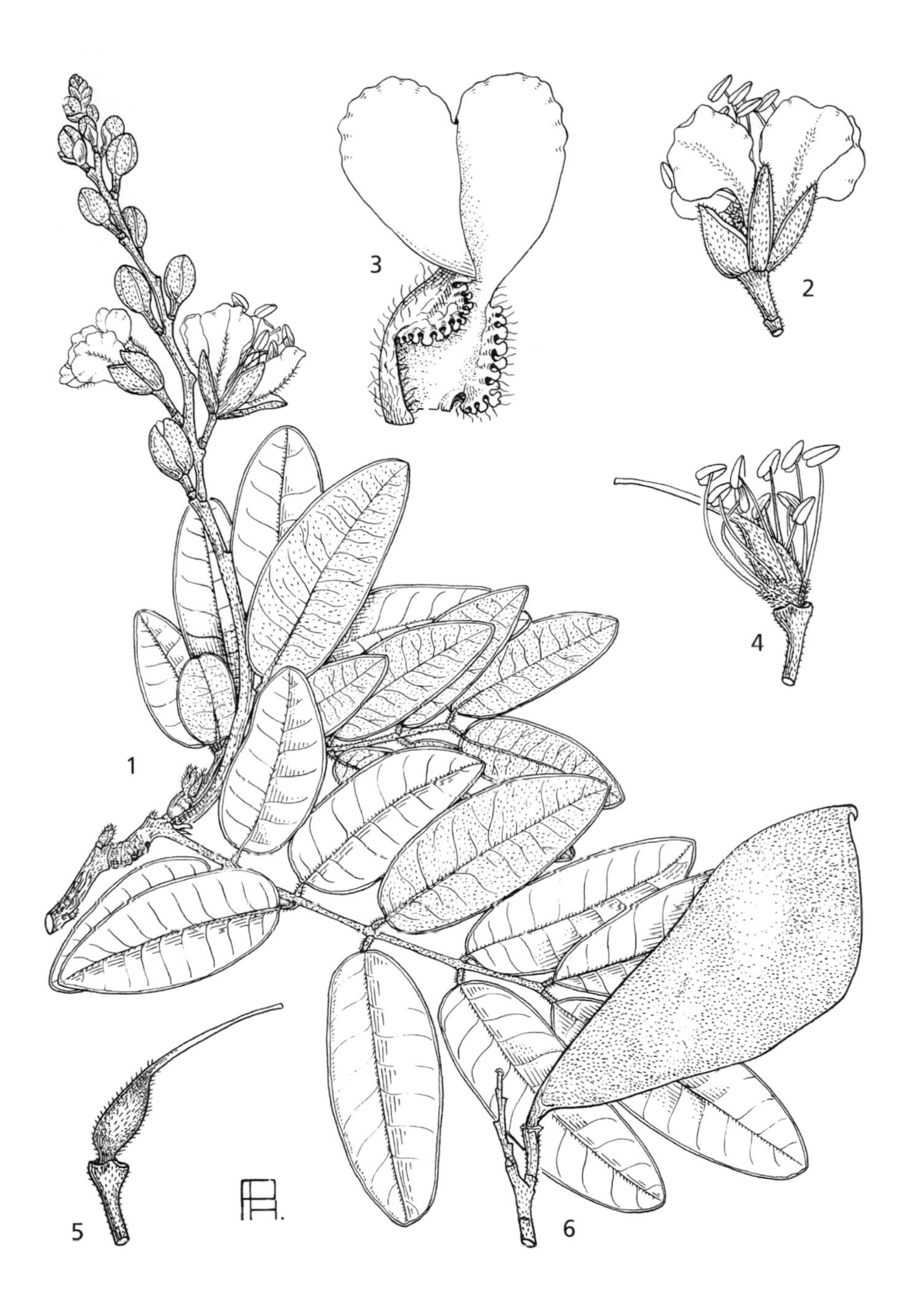

Fig. 3.2.**10**. BAIKIAEA PLURIJUGA. 1, flowering branch (× $^2/_3$), from *De Winter & Giess* 7026; 2, flower (× 1); 3, standard (× 2); 4, flower with sepals and petals removed (× 1); 5, gynoecium (× 1), 2–5 from *De Winter* 3816; 6, pod (× $^2/_3$), from *De Winter* 3762. Drawn by Pat Halliday. Reproduced with permission from Flora of Southern Africa.

caducous. Inflorescences of terminal or axillary racemes or panicles. Flowers distichous on the inflorescence axis in racemes or panicles; bracts and bracteoles small, early caducous. Hypanthium short. Sepals 4, narrowly imbricate, fulvous-sericeous inside. Petals 5, subequal, imbricate, clawed, lamina crinkled. Stamens usually 10 but sometimes fewer; filaments tomentose below, one free, the rest connate below into a short tube; anthers dorsifixed, dehiscing by longitudinal slits. Ovary stipitate, at least the stipe tomentose; ovules several; style elongate, coiled when young; stigma capitate. Pods flattened, woody, short, apparently indehiscent at least in the Zambian species, smooth or warty outside. Seeds 1–4, without areole but with a small basal aril.

About 12 species, mainly in the Guineo-Congolian forests of Africa.

Tessmannia burttii Harms in Repert. Spec. Nov. Regni Veg. **43**: 110 (1938). —White, F.F.N.R.: 128, fig.21H (1962). —Brenan in F.T.E.A., Legum.-Caesalp.: 107 (1967). Type: Zambia, Mbala (Abercorn), *B.D. Burtt* 6004 (K holotype; BM, BR, EA). FIGURE 3.2.**11**.

Tree to 13 m; bark grey or dark brown, smooth, cracking vertically. Branchlets brown, soon grey, striate, glabrous. Leaves pinnate; stipules not seen, probably early caducous; petiole 1–4 cm long, glabrous; rachis 2.5–7 cm long, glabrous; leaflets alternate, (3)4–7, 3.5–7.5 × 1.2–2.5 cm, ovate to rhombic to elliptic, often slightly asymmetric, base somewhat asymmetrically cuneate, apex acute to acuminate, the extreme tip rounded or retuse, glabrous; venation prominulous beneath, less so above, lateral nerves joined into a submarginal nerve. Inflorescences axillary corymbose racemes, sometimes paired, the axis rufous-pubescent; bracts and bracteoles not seen, apparently early caducous. Flowers 2.5–3 cm long; pedicels up to 3 cm long, rufous-pubescent. Calyx lobes 4, 10–12 mm long, one c. 8 mm wide, two c. 5 mm wide, one c. 2.5 mm wide, elliptic to obovate, rufous-pubescent outside, densely sericeous inside except on the overlapping margins. Petals 5, pink, c. 5 mm wide and 15–16 mm long (25–30 mm according to Harms) including a claw 5 mm long, elliptic, the dorsal petal with a thickened and grooved claw, glabrous, the rest pubescent on the midrib inside. Stamens 7–10, c. 25 mm long, the one opposite the dorsal petal free and almost glabrous, the rest fused and densely pubescent in the basal 5 mm; anthers dorsifixed. Ovary c. 10-ovulate, c. 7 × 3 mm, obovate, densely rufous-pubescent particularly on the margins, verrucose in the centre, stipe c. 4 mm long, rufous-pubescent; style c. 15 mm long, pubescent in the lower third; stigma capitate, bilobed. Pod up to 5 × 2.5 cm, asymmetrically obovate to rhombic, rufous-pubescent and with numerous resiniferous warts, usually 1-seeded. Seed c. 12 × 10 mm, ovoid, somewhat flattened.

Zambia. N: Mbala Dist., Kapata Village, Lunzua Gorge just above the falls, L. Tanganyika coast, fl. 20.v.1936, *B.D. Burtt* 6004 (BR, EA. K). W: Mwinilunga Dist., Zambezi R., c. 6 km NW of Kalene Hill Mission, fr. 20.xi.1952, *Holmes* 895 (FHO, K).

Also in Tanzania and Congo. In riverine woodland and termite mound thickets; 850–1400 m.

Conservation notes: Known from a few scattered localities in riverine woodland; Vulnerable (B1) globally.

The plant must be spectacular when flowering and it is surprising that there are not more collections (only 5 are known); it seems likely that it is truly rare, and that the flowers do not last long. The leaflets of the flowering specimen are smaller than those of the fruiting ones; they may be immature. The distinctive venation pattern is, however, similar in both. Harms's larger petal measurement may reflect the age of the flower he dissected.

I cannot distinguish the 'pellucid gland dots', said to be characteristic of the genus, on the leaves of this species. Small sessile glands can sometimes be seen; the pellucid glands may be more obvious in fresh material.

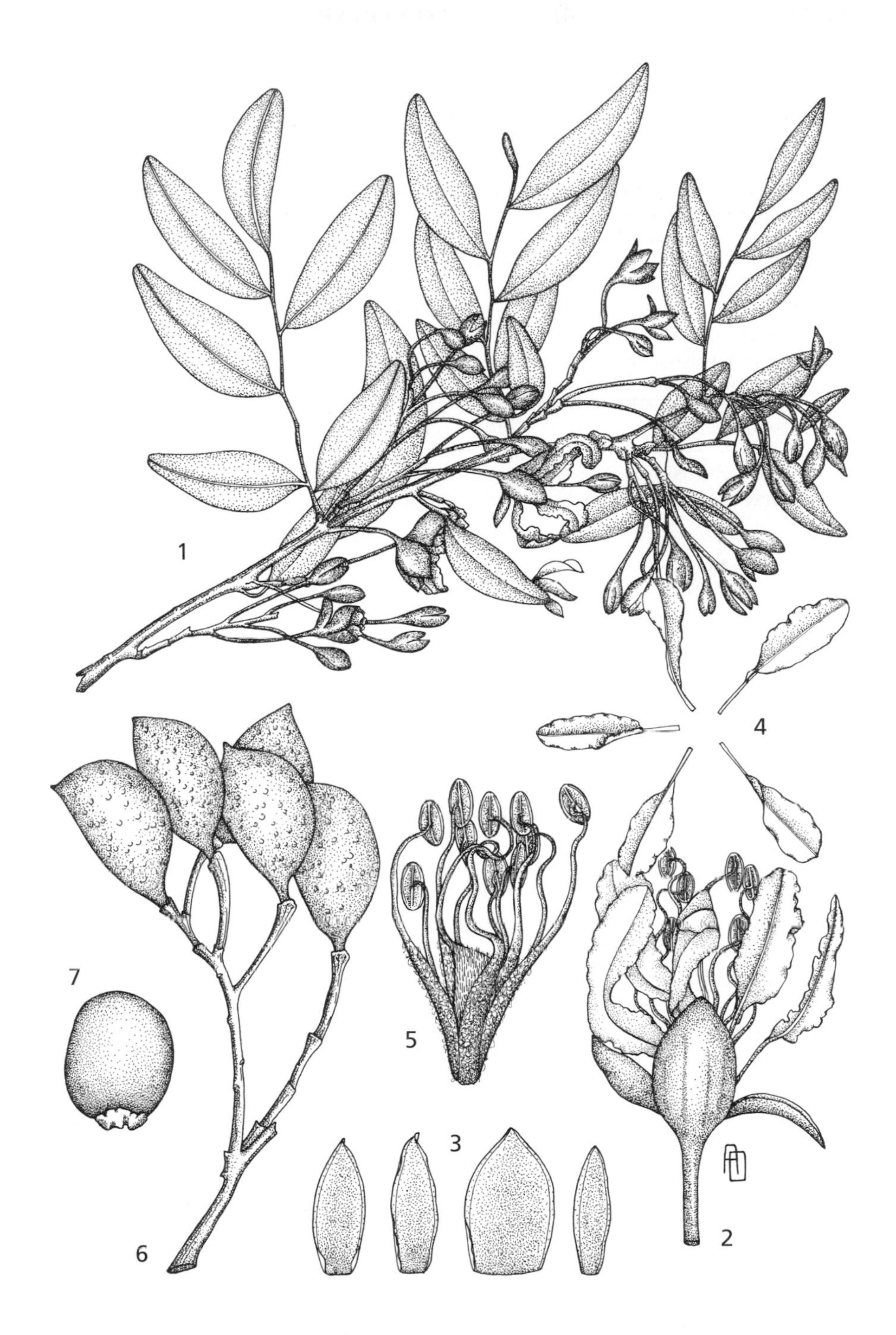

Fig. 3.2.**11**. TESSMANNIA BURTTII. 1, flowering branch (× $^1/_2$); 2, flower (× $1^1/_2$); 3, sepals (× $1^1/_2$); 4, petals (× 1); 5, stamens and gynoecium (× $2^1/_2$), 1–5 from *B.D. Burtt* 6004; 6, pods (× $^1/_2$); 7, seed (× $1^1/_2$), 6 & 7 from *Fanshawe* 3586. Drawn by Ann Davies.

12. COPAIFERA L.

Copaifera L., Sp. Pl., ed.2: 557 (1762). —Léonard in Bull. Jard. Bot. État **19**: 391–398 (1949).

Unarmed evergreen trees or occasionally shrubs or subshrubs. Leaves simply pinnate; leaflets opposite or alternate, shortly to conspicuously petiolulate, with a conspicuous marginal nerve, often with pellucid gland dots; stipules small, falling very early. Inflorescences of groups of spikes, or panicles, condensed when young; flowers distichous on the inflorescence axis; bracts small, caducous; bracteoles 2, small, caducous, not enclosing the bud. Hypanthium absent. Calyx lobes 4, almost valvate but with the margins slightly imbricate. Petals absent. Stamens usually 10 but sometimes 8; filaments alternately long and short in the bud; anthers dorsifixed, dehiscing by longitudinal slits. Ovary stipitate or sessile; ovules usually 2; style elongated, filiform; stigma terminal. Pods suborbicular or obliquely elliptic, generally coriaceous and very tardily dehiscent. Seeds 1(2), without areole, usually with a variously developed fleshy aril.

A tropical genus with about 25 species in South America and 5 in Africa. Generic limits between *Copaifera* and *Guibourtia* are under study and may change.

Copaifera baumiana Harms in Warburg, Kunene-Sambesi Exped. Baum: 246, fig.1A–C (1903). —Harms in Engler, Pflanzenw. Afrikas **3**(1): 446 (1915). —Baker, Legum. Trop. Afr.: 751 (1930). —Léonard in F.C.B. **3**: 305 (1952). —Torre & Hillcoat in C.F.A. **2**: 243 (1956). —White, F.F.N.R.: 121 (1962). —Schreiber in Merxmüller, Prodr. Fl. SW Afrika, fam. 59: 18 (1967). Type from Angola. FIGURE 3.2.**12**.

Shrub, or subshrub from a woody rootstock, 1–3(4) m tall. Branchlets brown, smooth with fine longitudinal cracks, very sparsely puberulous, glabrescent. Leaves imparipinnate; stipules 7–9 × 1 mm, very narrowly ovate, caducous; petiole 0.5–1 cm long, sparsely pubescent; rachis 6–9.5 cm long, sparsely pubescent; leaflets subopposite to alternate, in 4–5 pairs, 2.7–5.2 × 0.9–1.6 cm, elliptic to rhombic-elliptic, often asymmetric, apex rounded or retuse, base cuneate, glabrous; venation slightly prominent on both surfaces, marginal vein prominent, areolae often with dark patches (?resin) in older leaves. Inflorescences axillary and terminal, paniculate; axis sparsely puberulous to pubescent; bracts c. 2 mm in diameter, suborbicular, concave, pubescent; bracteoles c. 1 mm long, very broadly ovate, pubescent; bracts and bracteoles falling at anthesis. Young inflorescence branches with the distichous buds tightly packed into short bracteate spikes. Flowers 5–7 mm long, distichously arranged on the axis, sessile. Calyx lobes 4, c. 3 × 1–1.5 mm, ovate, thinly puberulous to pilose outside, mainly towards the apex and on the margins, glabrous to pilose inside. Petals absent. Stamens 8–10; filaments c. 6 mm long, glabrous; anthers dorsifixed. Ovary 1–2-ovulate, c. 2 mm diameter, subcircular, pubescent particularly on the margins; stipe very short; style c. 3.5 mm long, glabrous; stigma capitate, bilobed, papillose. Pod 1-seeded, 2.5–3 cm in diameter, suborbicular, flattened, slightly asymmetric, glabrous, with numerous resin-filled vesicles in the pod wall, tardily dehiscent. Seed c. 15 × 12 mm, broadly ellipsoidal, somewhat flattened.

Zambia. B: Mongu-Lealui, Mongu, fl. & fr. 10.xi.1952, *White* 2039 (BM, FHO, K). W: Mwinilunga Dist., c. 25 km from Mwinilunga on Matonchi road, fl. 21.xii.1969, *Simon & Williamson* 1906 (K, LISC, SRGH). S: Namwala, fl. 20.x.1959, *Fanshawe* 5254 (K).

Also in Angola and Congo. Grassland, shrubland and woodland on Kalahari sands, sometimes forming low thickets; 750–1400 m.

Conservation notes: Fairly widespread and apparently often common where it occurs; Lower Risk, Least Concern.

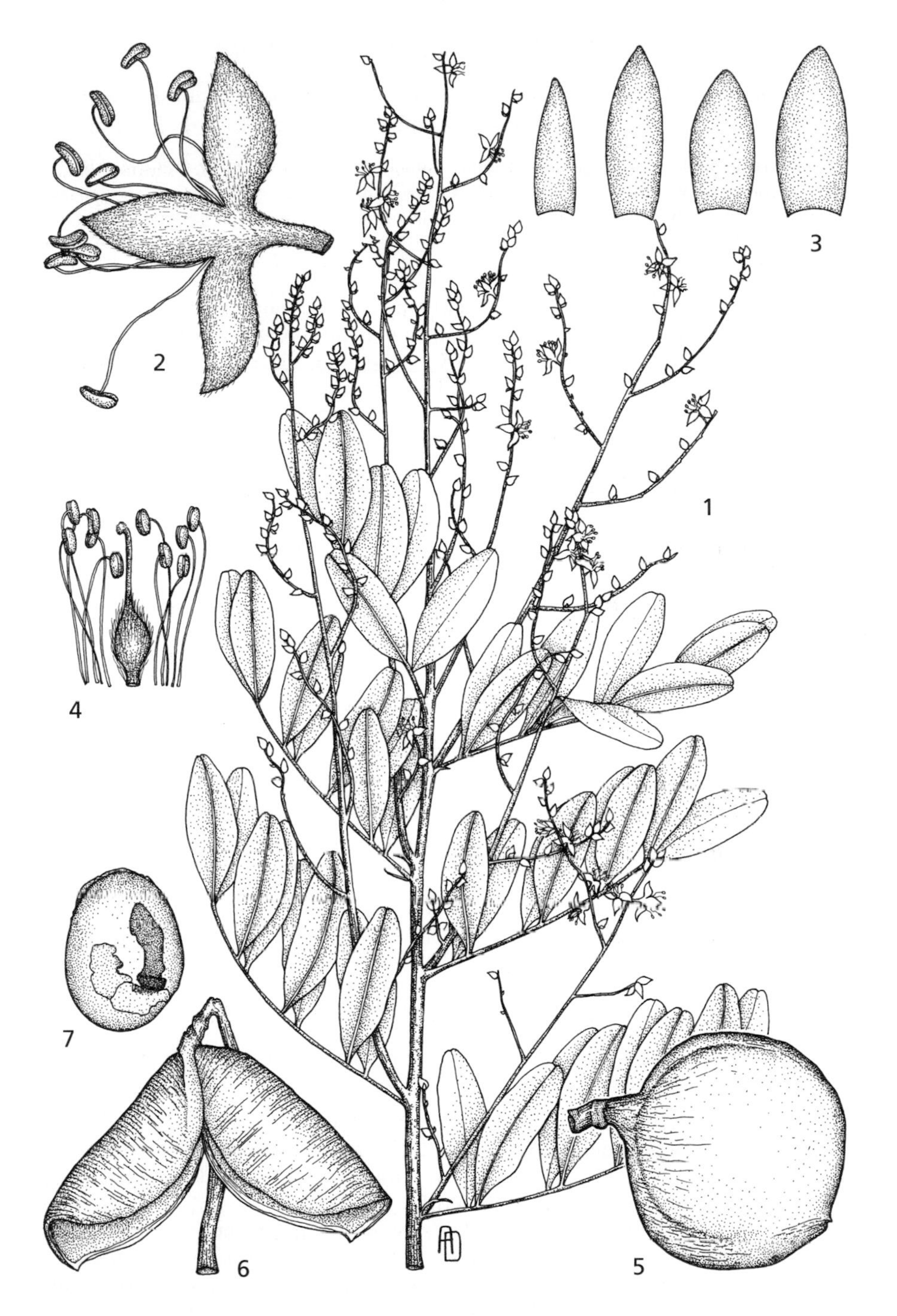

Fig. 3.2.**12**. COPAIFERA BAUMIANA. 1, flowering branch (× $^1/_2$); 2, flower (× 6); 3, sepals (× 5); 4, stamens and gynoecium (× 4), 1–4 from *Drummond & Cookson* 6223; 5, pod (× $1^1/_2$); 6, dehisced pod (× $1^1/_2$); 7, seed (× $1^1/_2$), 5–7 from *Angus* 2250. Drawn by Ann Davies.

13. SCORODOPHLOEUS Harms

Scorodophloeus Harms in Bot. Jahrb. Syst. **30**: 77 (1901). —Léonard in Bull. Jard. Bot. État **21**: 418–420 (1951).

Shrubs or trees. Leaves paripinnate, with 2–many leaflets; stipules linear, caducous; leaflets alternate, sessile, asymmetric, midrib subcentral. Inflorescence racemose, usually simple. Hypanthium present, up to 5 mm long. Sepals 4, subequal. Petals white, 4–5, subequal. Stamens 9–10. Ovary 2-ovulate; style attached to the side of the hypanthial cup. Pods without a lateral nerve; dorsal suture winged.

A genus of at least three species, one in Guineo-Congolian forests and two in the East African coastal forests of the Zanzibar-Inhambane Region. It is possible that some of the plants described as species of *Cynometra* in F.T.E.A. may in fact belong to this genus, but without flowers it is impossible to be sure. Flowering in the *Cynometra* group appears to be irregular, and the flowering period to be very short, so there are several taxa in the F.T.E.A. area that are poorly known.

Two sterile specimens from the extreme north of Mozambique may represent the East African species *Scorodophloeus fischeri* (Taub.) J.Léonard. The 4–5 alternate leaflets, with the distal ones much larger than the proximal, strongly suggest this species but flowering material is needed for confirmation. Mozambique. N: between Mocímboa da Praia and Diaca, 15.ix.1948, *Pedro & Pedrogão* 5222 (LISC); Nangade towards Palma, 27.ix.1948, *Pedro & Pedrogão* 5404 (LISC).

Scorodophloeus torrei Lock in Kew Bull. **61**: 257 (2006). Type: Mozambique, Angoche, ix.1970, *Patacas* s.n. (LISC holotype). FIGURE 3.2.**13**.

Shrub, or tree to 20 m; bark (from photograph with *Torre & Correia* 14585) pale, smooth on main branches. Young twigs terete, red-brown pubescent, glabrescent; older twigs grey, lenticellate. Leaves paripinnate, 30–70 mm long, 10–16-foliolate; stipules c. 3 mm long, very narrowly triangular to linear, red-brown pubescent, caducous; petiole and rachis pubescent, ridged; leaflets alternate (the basal and upper pair sometimes opposite), 11–26 × 6–12 mm, asymmetric, rhombic-ovate, apex obtuse or rounded to emarginate, base asymmetric, proximal side rounded, distal side cuneate, glabrous; midrib subcentral, venation brochidodromous. Inflorescences simply racemose, terminal on young flush growth, 7–10-flowered; bracts and bracteoles 3–4 mm long, linear-spathulate, caducous at time of anthesis; pedicels 4–7 mm long, sparsely puberulous below. Hypanthium 3–4 mm long, narrowly funnel-shaped, glabrous. Sepals 4, 4 × 1.2 mm, narrowly triangular, reflexed at anthesis. Petals 4(5), narrowly obovate, tapering to a basal claw, 2 broader 10–11 × 3.5 mm, 2 narrower 10–11 × 2.5 mm. Stamens 9; filaments 10–12 mm long. Pods c. 7.5 × 4 cm, broadly oblong, somewhat asymmetric, glabrous, with narrow flanges along each side of the dorsal suture, elastically dehiscent.

Mozambique. N: Angoche (Antonio Enes), Farol de Saugage, fl. ix.1970, *Patacas* s.n. (LISC holotype); same locality, fr. 22.i.1968, *Torre & Correia* 17320 (LISC). Z: Maganja da Costa, Floresta de Gobene, near Raraga Beach, 15 km from Maganja da Costa, fr. 10.i.1968, *Torre & Correia* 17006 (LISC).

Not known elsewhere. Coastal mixed forest, sandy soils; 0–80 m.

Conservation notes: Although little known, Vulnerable is probably appropriate. Restricted to a rather small part of the Mozambique coast. The coastal forests are poorly known and probably under pressure from clearance for agriculture and exploitation for timber and firewood. Specimens of this plant have lain in the herbarium at LISC for several years, and their number suggest that it is not uncommon where it occurs. There are no recent collections.

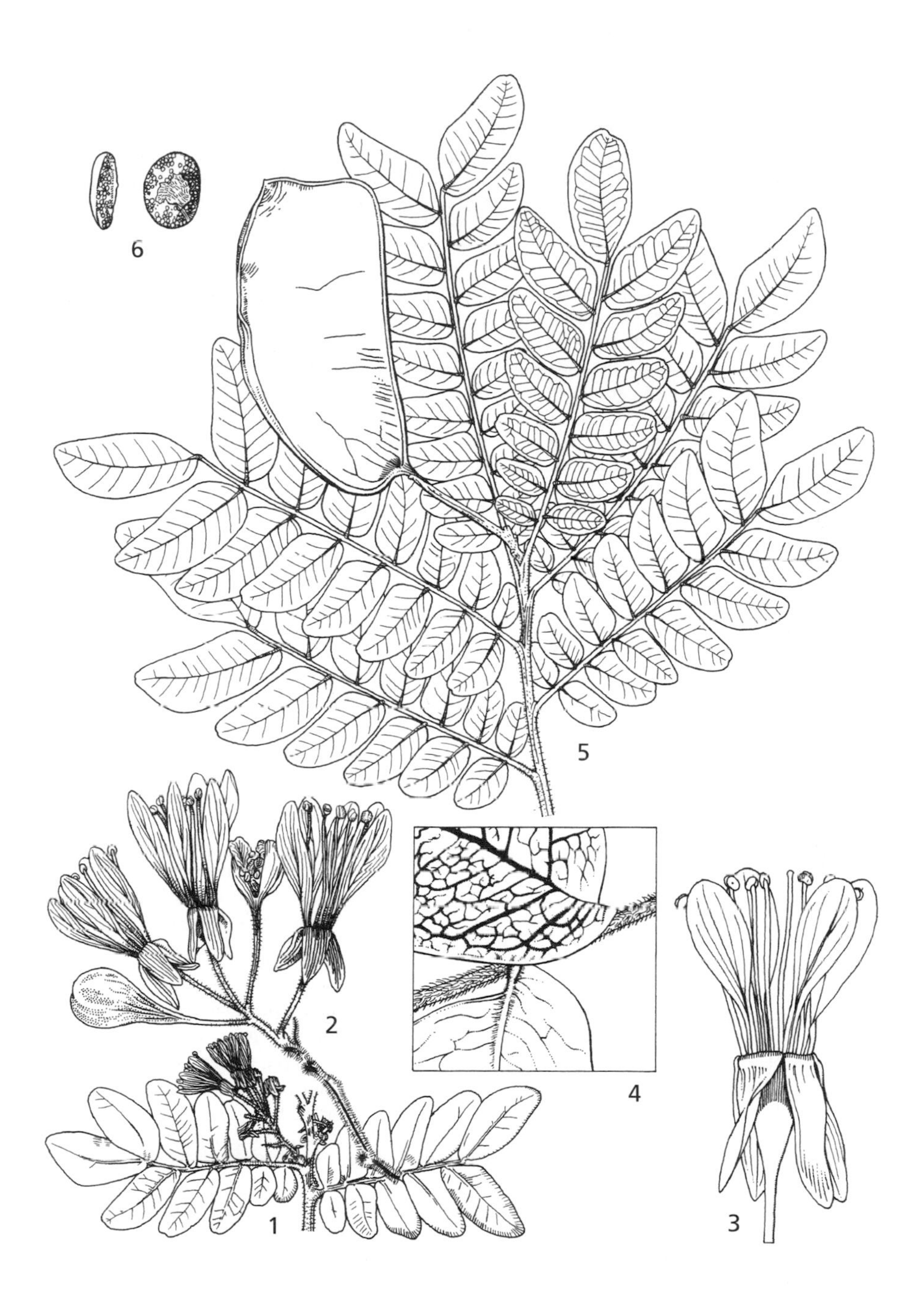

Fig. 3.2.**13**. SCORODOPHLOEUS TORREI. 1, flowering branch (× ½); 2, inflorescence (× 2); 3, flower (× 7); 4, base of leaflets and rachis (× 6); 1–4 from *Patacus* s.n.; 5, fruiting branch (× ½), from *Torre & Correia* 17006; 6, seed (× ⅔), from *Torre & Correia* 14543. Drawn by Emmanuel Papadopoulus.

The type species of the genus, *S. zenkeri* Harms, has bark that smells strongly of garlic, hence the generic name. No collector of the present species has noted such a scent.

14. MICKLETHWAITIA G.P. Lewis & Schrire

Micklethwaitia G.P. Lewis & Schrire in Kew Bull. **59**: 166 (2004)
Brenaniodendron J. Léonard in Bull. Jard. Bot. Belg. **67**: 16 (23.iii.1999) non *Brenandendron* H. Rob. (3.iii.1999).

Small trees, deciduous. Leaves alternate or crowded on short shoots which later grow out, paripinnate; leaflets opposite, with short twisted petiolules and with a few glands beneath near the base; stipules intrapetiolar, early caducous. Inflorescence racemose, leaf-opposed on short shoots, rather few-flowered. Hypanthium shallowly cupular, much shorter than the sepals. Sepals 4, reflexed at anthesis. Petals 4–5, equalling or slightly longer than the stamens. Stamens 8–9(10), all equal in length; anthers 0.4–0.5 times the length of the filament, narrowly oblong-elliptic. Ovary 2-ovulate, the stipe partially fused to the walls of the hypanthium. Pod asymmetrically narrowly obovate, at least 6 cm wide at maturity, ridged along each side of the upper suture, elastically dehiscent, apex long-acuminate.

One species in the coastal forests of Mozambique.

Micklethwaitia carvalhoi (Harms) G.P. Lewis & Schrire in Kew Bull. **59**: 166 (2004). Type: Mozambique, Mossuril (Mussoril) to Cabeceira (Cabessira), 1884–1885, *M.R. Carvalho* (COI holotype). FIGURE 3.2.**14**.

Cynometra? carvalhoi Harms in Bot. Jahrb. Syst. **26**: 261 (1899); in Engler, Pflanzenw. Afrikas **3**(1): 435 (1915). —Lewis in Kew Bull. **51**: 595–599, fig.1 (1996).

Brenaniodendron carvalhoi (Harms) J. Léonard in Bull. Jard. Bot. État **67**: 18 (1999). — M. Coates Palgrave, Trees Sthn. Africa: 346 (2002).

Deciduous tree 6–10 m tall; bark grey, smooth. Branchlets brown becoming grey, pubescent, strongly lenticellate. Leaves paripinnate, in 3(4) pairs; stipules intrapetiolar, early caducous, leaving a low rim between the stem and the petiole; petiole 1.4–2.1 cm long, almost glabrous to pubescent, glabrescent; rachis 3–8 cm long, channelled on the upper surface, glabrous to pubescent; leaflets up to 7.6 × 4 cm, the terminal pair the largest, elliptic to obovate, slightly asymmetric, glabrous to pubescent, the distal side of the midrib pubescent near the base, apex bluntly acuminate, base asymmetrically cuneate with the proximal side expanded, venation slightly prominent on both surfaces; 1–3 resin vesicles near the base of each leaflet. Flowers 8–12, in racemes that are leaf-opposed and appear terminal on short shoots; axis usually unbranched, terete, glabrescent; bracts and bracteoles very early caducous, the bracts c. 2 × 0.5 mm, spathulate, with ciliate margins; bracteoles c. 1.5 mm long, spathulate; pedicels 1.5–2 cm long, terete. Sepals 4, 4.5 × 2 mm, ovate, reflexed at anthesis, glabrous. Petals white, 4–5, 7 × 1.5 mm, narrowly elliptic to narrowly obovate, with a strong midvein. Stamens 8–10; filaments 5.6–6 mm long, glabrous; anthers 2 × 3 mm. Ovary 2-ovulate, asymmetrically obovate-elliptic, c. 3.5 mm long, glabrous; stipe very short, glabrous; style c. 7 mm long, glabrous; stigma capitate. Pod 10–14 × 3.5–4.5 cm, asymmetrically obtriangular, flattened, glabrous, ridged along each side of upper margin, explosively dehiscent; stipe c. 8 mm long. Seed c. 2.8 × 2.3 × 0.8 cm, very broadly ellipsoidal, flattened, dull brown with a darker rim.

Mozambique. N: Pemba Dist., Marunga, Nangororo, 12°54'S, 40°22'E, fl. xi/xii.1960, *Gomes e Sousa* 4528 (COI, K, WAG); 5 km from Nangororo Sisal Farm, 13°05'S, 40°25'E, fr. 12.iii.1961, *Gomes e Sousa* 4657 (COI, K).

Not known elsewhere. Open forest on red or black clay soils; 10–150 m.

Conservation notes: Restricted distribution; probably Vulnerable.

A distinctive plant known only from coastal N Mozambique. Most of the recent material comes from one small area, but the type and several other collections are from c. 150 km further south.

Fig. 3.2.**14**. MICKLETHWAITIA CARVALHOI. 1, leaf (× ²⁄₃); 2, twisted petiole base (× 2), 1 & 2 from *Gomes e Sousa* 4528; 3, flowering branchlet (× ²⁄₃); 4, flower (× 3); 5, petal (× 4); 6, stamen (× 4); 7, stigma (× 10), 3–7 from *Montenegro* 3; 8, young pod (× ²⁄₃), from *Montenegro* 3; 9, twisted valve of dehisced pod (× ²⁄₃); 10, seed (× 1), 9 & 10 from *Gomes e Sousa* 4640. Drawn by Eleanor Catherine. From Kew Bulletin.

The genus name was recently changed from *Brenaniodendron* since the nomenclatural Committee for Spermatophyta voted that *Brenaniodendron* J. Léonard should be considered as a later homonym of *Brenandendron* H. Rob., a genus in the Compositae (see Lewis & Schrire 2004).

A note on *Montenegro* 5 (K) ('This plant has not flowered the last two years') suggests that flowering may be irregular, as in some of the East African species of *Cynometra*.

15. TAMARINDUS L.

Tamarindus L., Sp. Pl.: 34 (1753); Gen. Pl. ed.5: 20 (1754). —Léonard in Mém. Acad. Roy. Sci. Belg. **30**(2): 200 (1957).

Tree, evergreen or semi-deciduous, unarmed. Leaves paripinnate; leaflets opposite, in rather numerous pairs, almost sessile, asymmetric at base, with 1–2 secondary nerves from base on the proximal side of main nerve; translucent gland-dots absent; stipules free, asymmetrically lanceolate, very early caducous. Flowers in lax terminal and axillary racemes; bracteoles 2, well developed, valvate, completely enclosing the young flower buds but early caducous. Hypanthium shortly elongate-turbinate. Sepals 4, imbricate. Upper 3 petals well developed, lower 2 minute, setiform. Stamen-filaments connate to about half-way into a pubescent band terminating in 3 upcurved anther-bearing filaments alternating with 5 sterile teeth. Ovary ± pubescent, long-stipitate, with the stipe adnate to one side of the hypanthium; ovules 8–14; style elongate, gradually enlarged into the stigma. Pod cylindrical with irregular constrictions, indehiscent, with a dry outer shell and pulpy inner layer. Seeds ± compressed, with a continuous-margined areole on each face.

A single species, probably native to Madagascar and perhaps elsewhere. Very widely planted and often completely naturalised.

Tamarindus indica L., Sp. Pl.: 34 (1753). —Baker, Legum. Trop. Afr.: 702 (1930). —Brenan, Check-list For. Trees Shrubs Tang. Terr.: 106 (1949). —Léonard in F.C.B. **3**: 436 (1952). —Torre & Hillcoat in C.F.A. **2**: 217 (1956). —White, F.F.N.R.: 128 (1962). —Gomes e Sousa, Dendrol. Moçamb. Estudo Geral **1**: 263, fig.62 (1966). —Brenan in F.T.E.A., Legum.-Caesalp.: 151, fig.32 (1967). —M. Coates Palgrave, Trees Sthn. Africa: 330 (2002). Type uncertain. FIGURE 3.2.**15**.

Tree to 25 m; crown rounded; bark rough, grey or grey-black. Young branchlets pubescent to puberulous. Leaves: stipules free, narrowly ovate, small, early caducous; petiole with rachis 5–12(16) cm long, pubescent; leaflets in 10–18 pairs, (0.8)1.2–3.2 cm long, narrowly oblong, rounded and asymmetric at the base, rounded to truncate or sometimes emarginate at the apex, pubescent or occasionally glabrous except for a tuft of yellowish hairs at the base, venation raised-reticulate on both surfaces. Racemes 1–15(22) cm long; axis pubescent to densely so; pedicels 3–14 mm long, glabrous to pubescent. Flower buds red. Hypanthium 3–5 mm long. Sepals pale yellow inside, reddish outside, 8–12 mm long, elliptic, acute, imbricate. Petals yellow with red veins, large ones 10–13 mm long, elliptic to obovate-elliptic. Pods 3–14 cm long, 2–3 cm in diameter, cylindrical with irregular constrictions, straight or curved, brown-scurfy, 1–10-seeded. Seeds chestnut-brown, 11–17 × 10–12 mm, rhombic to trapezioid.

Fig. 3.2.**15**. TAMARINDUS INDICA. 1, flowering branchlet (× ²/₃); leaflet base, lower surface (× 6); 3, leaflet, upper surface, showing venation (× 4); 4, flower bud with bract (× 4); 5, flower bud with bracteoles (× 4); 6, older flower bud after fall of bract and bracteoles (× 4); 7, flower (× 2); 8, sepal (× 2); 9, one of three upper large petals (× 2); 10, one of two lower minute petals (× 2); 11, stamens (× 2); 12, gynoecium, with ovary opened (× 4), 1–12 from *Semsei* in *For. Herb.* 2867; 13, mature pod, breaking up (× ²/₃); 14, detail of pod surface (× 4); 15, seed, showing areole (× 2), 13–15 from *Hughes* 5. Drawn by Lura Ripley. From F.T.E.A.

1
2
3
4
5
6
7
8
9
10
11
12
13
14
15
LMR

Zambia. C: Bank of Zambezi R. at Chiawa, up river from Luangwa (Feira), fr. 28.iv.1957, *Angus* 1563 (FHO, K). E: Chikowa Mission to Jumbe, km 12 (mile 8), fr. 13.x.1958, *Robson & Angus* 87 (K, LISC). S: Gwembe, fl. 11.xi.1955, *Bainbridge* 106/55 (K, FHO). **Zimbabwe.** N: Guruve Dist., Mushumbi Pools, fl. 18.i.1998, *Poilecot* 7655 (G, K). W: Hwange (Wankie) Dist., banks of Zambezi R., Deka R. valley, fr. 26.iv.1959, *Noel* CAH 3671 (K, SRGH). **Malawi.** N: Karonga Dist., 26 km (17 miles) N of Chilumba, at Ngala, fr. 4.vii.1970, *Pawek* 3571 (K). S: Chikwawa Dist., Lengwe Game Reserve, fl. 7.iii.1970, *Brummitt* 8952 (K). **Mozambique.** N: Malema, Mutuali, fr. 29.v.1947, *Pedro* 3287 (LISC). Z: Morrumbala, at limits of Posto do Derre, fr. 10.vi.1949, *Andrada* 1564 (COI, LISC). T: Near Posto do G.P.Z., fl. 24.i.1972, *Macedo* 4663 (K, LISC). MS: between Mutarara-a-Velha and Sinjal, 35.8 km from Mutarara-a-Velha, fr. 18.vi.1949, *Barbosa & Carvalho* 3129 (K, LISC). GI: c. 5 km N of Cheline, between Inhambane and Vilankulo (Vilanculos), fr. 8.v.1971, *Edwards & Vahrmeyer* 4240 (K, PRE). M: Inhaca Is., Ilha dos Portugueses, E end, st. 23.i.1962, *Mogg* 29829 (K).

Widespread in tropical Africa; believed to be native to SE Madagascar; 20–1000 m.

Conservation notes: Lower Risk, Least Concern. Very widespread as a naturalised or cultivated tree. Locally abundant in its Madagascar stations.

Specimens from the Flora area are mainly from low altitudes and in riverine vegetation. This is the precise environment in which it grows apparently wild in Madagascar and it is possible that its native range extends into Africa. Planted trees may be found in frost-free areas but collections specifically stated to be from cultivated trees are few. The fruit pulp is sweet-sour and used for jams and chutneys; it is also mildly laxative.

16. AFZELIA Sm.

Afzelia Sm. in Trans. Linn. Soc. **4**: 221 (1798) nom. conserv. —Léonard in Reinwardtia **1**: 61–66 (1950); in Mém. Acad. Roy. Sci. Belg. **30**(2): 106 (1957).

Trees, evergreen or deciduous, unarmed. Leaves paripinnate; leaflets opposite or subopposite, without gland dots, but sometimes with a dot-like gland near the base of the leaflet; petiolules twisted; stipules intrapetiolar, upper parts early caducous. Flowers in racemes or panicles, spirally arranged on the inflorescence axis; bracts small, caducous; bracteoles 2, small, caducous. Hypanthium present, generally elongate. Sepals 4, subequal, differentiated into 2 outer and 2 inner, imbricate. Corolla of a single large petal, with up to 4 represented as minute rudiments. Fertile stamens 7–9, filaments long; anthers dorsifixed, dehiscing by longitudinal slits; 2 staminodes often present. Ovary stipitate; ovules usually 10–15; style elongate, filiform; stigma terminal, small. Pods oblong or reniform, compressed, with thick woody valves, tardily dehiscent. Seeds up to 10, without areole, lying in compartments in a white pithy endocarp, with a yellow or red fleshy aril partially enveloping the seed.

A genus of about 15 species in the tropics of the Old World. The African species need revision on a continent-wide basis. Some species yield valuable timber.

Leaflet apex obtuse, rounded or retuse; inflorescence usually a simple raceme; pod usually ± straight, the seeds set at right angles to the long axis; aril cup-shaped, less than half seed length . . . . **1.** *quanzensis*

Leaflet apex distinctly (sometimes bluntly) acuminate; inflorescence usually paniculate (but only known in fruit); pod curved into a semi-circle, the seeds set obliquely to the long axis; aril deeply bilobed, with at least one lobe more than half seed length . . . . **2.** *peturei*

1. **Afzelia quanzensis** Welw. in Ann. Consel. Ultram., Parte Não Official, Sér. 1 [Apontamentos Phytogeographicos] **1**: 586 (1859). —Oliver in F.T.A. **2**: 302 (1871) as *cuanzensis*. —Sim, For. Fl. Port. E. Afr.: 48, fig.45 (1909). —Baker in J. Linn. Soc., Bot. **40**: 63 (1911) as *cuanzensis*. —Harms in Engler, Pflanzenw. Afrikas **3**(1): 458 (1915) as *cuanzensis*. —Eyles in Trans. Roy. Soc. S. Afr. **5**: 366 (1916) as *cuanzensis*. —Baker, Legum. Trop. Afr.: 701 (1930). —Miller in J. S. Afr. Bot. **18**: 27 (1952). —Pardy in Rhod. Agric. J. **49**: 82 (1952). —Léonard in F.C.B. **3**: 354, fig.27B (1952) as *cuanzensis*. —Torre & Hillcoat in C.F.A. **2**: 215 (1956). —Coates Palgrave, Trees Central Africa: 61 (1957). —Topham in Burtt Davy & Hoyle, Check List For. Trees Shrubs Nyasaland Prot.: 34 (1958). —White, F.F.N.R.: 98, fig.21 (1962). —Gomes e Sousa, Dendrol. Moçamb. Estudo Geral **1**: 256, fig.57 (1966). —Brenan in F.T.E.A., Legum.-Caesalp.: 125, fig.22 (1967). —Drummond in Kirkia **10**: 246 (1977). —Ross in F.S.A. **16**(2): 42, fig.9 (1977). —F. & J.A. Venter, Indig. Trees: 166 (1996). —M. Coates Palgrave, Trees Sthn. Africa: 331 (2002). Type: Angola, Cuanza Norte, R. Cuanza between Sansamanda & Quisonde, *Welwitsch* 594 (LISU holotype; BM, K). FIGURE 3.2.**16**.

Afzelia petersiana Klotzsch in Peters, Naturw. Reise Mossambique **6**(1): 19 (1861). Type: Mozambique, Mocímboa (Mossimboa), *Peters* s.n. (B† holotype).

Afzelia attenuata Klotzsch in Peters, Naturw. Reise Mossambique **6**(1): 20 (1861). —Baker, Legum. Trop. Afr.: 701 (1930). Type: Mozambique, Inhambane, *Peters* s.n. (B† holotype; K).

Pahudia quanzensis (Welw.) Prain in Sci. Mem. Off. Med. Dept. Gov. India **12**: 16 (1901) as *quangensis*. —Baker in J. Linn. Soc. Bot. **40**: 62 (1911) as *quangensis*. —Eyles in Trans. Roy. Soc. S. Afr. **5**: 366 (1916).

Tree to 25 m tall; bark grey, irregularly fissured, later flaking to leave paler patches. Branchlets brown, sparsely pubescent, glabrescent. Leaves paripinnate; stipules c. 9 × 2 mm, narrowly elliptic, upper parts early caducous (only seen in seedling material); petiole 2.5–4(6) cm long, sparsely pubescent, glabrescent; rachis (7)12–19 cm long, almost glabrous; leaflets opposite to subopposite, in (3)4–6(7) pairs, 4–8.5 × 2.5–4.8 cm, broadly ovate to broadly elliptic, almost symmetrical, apex rounded or emarginate, base cuneate, glabrous, usually with a small dot-like gland at proximal side of base either on lower surface in angle between margin and midrib or on the margin itself; venation slightly prominent on both surfaces, marginal vein prominent. Inflorescences appearing leaf-opposed, usually simple, racemose, 4–7-flowered; axis sparsely to densely puberulous; bracts c. 6 × 5 mm, broadly ovate, puberulous, early caducous; bracteoles paired, 6 × 4 mm, broadly ovate, puberulous, caducous; flowers spirally arranged on the axis; pedicel c. 5 mm long. Hypanthium c. 15 mm long, puberulous like the pedicel. Calyx lobes 4, puberulous outside, glabrous inside, margins sparsely glandular-pilose particularly towards the tip, the outer two c. 1.7 × 0.9 cm, broadly triangular, somewhat auriculate at the base, the inner two c. 2 × 1.2 cm, broadly ovate, cuneate at the base. Petal entirely red or green with a red median line, c. 3.5 cm long, claw c. 2 cm long, pilose near the base, abruptly expanded into a broad bilobed lamina c. 2.5 cm wide. Fertile stamens 7–9; staminodes usually 2; filaments c. 35 mm long, pilose near the base, otherwise glabrous; anthers dorsifixed. Ovary c. 3.5 × 1.5 mm, oblong, pilose particularly on the margins, 10–15-ovulate; stipe c. 3 mm long; style c. 25 mm long, glabrous; stigma capitate, papillose. Pod 10–20 × 5–8 cm, straight, oblong, flattened, apiculate, the pedicel attached asymmetrically at the base, glabrous, tardily dehiscent, 5–13-seeded. Seeds arranged at right angles to the long axis of the pod, smooth, black, 2.6–3.5 × 1.2–1.7 cm, ellipsoid; aril orange or red, 0.8–1.3 cm long, cup-shaped. Germination epigeal; first true leaves opposite, paripinnate, in 4 pairs, elliptic, apex acuminate, base cuneate.

Caprivi Strip. Katima Mulilo, fl. 21.x.1970, *Vahrmeijer* 2208 (K, PRE). **Botswana.** N: Chobe Dist., Kasane, banks of Chobe R., fl. 10.x.1954, *Story* 4816 (K, PRE). **Zambia.** B: Kalabo Dist., Kalabo Boma, st. 16.ii.1962, *White* 2084 (FHO, K). N: by L. Bangweulu (Bangweolo), N of Samfya Mission, fl. & fr. 7.x.1947, *Brenan & Greenway* 8049 (FHO, K). W: Solwezi Dist., near Mbulungu stream, W of Mutanda Bridge, fr.

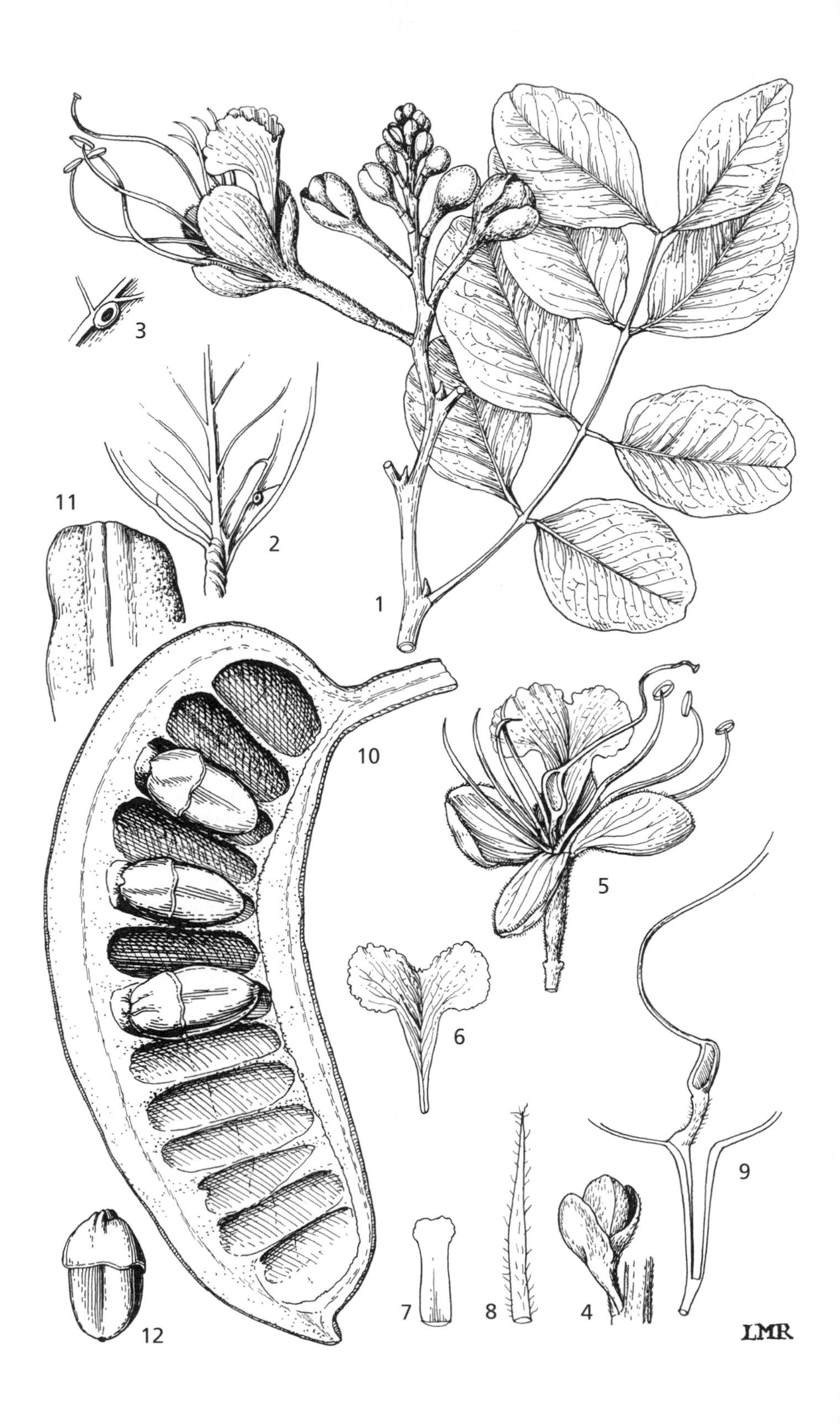
1
2
3
4
5
6
7
8
9
10
11
12
LMR

14.vii.1930, *Milne-Redhead* 702 (K). C: c. 15 km S of Lusaka, near Mt. Makulu Res. Station, fl. 1.x.1957, *Angus* 1724 (FHO, K). E: Chipata (Fort Jameson), fl. 2.x.1966, *Mutimushi* 1584 (K). S: c. 3 km E of Mapanza, fl. 26.ix.1954, *E.A. Robinson* 889 (K). **Zimbabwe.** N: Makonde (Lomagundi), fl. x.1920, *Eyles* 2717 (K, SRGH). W: Matobo Dist., Matopos, fl. 27.x.1951, *Plowes* 1297 (K, SRGH). C: Goromonzi Dist., Enterprise, near Harare (Salisbury), fl. xi.1917, *Willoughby* in *Eyles* 872 (BM). E: Mutare Dist., SE side of Palmerston, NW of Mutare (Umtali) Golf Course, fl. 25.x.1962, *Chase* 7879 (BM, FHO, K, LISC, SRGH). S: Runde (Lundi) R., fl. 13.x.1951, *Mylne* 46/51 (FHO). **Malawi.** N: Mzimba Dist., Mbawa, fl. 15.x.1952, *Jackson* 981 (K). C: Salima Dist., dunes N of Lake Nyasa Hotel, fr. 30.vii.1951, *Chase* 3908 (BM). S: Mangochi (Fort Johnson), Funwe Hill, st. 19.xi.1954, *Jackson* 1385 (BM, FHO, K). **Mozambique.** N: between Quinhandoa (Kinyatugo) and Messalo (Msalu) R., fl. 28.ix.1906, *Stocks* 36 (K). Z: Lugela Dist., Lugela–Mocuba, Namagoa Estate, fl. x. [no year], *Faulkner* PRE 12 (BM, COI, K, LISC). T: Tete, Cabora Bassa, R. Mecangádzi, fl. 22.x.1973, *Correia, Marques & Diniz* 3558 (K). MS: Beira, fl., n.d., *Honey* 648 (K). GI: Inhambane, Pomene, some km past airstrip, fl. 24.ix.1980, *Jansen, de Koning & Zunguze* 7529 (K). M: Santaca, 90 km S of Maputo (Lourenço Marques), fl. 15.x.1947, *Gomes e Sousa* 3634 (COI, K, LISC).

East African coastal region from Somalia to S Mozambique; widespread in southern Africa, northwards to southern Congo and Angola. Dry forests, woodland and scrub forest; up to c. 1350 m.

Conservation notes: A very widespread species, unlikely to be under any immediate threat, although there may be pressure on large individuals owing to their timber value.

The timber is excellent ('Pod Mahogany') and much used for furniture and for similar uses.

2. **Afzelia peturei** De Wild. in Bull. Séances Inst. Roy. Colon. Belge **6**: 203, figs.1–3, figs.1–6 (1935). —Léonard in Reinwardtia **1**: 65 (1950); in F.C.B. **3**: 353, fig.27E (1952). Type from Congo.

Tree to 10(20) m. Branchlets glabrous, brown, cracking longitudinally. Leaves paripinnate, in 4–6 pairs, glabrous; petiole 2–2.5 cm long; rachis 7–14 cm long; petiolules 2–3 mm long; leaflets somewhat shiny, 7.7–5.8 × 3–3.9 cm, ovate to elliptic, apex bluntly acuminate, base rounded to slightly cordate, glabrous, with a gland beneath near the base on the proximal side; midrib prominent beneath, impressed above; venation more prominent above than beneath. Young inflorescences and flowers unknown. Inflorescence axis branched, puberulous. Fruit brown, c. 10.5 × 5 cm, reniform, apiculate, hard, woody. Seeds shiny black, c. 2 cm long, 1 cm in diameter, ellipsoidal; aril orange, bilobed, the longest lobe almost as long as the seed.

Zambia. W: Mwinilunga Dist., Kalene Hill, fr. 19.v.1969, *Mutimushi* 3377 (FHO, K).

Otherwise known only from Congo. The type is from Katanga, about 150 km N of Mwinilunga. The single collection from the Flora area is recorded as '30 ft. [9 m] tree, 8" [20 cm] dia. of miombo woodland by river (seen also growing in mushitu to 50 ft.[15 m] high)'.

Fig. 3.2.**16**. AFZELIA QUANZENSIS. 1, flowering branchlet (× ²⁄₃); 2, base of leaflet, lower surface, showing gland (× 4); 3, gland (× 8); 4, flower bud, showing bracteoles (× 2), 1–4 from *Hornby* 340; 5, flower (× ²⁄₃); 6, large petal (× ²⁄₃); 7, one of four small petals (× 8); 8, one of two staminodes (× 8); 9, longitudinal section of gynoecium and hypanthium (× 1), 5–9 from *Milne-Redhead & Taylor* 7061; 10, pod, dehisced (× ²⁄₃); 11, part of pod showing suture (× ²⁄₃); 12, seed, showing aril (× ²⁄₃), 10–12 from *Richards* 6348. Drawn by Lura Ripley. From F.T.E.A.

Conservation notes: Apparently of restricted distribution and habitat, with few collections made. There are also problems of species definition (see note below). Best treated as Data Deficient.

The kidney-shaped fruit with obliquely-arranged seeds with lobed arils immediately distinguishes this taxon from *A. quanzensis.*

Mutimushi 3377 is treated here as *A. peturei* rather than *A. bipindensis* Harms or *A. bella* Harms. It shows the short petiolules that Léonard uses as a key character, although the leaflets are somewhat beyond the size range in the key. The inflorescence and flowers of *A. peturei* are still unknown. The leaflets are larger than those of the type but can be matched in other material from Katanga (e.g. *Schmitz* 4859 (BR)). The pod of *Mutimushi* 3377 is somewhat immature; the descriptions of the fruit and seeds are taken from *Schmitz* 4859.

Afzelia bipindensis differs from *A. bella* in usually being a tree rather than a shrub, with more leaflets and smaller flowers and pods. Both are forest species according to Léonard, but *A. bipindensis* extends into drier regions than *A. bella.* The two species are not easy to distinguish in the herbarium and many of the key characters given by Léonard overlap. The whole genus is in need of revision.

17. **CRYPTOSEPALUM** Benth.

Cryptosepalum Benth. in Bentham & Hooker, Gen. Pl. **1**: 584 (1865). —Léonard in Mém. Acad. Roy. Sci. Belg. **30**(2): 270 (1957). —Duvigneaud & Brenan in Kew Bull. **20**: 1–23 (1966).

Unarmed trees, shrubs, or subshrubs with mostly annual shoots from a woody rootstock. Leaves paripinnate, with leaflets in 1–many pairs (elsewhere occasionally 1-foliolate); stipules intrapetiolar, linear, caducous; leaflets sessile or subsessile, opposite, occasionally with single glands near base of leaflet. Flowers spirally arranged in terminal or axillary racemes; bracteoles well developed, petaloid (pink or white), opposite, valvate, enveloping the flower bud, persistent. Hypanthium short, cupular. Sepals small, 0–6. Petals (0)1(3), well developed and elliptic when present. Stamens 3–6(8), normally 3 in the Flora area; staminodes sometimes present. Ovary stipitate, with stipe adnate to the side of the hypanthium; ovules 1–5; stigma terminal, abruptly enlarged. Pods compressed, woody, glabrous, dehiscing with twisting valves, without longitudinal nerves, but upper suture with a longitudinal wing-like ridge on either side projecting in a plane at right-angles to the valve. Seeds compressed, elliptic, without areoles.

11 species in tropical Africa.

An ecologically important component of 'mavunda' woodland in areas of higher rainfall miombo in W & N Zambia and N Malawi, but also occurring more widely in miombo as subshrubs with annual stems. This account is based on the erudite analysis by Duvigneaud & Brenan (1966). Additional material accrued since then adds virtually nothing. About half the genus occurs in the forests of west and central Africa and presents no undue problem, but in the deciduous woodlands of the Zambesian region there is an extraordinary complex pattern of variation that defies the normal criteria for defining taxa. The following account should be used with an understanding that transitions and anomalies are the norm rather than the exception. Nevertheless, Duvigneaud & Brenan clearly point out the main geographical elements of each of the two broadly defined species recognised in the Flora area.

Shrub or small tree, sometimes suffruticose but then stems branched above and bearing several inflorescences, or at least with clustered lateral inflorescences; stems apparently persisting for at least 2 seasons . . . . **1.** *exfoliatum*

Suffruticose, with a thickened rootstock and simple, erect, annual stems arising at or near ground level and each ending in a single inflorescence . . . . **2.** *maraviense*

1. **Cryptosepalum exfoliatum** De Wild. in Ann. Mus. Congo, sér. 4, Bot. **1**: 41, fig.3, figs. 1–8 (1902). —Baker, Legum. Trop. Afr.: 743 (1930). —Léonard in F.C.B. **3**: 493 (1952). —White, F.F.N.R.: 121 (1962). —Duvigneaud & Brenan in Kew Bull. **20**: 8 (1966). —Brenan in F.T.E.A., Legum.-Caesalp.: 199 (1967). Type: Congo, Katanga, 1900, *Verdick* (BR holotype).

Shrub or tree 0.2–18(30) m tall, sometimes suffrutcose but then branched above or at least with several often clustered lateral inflorescences, and with stems apparently persisting for at least two seasons; bark grey to brown. Leaves with rachis 1.5–9.5 cm long; leaflets in 2–11 pairs, 0.5–7 × 0.15–3.3. cm, ± asymmetrically oblong to elliptic, emarginate or rounded to subacute at apex, asymmetric at base, glabrous to hairy. Racemes terminal or lateral, solitary or clustered up to 3 together, 2–10 cm long, glabrous to pubescent. Flowers white or pale pink, sweetly scented. Bracteoles 5–8 × 3–6 mm, elliptic. Sepals 1–4, up to 2 mm long, triangular. Petals 1(2), 5–10 mm long. Stamens 3. Pods 4.5–6 × 2–2.5 cm. Seeds 1.3–1.5 × 0.8–1 cm.

Five subspecies and three varieties in Angola, a small part of the Katanga, W and N Zambia into SW Tanzania and also in N Malawi.

1. Shrubs or trees mostly 1.5–18 m tall 2
– Stems numerous from a woody rootstock, short but branched or at least with lateral inflorescences 4
2. Leaflets beneath glabrous except sometimes at the extreme base, in 3–5(6) pairs per leaf; inflorescence axis and pedicels glabrous .... **ia)** subsp. *exfoliatum* var. *fruticosum*
– Leaflets beneath ± pubescent, usually rather densely so; well-developed leaves with 5–8 pairs of leaflets; inflorescence axis and pedicels pubescent 3
3. Leaflets 40–50 × 25–30 mm; leaves each with 2–4 pairs of leaflets .... **ib)** subsp. *exfoliatum* var. *pubescens*
– Leaflets 10–30 × 3–12 mm; well-developed leaves with 5–8 pairs of leaflets .... **ii)** subsp. *pseudotaxus*
4. Main nerve of leaflets marginal; stems with spreading pubescence; leaflets softly appressed-pubescent all over, with 8–10 pairs per leaf .. **v)** subsp. *craspedoneuron*
– Main nerve of leaflets central to somewhat excentric but not marginal 5
5. Stem with ± dense spreading pubescence; leaflets up to 6–9 pairs per leaf, ± pubescent on surface or on midrib and margins only, the largest 7–18 mm wide .... **iii)** subsp. *suffruticans*
– Stem usually with appressed puberulence; leaflets 2–4 pairs per leaf, ± appressed puberulous on surface or glabrous .... **iv)** subsp. *puberulum*

i) Subsp. **exfoliatum** P.A. Duvign. & Brenan in Kew Bull. **20**: 8 (1966). —Brenan in F.T.E.A., Legum.-Caesalp.: 199 (1967).

Shrub or small tree 1.5–7.5 m high. Branchlets glabrous to densely pubescent. Leaflets large, up to 25–75 × 10–33 mm.

ia) Var. **fruticosum** (Hutch.) P.A. Duvign. & Brenan in Kew Bull. **20**: 8 (1966). —Brenan in F.T.E.A., Legum.-Caesalp.: 199 (1967). Type: Zambia, Mbala Dist., 29 km NW of Mbala (Abercorn), *Hutchinson & Gillett* 3947 (K holotype, BM, BR, LISC).

Cryptosepalum fruticosum Hutch. in Bull. Misc. Inform., Kew **1931**: 250, fig. p. 238 (1931). —Brenan, Check-list For. Trees Shrubs Tang.Terr.: 100 (1949). —Léonard in F.C.B. **3**: 494 (1952).

Branchlets glabrous. Leaflets glabrous, or pubescent only at extreme base of lower side. Inflorescence axes and pedicels glabrous.

Zambia. N: Mporokoso, fr. 11.x.1958, *Fanshawe* 4918 (K, LISC); Mbala–Mpulungu, fl. 31.vii.1949, *Greenway & Hoyle* 8354 (EA, FHO, K).

Also in Ufipa Dist. of SW Tanzania. Higher rainfall miombo woodland, sometimes mixed with genera such as *Monotes, Marquesia* and *Pterocarpus*; 800–1650 m.

An attractive small tree commonly collected in the Mbala and Mporokoso Districts of Zambia and just over the border in Tanzania. *Bullock* 1406 (K) from Mbala Dist., Luchechi R., fl. 9.xi.1949, is somewhat intermediate with var. *pubescens*, having pubescent stems, 4 pairs of leaflets 12–30 × 7–16 mm and glabrous except on the margins, and glabrous inflorescence axes.

Var. *exfoliatum* is known only from Katanga in Congo. It has leaflets sparsely pubescent on the midrib and near the base and the inflorescence axes and pedicels with spreading pubescence.

ib) Var. **pubescens** P.A. Duvign. & Brenan in Kew Bull. **20**: 9 (1966). —Brenan in F.T.E.A., Legum.-Caesalp.: 200 (1967). Type: Zambia, Kalambo Falls, *Bullock* 3002 (K holotype; BM).

Branchlets with dense spreading pubescence. Leaflets rather densely pubescent beneath and ± pubescent above. Inflorescence axes and pedicels densely spreading pubescent.

Zambia. N: Mbala Dist., Kalambo Falls, fl. 10.ix.1969, *Sanane* 894 (K). W: Mwinilunga Dist., N of Mwanamitowa stream S of Mwinilunga, st. ix.1934, *Trapnell* 1573 (K).

Also in Ufipa Dist. of SW Tanzania. Miombo woodland; 1050–1200 m.

Grows within the range of var. *fruticosum*, with one record further west near Mwinilunga. The two varieties have been collected together: Mbala Dist., Mpulungu, fl. ix.1956, *Clayphan* 19 (var. *fruticosum*) and 19A (var. *pubescens*).

ii) Subsp. **pseudotaxus** (Baker f.) P.A. Duvign. & Brenan in Kew Bull. **20**: 10 (1966). Type from Angola.

Cryptosepalum pseudotaxus Baker f. in J. Bot. **66**: 148 (1928); Legum. Trop. Afr.: 745 (1930). —Milne-Redhead in Hooker, Icon. Pl. **32**: fig.3196 (1933). —Léonard in F.C.B. **3**: 492 (1952). —Duvigneaud in C.F.A. **2**: 237 (1956). —White, F.F.N.R.: 122 (1962) in part excl. *White* 3446.

Cryptosepalum arboreum Baker f. in J. Bot. **66**: 148 (1928); Legum. Trop. Afr.: 745 (1930). —Duvigneaud in C.F.A. **2**: 237 (1956). Type from Angola.

Small to medium-sized trees, mostly 5–18(30) m tall. Branchlets densely spreading pubescent. Leaflets up to 11–27 × 3–12 mm, mostly ± densely pubescent beneath and ± pubescent above. Inflorescence axes densely hairy.

Zambia. B: Zambezi Dist., Zambezi (Balovale), fl. vii.1952 & fr. ix.1952, *Gilges* 182 (K). N: S Isoka, fl. & fr. 26.ix.1938, *Greenway* 5785 (EA, K). W: Mwinilunga Dist., L. Chibesha, fl. 21.v.1969, *Mutimushi* 3265 (K, NDO). S: Livingstone, fr. 19.xii.1952, *Angus* 967 (K). **Malawi.** N: Chitipa Dist., 17.5 km (11 miles) S of Chisenga, fl. & fr. 11.ix.1977, *Pawek* 12979 (K, MAL, MO, SRGH).

Also in Angola and possibly in Congo. Locally dominant on Kalahari sands, forming 'mavunda' woodland, but also in miombo and higher rainfall mixed woodlands with *Guibourtia, Copaifera, Erythrophleum, Isoberlinia,* etc. in sandy places and on rocky ridges; 550–1700 m.

Duvigneaud & Brenan (1966) recognised two rather distinct forms. The populations on Kalahari sands in Angola, Western, Barotse and Southern Provinces of Zambia tend to have small leaflets, the largest 11–20 × 3–8.5 mm. Those from the Mporokoso, Mpika and Isoka Districts of Zambia and from the Northern Province of

Malawi tend to have leaflets 17–27 × 8–12 mm. Both forms occur in Angola and some of the Malawi material seemed somewhat intermediate.

iii) Subsp. **suffruticans** (P.A. Duvign.) P.A. Duvign. & Brenan in Kew Bull. **20**: 11 (1966). Type from Angola.

Cryptosepalum suffruticans P.A. Duvign. in Bol. Soc. Brot. Sér. 2, **29**: 85 (1955); in C.F.A. **2**: 238 (1956).

Cryptosepalum pseudotaxus sensu White, F.F.N.R.: 122 (1962) in part as regard *White* 3446, non Baker f.

Subshrub up to 40 cm tall. Branchlets ± densely spreading pubescent. Leaflets in 6–9 pairs, 11–30 × 4–12(18) mm, ± pubescent on both surfaces or sometimes hairy only along the midrib and margins; main nerve median or a little to one side. Inflorescence axes densely spreading pubescent.

Zambia. B: Mankoya Dist., 6.5 km W of Luampa R., fr. 25.ii.1952, *White* 2039B (FHO, K). W: Mwinilunga Dist., edge of Mayowa Plains, fl. & imm.fr. 4.x.1952, *White* 3446 (FHO, K).

Also in Angola. *Isoberlinia* woodland, with *Brachystegia* or *Uapaca* and *Syzygium*, on Kalahari sands; 1000–1350 m.

This subspecies, with its densely spreading-pubescent stems and its comparatively small leaflets in up to 9 pairs per leaf, is the suffruticose counterpart to subsp. *pseudotaxus*. It was included under *C. pseudotaxus* by White (1962).

iv) Subsp. **puberulum** P.A. Duvign. & Brenan in Kew Bull. **20**: 12 (1966). Type: Zambia, Itabu R., Kalungwishi, *Fanshawe* 4853 (K holotype).

Subshrub 15–40 cm tall. Branchlets appressed puberulous. Leaflets in 2–4 pairs, 11–40 × 5–16 mm (mature leaflets at least 14 mm wide), minutely puberulous to glabrous beneath; main nerve median or slightly to one side. Inflorescence axis puberulous like the branchlets, or glabrous.

Zambia. N: Mporokoso Dist., Msanka-Lufubu divide, fl. & imm.fr. 29.ix.1956, *Richards* 6297 (K); Mbala Dist., Chinakila, fl. 5.x.1956, *Richards* 6353B (K).

Known only from N Zambia. Miombo woodland, sometimes in rocky places, said to be locally common; 900–1400 m.

This is a rather distinctive variant, but *Richards* 6317 (K), from the Nsama–Mporokoso road, collected in late flower on 27.ix.1956, may indicate some similarity to subsp. *exfoliatum*. The plant is described as a low bush to 3 feet, the stems are more branched than usual for the suffruticose subspecies, the branchlets are nearly glabrous, the few small hairs seen by Brenan on the basal part of young lateral stems near where they leave the parent stem more spreading than appressed (not evident on the specimen now at Kew); the glabrous leaves look like those of subsp. *puberulum* and subsp. *exfoliatum* var. *fruticosum*; the inflorescence axes are glabrous.

v) Subsp. **craspedoneuron** P.A. Duvign. & Brenan in Kew Bull. **20**: 12 (1966). Type: Zambia, Kawambwa Dist., between Luongo R. pontoon and Kawambwa, *Brenan & Greenway* 8114 (K holotype).

Subshrub 20–40 cm tall. Branchlets ± densely spreading pubescent. Leaflets in 6–10 pairs, older ones 15–25 × 7–9 mm, densely pubescent on both surfaces; main nerve submarginal only 0.75–1.5 mm from the proximal margin. Inflorescence axes densely spreading pubescent like the branchlets.

Zambia. N: Kawambwa Dist., between the Luongo R. pontoon and Kawambwa, fl. 15.x.1947, *Brenan & Greenway* 8114 (K).

Known only from the type collection. Miombo woodland; c. 1350 m.

This is the only element of *C. exfoliatum* to have a submarginal nerve to the leaflets, a feature not uncommon in *C. maraviense* from the same general region.

2. **Cryptosepalum maraviense** Oliv. in F.T.A. **2**: 304 (1871). —Baker, Legum. Trop. Afr.: 744 (1930). —Hutchinson, Botanist Sthn. Africa: 506 (1946). —Brenan, Check-list For. Trees Shrubs Tang. Terr.: 100 (1949). —Léonard in F.C.B. **3**: 487, t.34 & fig.42A (1952). —White, F.F.N.R.: 122, fig.25 (1962). —Duvigneaud & Brenan in Kew Bull. **20**: 12 (1966). —Brenan in F.T.E.A., Legum.-Caesalp.: 200, figs.43 & 44 (1967). Type: Mozambique, Maravi country W of L. Malawi, *Kirk* s.n. (K holotype, BR frag.) FIGURE 3.2.**17**.

Stems annual from a woody rootstock, erect, tufted, directly from the rootstock or from the basal part of the previous season's growth, 4–40 cm high, simple and with a terminal inflorescence. Leaves with rachis 3–14 cm long; leaflets in 3–16(18) pairs, 0.6–8 × 0.2–2.7 cm, ± asymmetrically oblong-lanceolate, oblong-elliptic or oblong, the upper ones sometimes slightly obovate, rounded to subacute or occasionally acute at apex, asymmetric at base, usually glabrous, sometimes ± pubescent. Racemes terminal, single, 2–12(16) cm long, glabrous to ± pubescent. Bracteoles 5–15 × 2.5–8 mm, elliptic. Sepals 1–6, small or very small. Petal 1, 7–9 mm long, sometimes with 1(2) smaller ones 3.5–5 × 1–1.5 mm. Stamens usually 3, rarely with up to 2(3) further ones that are generally smaller. Pods mostly 2.5–5 × 1.5–2.7 cm, 1–2-seeded. Seeds 1.2–1.3 × 0.7–1 cm.

Eight minor variants are recognised in miombo woodlands from Angola and Congo to Tanzania, Mozambique and Zimbabwe.

The species is defined in a very broad sense to include all the material of the genus with annual stems terminating in a single inflorescence. There is much variation in height of the stems, indumentum and leaf characters. The differences occur in numerous uncorrelated combinations, but Duvigneaud & Brenan (1966) devised a system of informal groups which have some geographical basis, and emphasise some of the main nodes of the variation. Most of the forms overlap the range of one or sometimes even several of the others. The authors emphasize that they are only attempting to separate tendencies, to which numerous exceptions exist and where frequent doubtful specimens are to be expected. Nonetheless, material collected in the last 40 years all falls relatively easily into the pattern they recognised and affirms their analysis.

They divided the species complex into three classes on size: plants > 15 cm tall, 5–15 cm tall and stems (up to base of the inflorescence) < 4.5 cm (even in fruit). The largest plants (Variants A–D) extend down from Tanzania into N Zambia, Malawi, Mozambique and E Zimbabwe. Among these, a variant with an excentric main nerve to the leaflets (Variant B) occurs SW of Lake Tanganyika around Mporokoso, one with few large leaflets (Variant C) ranges from the eastern side of Lake Tanganyika south to N Zambia (intergrading eastwards with Variant A) and another that is uncommonly

---

Fig. 3.2.**17**. CRYPTOSEPALUM MARAVIENSE. 1, habit (× 1); 2, lower surface of leaflet, showing venation (× 1), 1 & 2 from *Heusghem* in *Delevoy* 1296; 3, same, showing variation (× 1), from *Quarré* 3384; 4, flower bud, with bract and bracteoles (× 5); 5, flower opened out (× 5); 6, section of flower (× 5), 4–6 from *Heusghem* in *Delevoy* 1296; 7, pod (× 1); 8, seed (× 2), 7 & 8 from *Schmitz* 3084. Drawn by J.M. Lerinckx. Reproduced from Flora du Congo Belge with permission of the National Botanic Garden of Belgium.

4
1
5
6
3
2
8
7
J.M. LERINCKX DEL.

hairy (Variant D) from L. Mweru to Luwingu in N Zambia. The middle-sized plants are mostly aggregated into Variant E, which is widespread and the common form over much of Zambia and adjoining Katanga in Congo. Variant H is similar, except the main nerve of the leaflets is very excentric, like Variant B, and occurs in a small area slightly further south across the intruding 'finger' of the Katanga. The smallest plants, with stems mostly less than 4.5 cm tall, occur in small areas: Variant F around Solwezi in W Zambia (with an apparently disjunct population around Iringa in S Tanzania) and one with more numerous small leaflets, Variant G, in N Zimbabwe.

1. Annual shoots (measured from base of stem to end of inflorescence) usually more than 15 cm high, sometimes as much as 35 cm or more; basal scales on annual shoots usually with a tendency to be concentrated near base of stem; indumentum on stems variable, spreading, appressed or absent; leaflets variable, glabrous to pubescent all over 2
– Annual shoots less than 15(17) cm high; basal scales on annual shoots usually with a tendency to be spread out over the stem below the lowest leaf; indumentum on stems usually of spreading hairs, sometimes absent; leaflets glabrous or subglabrous except for a few hairs on margin and sometime midrib 5
2. Petioles 0–3.5 mm long; stems tall, spreading hairy; leaflets pubescent all over on both surfaces (L. Mweru to Luwingu in N Zambia) .... **iv)** Variant **D**
– Petioles 5 mm or more long, vary rarely as short as 3 mm; indumentum on stem and leaflets variable 3
3. Leaflets, or at least some of them, up to 15 mm or more wide, in not more than 7 pairs; stems spreading hairy to glabrous (N Zambia) .... **iii)** Variant **C**
– Leaflets usually up to 13 mm or less wide, variable in number, but in up to 13 pairs; stem often appressed hairy, but sometimes with spreading hairs or glabrous ... 4
4. Main nerve of leaflet not very strongly excentric (Mozambique, Malawi, N Zambia) .... **i)** Variant **A**
– Main nerve of leaflet very strongly excentric (Zambia, Mporokoso Dist.) .... **ii)** Variant **B**
5. Main nerve of leaflet strongly excentric or marginal, less than 1 mm away from margin (NW Zambia along Katanga border region) .... **viii)** Variant **H**
– Main nerve of leaflet not very strongly excentric 6
6. Annual shoots usually 5 cm or more long from base of shoot to base of inflorescence (variable widespread complex) .... **v)** Variant **E**
– Annual shoots usually very short, usually 4.5 cm or less from from base of shoot to to base of inflorescence 7
7. Leaflets in up to 7–9 pairs per leaf, the widest leaflets (7)10–15 mm wide (Solwezi Dist. in W Zambia) .... **vi)** Variant **F**
– Leaflets up to 10–11 pairs per leaf, widest leaflets 6–9 mm wide (Zimbabwe) .. .. **vii)** Variant **G**

i) Variant **A**

Cryptosepalum maraviense Oliv. in F.T.A. **2**: 304 (1871) sensu stricto.

Cryptosepalum pulchellum Harms in Engler, Bot. Jahrb. Syst. **30**: 319, fig. p. 320 (1901). Type: Tanzania, Rungwe Dist., Kirira valley, Untali, *Goetze* 1472 (B† holotype; BR, P).

Shoots usually more than 15 cm tall; basal scales usually aggregated at base. Stems usually appressed-pubescent, sometimes spreading hairy or glabrous. Leaflets in up to 13 pairs, but often less than 8 pairs, up to 13 mm wide or less, the main nerve not very strongly excentric.

Zambia. N: Mbala Dist., Chimbwa area, 21.viii.1955, *Nash* 161 (BM). **Zimbabwe.** E: Nyanga Dist., Holdenby, above Nyakombe R., fl. 12.xi.1964, *Masterson* 456 (K,

SRGH). **Malawi.** N: Nkhata Bay Dist., near Chikwina, c. 20 km ENE of Mzuzu, fl. 4.ii.1992, *Goyder, Paton & Tawakali* 3614 (K, MAL). C: Nkhota Kota Fly Post, fr. 19.xi.1963, *Salubeni* 138 (K, MAL). **Mozambique.** N: between Maniamba and Lichinga (Vila Cabral), fl. 11.x.1942, *Mendonça* 778 (BM, K, LISC). Z: between Mualama and Naburi, 9.4 km from Mualama, fl. 5.x.1949, *Barbosa & Carvalho* 4308 (K, LISC). T: Maravi country W of L. Malawi, fl. [no date], *Kirk* s.n. (K).

Extends from S Tanzania south into the Zambesia Province of Mozambique and E Zimbabwe and west into Malawi and just across to Mbala from the Tanzanian side of the border. Miombo woodland; 450–1550 m.

Overlaps and intergrades with the most widespread element, Variant E, and abuts or possibly intergrades with Variant C in the drainage area of the Luangwa R. in N Zambia.

ii) Variant **B**

Shoots usually more than 15 cm tall; basal scales usually aggregated at base. Stems usually appressed-pubescent, sometimes spreading hairy or glabrous. Leaflets in up to 13 pairs, but often less than 8 pairs, up to 13 mm wide or less, the midrib very excentric.

Zambia. N: Mporokoso Dist., Chishi (Chisi) woodlands, fl. & fr. 24.ix.1956, *Richards* 6272 (K); Mporokoso–Chiengi Road, 113.5 km (71 miles), fl. 27.x.1949, *Hoyle* 1325 (FHO, K).

Known only from around Mporokoso. *Gilges* 118 (K) from Barotseland, Zambezi (Balovale) cited by Duvigneaud & Brenan (1966), is better ascribed to the ubiquitous Variant E.

iii) Variant **C**

Cryptosepalum busseanum Harms in Bot. Jahrb. Syst. **33**: 156 (1902). Type: Tanzania, Songea Dist., Madjanga-Kwa-Bagaya, *Busse* 633 (B† holotype; BM, BR, EA, K).

Cryptosepalum boehmii Harms in Bot. Jahrb. Syst. **33**: 156 (1902). Type: Tanzania, Mpanda Dist., between Kabombue and Ndani, *Boehm* 17a (B† holotype).

Shoots usually more than 15 cm tall; basal scales usually aggregated at base. Stems spreading hairy to glabrous. Leaflets in not more than 7 pairs, up to 15 mm or more wide, with the main nerve ± median.

Zambia. N: Mporokoso Dist., 24 km beyond Kalungwishi Ferry by road to Mporokoso, fl. 16.x.1947, *Brenan & Greenway* 8123 (FHO, K); Mbala Dist., Kambole Escarpment, fl. 23.viii.1956, *Richards* 5934 (K).

Principally from the Buha region of NW Tanzania (northern limit of the genus) down L. Tanganyika into the Mbala and Mporokoso Districts of Zambia. It occurs in drier woodlands to the east across the drainage basin of the Luangwa Valley in N Zambia, but these records, including for example *Hutchinson & Gillett* 3788 (K) from between Mpika and the Chambesi R., form a transition with Variant A to the east. Duvigneaud & Brenan (1966) cite other odd specimens from SE Tanzania, N Malawi and N Mozambique, but certainly *Hornby* 2174 (K) from Metonia in N Mozambique, and *B.D.Burtt* 5853 (EA, K), from west of Nkhota Kota in central Malawi, now seem more conformable in Variant A. The collections from N Malawi have been much enhanced in the last 40 years.

iv) Variant **D**

Cryptosepalum pulchellum var. *rhodesicum* R.E. Fr. in Wiss. Ergebn. Schwed. Rhod.-Kongo-Exped. **1**: 69 (1914). Type: Zambia, Luwingu Dist., Malolo near Luwingu, *R.E. Fries* 1109 (UPS holotype, K photo).

Shoots usually more than 15 cm tall; basal scales usually aggregated at base. Stems spreading hairy. Leaflets in 9–12 pairs, less than 15 mm wide, pubescent all over on both surfaces, with the main nerve ± median; petiole unusually short, 0–3.5 mm long.

Zambia. N: Luwingu, near Government Rest House, fl. & fr. 15.x.1947, *Brenan & Greenway* 8110 (K); Luwingu, fl. 30.v.1964, *Fanshawe* 8725 (K, NDO).

Known only from the Northern Province of Zambia between L. Mweru and Luwingu. Miombo woodland; 1000–1450 m.

A very distinct variant with short petioles and hairy leaves.

v) Variant **E**

Cryptosepalum dasycladum Harms in Bot. Jahrb. Syst. **30**: 319, fig. on p. 320 (1901). Types: Tanzania, Mbeya Dist., Mbozi hill, *Goetze* 1384 (B† syntype, BR) and Unyika, Kananda village, *Goetze* 1438 (B† syntype).

Shoots usually 5–15 cm tall; basal scales with a tendency to be spread out over the stem below the lowest leaf. Stems usually spreading hairy, rarely glabrous. Leaflets in 5–14 pairs, less than 15 mm wide, glabrous or subglabrous, with the main nerve not very excentric.

Zambia. B: Zambezi (Balovale), fl. viii.1952, *Gilges* 179 (K). N: Mbala Dist., Kawimbe road, fl. 9.vii.1970, *Sanane* 1262 (K). W: Kabompo Dist., between Kabompo and Mayowa Plains, fl. 5.x.1952, *White* 3457A (BM, FHO, K). C: Mutinondo Wilderness Area, fl. 13.vi.1998, *P.P. Smith* 1703 (K). **Zimbabwe.** N: Manyame (Hunyani) Mts., fl. & imm.fr. x.1927, *Henkel* 1386 (K, SRGH). **Malawi.** N: Livingstonia Escarpment, fl. 31.xii.1973, *Pawek* 7678 (K, MAL, MO). S: Lukulesi R. area, fl. & fr. 18.xii.1957, *Chapman* 502 (K). **Mozambique.** N: Metonia, fl. viii.1931, *Gomes e Sousa* 872 (K). T: between Furancungo and Ulongue (Vila Coutinho), fl. 28.ix.1942, *Mendonça* 510 (K, LISC).

Widespread from Angola to Tanzania and south to N Zimbabwe. Miombo woodland, most evident in places with rather impeded growth after recent fires, along dambos and in sand; 1000–1900 m, perhaps rather lower in Mozambique.

The most commonly collected and heterogenous variant, extending over much of the range, in places transitional to Variant A, which is just taller; elsewhere Duvigneaud & Brenan (1966) thought it might have introgressed with Variants C and H.

vi) Variant **F**

Shoots very short, usually 4.5 cm or less tall; basal scales with a tendency to be spread out over the stem below the lowest leaf. Stems usually spreading hairy. Leaflets in up to 7–9 pairs, the widest (7)10–mm wide, glabrous or subglabrous, with the main nerve not very excentric.

Zambia. W: Solwezi Dist., by road to Solwezi, 102 km from Nchanga, fr. 29.ix.1947, *Brenan & Greenway* 7986 (K); Ndola Dist., Mufulira, fr. 2.xi.1953, *Fanshawe* 471 (K, NDO).

Possibly also in S Tanzania. Miombo woodland; 1300–1400 m.

In Zambia known only around Solwezi and conspicuous by the very short stems, otherwise similar to Variant E.

vii) Variant **G**

Shoots usually 4.5 cm or less tall; basal scales with a tendency to be spread out over the stem below the lowest leaf. Stems usually spreading hairy. Leaflets in up to 10–11 pairs, the widest 6–9 mm wide, glabrous or subglabrous, with the main nerve not very excentric.

Zimbabwe. N: Chinoyi–Mwami (Sinoia–Miami) road, fl. 26.ix.1945, *Wild* 103 (K, SRGH).

Known only from N Zimbabwe. Miombo woodland; 1200–1350 m.

Generally shorter than Variant E, but otherwise essentially the same.

viii) Variant **H**

Cryptosepalum mimosoides sensu R.E. Fries, Wiss. Ergebn. Rhod.-Kongo Exped. **1**: 69 (1911) non Oliv.

Shoots usually 5–15 cm tall; basal scales with a tendency to be spread out over the stem below the lowest leaf. Stems usually spreading hairy, rarely glabrous. Leaflets in 10–16 pairs, less than 15 mm wide, glabrous or subglabrous, with the main nerve excentric or marginal.

Zambia. N: Kapalala, fl. 3.ix.1953, *Fanshawe* 299 (EA, K, NDO). W: c. 11 km N of Kapiri Mposhi by the Ndola road, fl. 24.ix.1947, *Brenan & Greenway* 7950 (K).

Also in adjoining parts of the Congo (Katanga). Miombo woodland; 1250–1400 m.

The leaflets have a markedly excentic main nerve like Variant D from a little further north, but the plants are smaller in stature.

18. BERLINIA Hook. f. & Benth.*

Berlinia Hook. f. & Benth. in Hooker, Niger Flora: 326 (1849). —Léonard in Mém. Acad. Roy. Sci. Belg. **30**(2): 180 (1957).

Trees, evergreen, unarmed. Leaves paripinnate; leaflets opposite or subopposite, in (1)2–5 pairs, petiolulate, equal- to unequal-sided at the base, with a marginal nerve, with or without translucent dots; stipules intrapetiolar, connate into a scale. Inflorescence of simple racemes or panicles; bracteoles 2, well developed, valvate, completely enclosing the flower buds, persistent. Hypanthium 7–20 mm long tubular. Sepals 5, subequal. Petals 5, the upper one much larger than the others and long-clawed (in the F.Z. area). Stamens 10; nine filaments shortly connate at the base, one free. Ovary shortly stipitate, inserted on the tube of the hypanthium, 2–8-ovulate. Pod dehiscing elastically into two obliquely transversely nerved valves; upper suture narrowly winged on each side. Seeds flattened, obovate-elliptic to elliptic or quadrate; testa thin; funicle short.

About 15 species, mainly in Guineo-Congolian forests; taxonomically difficult and currently under revision.

Leaflets in 2–4 pairs; pedicels 1.5–3 cm long; hypanthium glabrous or nearly so .... **1.** *orientalis*

Leaflets in 3–5 pairs; pedicels 5–7 cm long; hypanthium shortly velutinous .... **2.** *giorgii*

1. **Berlinia orientalis** Brenan in Kew Bull. **17**: 211, fig.2 (1963); in F.T.E.A. Legum.-Caesalp.: 143, fig.29 (1967). —Gomes e Sousa, Dendrol. Moçamb. Estudo Geral **1**: 265, fig.63 (1966). Type: Mozambique, mouth of Msalu R., *Allen* 35 (K holotype). FIGURE 3.2.**18**.

Tree to 20 m tall. Branchlets glabrous. Leaflets in 2–4 pairs, 4.5–16.5 × 2.5–7 cm, ovate to elliptic, cuneate to rounded at the base, acuminate with blunt tips at the apex, glabrous. Inflorescences paniculate, terminal, 3–11 cm long; pedicels 1.5–3 cm long, tomentellous; bracteoles 2–3 × 1–1.4 cm, obovate, tomentellous. Hypanthium 0.8–1.2 cm long, glabrous or nearly so. Sepals 5. Petals 5, the upper one white, shaded green in the centre, 3.5–4.5 cm long, the expanded portion 2–2.5 × 2.5–3.8 cm, the other petals 1–1.7 cm long, ± linear. Pods brownish, to 30 × 8 cm, oblong. Seeds up to 4 × 3.5 cm, elliptic, flattened.

Mozambique. N: Palma–Nangadi road, 30 km from Palma, fl. 23.x.1960, *Gomes e Sousa* 4575 (COI, K, LISC).

---

* I am grateful to Barbara Mackinder for comments on a draft of this account and for allowing me to see a draft of her account of the genus.

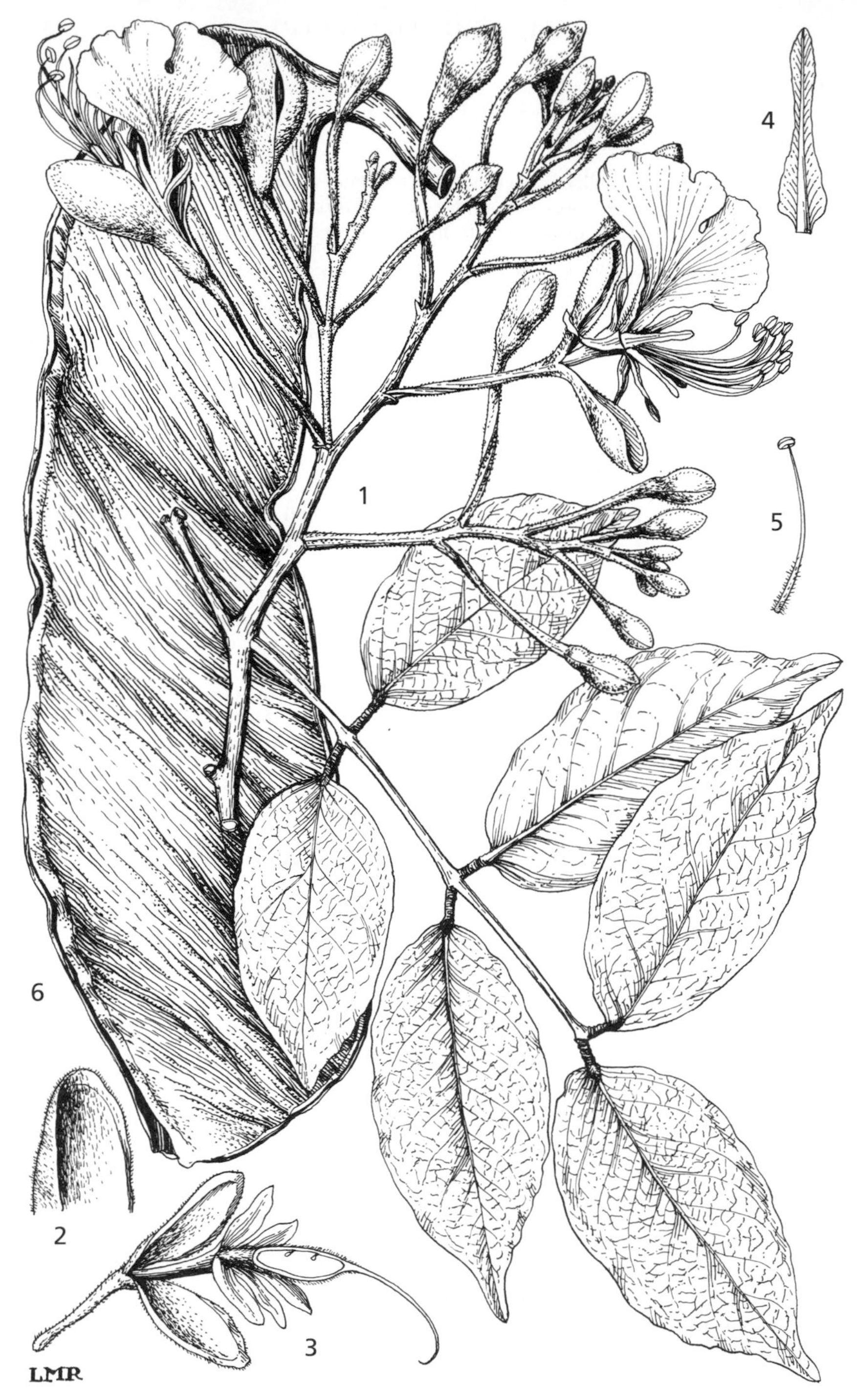

Fig. 3.2.**18**. BERLINIA ORIENTALIS. 1, flowering branchlet (× $^{2}/_{3}$), from *Eggeling* 6741; 2, part of bracteole to show indumentum (× 1); 3, flower, showing barcteoles, calyx and ovary, other parts removed ($^{2}/_{3}$), 2 & 3 from *Mendonça* 1005; 4, one of four smaller petals (× 2); 5, stamen (× $^{2}/_{3}$), 4 & 5 from *Eggeling* 6741; 6, immature pod (× $^{2}/_{3}$), from *Gomes e Sousa* 4575. Drawn by Lura Ripley. From F.T.E.A.

Also in S Tanzania. Open coastal forest on sandy clay soils; 10–150 m.

Conservation notes: Very conspicuous when in flower, and there are numerous collections from its apparently restricted range. Probably Lower Risk, Least Concern, but increased clearance and settlement in its area could change this to Vulnerable.

2. **Berlinia giorgii** De Wild., Pl. Bequaert. **3**: 140 (1925). —Baker, Legum. Trop. Afr.: 685 (1930). —Hauman in F.C.B. **3**: 394 (1952). Type from Congo.

Berlinia acuminata var. *pubescens* De Wild. in Ann. Mus. Congo. Belge, Bot. Sér. 5, **1**: 29 (1903). Type from Congo.

Berlinia gilletii De Wild., Pl. Bequaert **3**: 138 (1925). —Baker, Legum. Trop. Afr.: 634 (1930). Type from Congo.

Berlinia giorgii var. *gossweileri* Baker f. in J. Bot. **66**: 141 (1928). Type from Angola.

Berlinia giorgii var. *gilletii* (De Wild.) Hauman in F.C.B. **3**: 395 (1952).

Berlinia giorgii var. *vernicosa* Hauman in Bull. Séances Inst. Roy. Colon. Belge **23**: 482 (1952); in F.C.B. **3**: 394 (1952). Type from Congo.

Berlinia giorgii var. *pubescens* (De Wild.) Hauman in F.C.B. **3**: 395 (1952).

Berlinia sp. 1 sensu White, F.F.N.R.: 100 (1962).

Tree to 18 m tall. Branchlets minutely puberulous. Leaflets in 3–5 pairs, 10–16 × 7–9.5 cm, oblong, broadly elliptic or obovate-oblong, cuneate to rounded at the base, emarginate at the apex, sparsely and minutely puberulous on midrib and nerves beneath. Inflorescences paniculate, terminal, up to 35 cm long; pedicels 5–7 cm long, puberulous; bracteoles 3.2 × 1.6 cm, obovate-spathulate, shortly velutinous inside and outside, coriaceous. Hypanthium 1.2–1.5 cm long, shortly velutinous. Sepals 5. Petals 5, white, the upper one 6.5–7.5 cm long, the expanded portion 4.5–5 × 5.5–6 cm, the other petals c. 2 cm long, linear from an expanded base. Pods 30–35 × 7–8 cm, oblong, brown velutinous. Seeds 3 × 2 cm, flattened, elliptic.

Zambia. N: Mbala Dist., Inona R. on way to Mpulungu, fl. 7.v.1936, *B.D. Burtt* 5909 (BM, BR, K); Mbala (Abercorn) to Mpulungu, fr. 17.ix.1950, *Bullock* 3329 (K).

Also in Congo and northwards to Nigeria, although true distribution will only be clear after full revision of the genus. In riverine and gallery forests; c. 1400 m.

Conservation notes: Best treated as Data Deficient pending revision of the genus.

19. ISOBERLINIA Craib & Stapf

Isoberlinia Craib & Stapf in Bull. Misc. Inform., Kew, Add. Ser. **9**: 266 (1911); in Bull. Misc. Inform., Kew **1912**: 93 (1912). —Léonard In Mém. Acad. Roy. Sci. Belg. **30**(2): 173 (1957).

Unarmed trees or shrubs. Leaves paripinnate; leaflets in 2–5 pairs, opposite or subopposite, petiolulate, asymmetric at base, without translucent dots; stipules intrapetiolar, connate, free above. Inflorescences normally paniculate; bracteoles 2, well developed, valvate, completely enclosing the flower buds, persistent. Hypanthium tubular. Sepals (4)5(7), subequal. Petals 5(6), usually all well developed; upper petal longer or wider than the rest and often differently shaped. Stamens 10(14); filaments free. Ovary sessile to shortly stipitate, inserted on the tube of the hypanthium; ovules 4–8. Pod woody, dehiscing elastically into two flattened valves; upper suture unwinged. Seeds compressed, ± obovate-elliptic, with thin testa and short funicle.

A genus of about 6 species, some in equatorial forests but mainly in seasonal tropical woodlands, occurring in Africa both north and south of the Equator. White (F.F.N.R., 1962) took a very broad view and recognised only a single species in Zambia; here Brenan's taxonomic concepts as used in F.T.E.A. are followed.

Upper posterior petal bilobed or deeply emarginate at apex; ultimate branches of inflorescence comparatively short, 2.5–8 cm long; inflorescence comparatively dense and compact; bracts subtending pedicels 2–4 mm long .... **1.** *angolensis*
Upper posterior petal rounded or only slightly emarginate at apex; ultimate branches of inflorescence often elongate, 3.5–25 cm long; inflorescence rather dense to lax; bracts subtending pedicels 4–7 mm long .... **2.** *tomentosa*

1. **Isoberlinia angolensis** (Benth.) Hoyle & Brenan in Kew Bull. **4**: 78 (1949). Type: Angola, Cuanza Norte between Pungo Andongo & Candumbo, *Welwitsch* 568 (LISU syntype, BM).

Tree, sometimes shrubby and flowering when no more than 1 m tall, 1–15 m tall; bark grey to brown, fissured and scaly. Leaves: stipules connate, intrapetiolar, with upper parts normally absent; petiole with rachis 6.5–25 cm long; leaflets 3–4(5) pairs, (4)7–19 × 2–7.8 cm (larger on coppice shoots), ovate, elliptic or oblong, rounded to cuneate at the base, acute to obtuse or retuse at the apex, glabrous to shortly tomentose; lateral nerves (8)11–16 on each side of midrib. Panicles terminal or axillary, much branched, ± dense, ultimate raceme branches short, 3–8 cm long, ± shortly brown-tomentose to almost glabrous; bracts 3–4 mm long; pedicels of open flowers 3–4(15) mm long; bracteoles brown to greenish brown, 10.5–16 × 6–9 mm. Sepals white. Petals white; upper petal 9–12 × 6.5–8 mm, obovate or oblong-obovate, bilobed or deeply emarginate; other 4 petals 7–12 × 3–5 mm. Stamens 10. Pods large, 19–35 × 6–8 cm, rusty-tomentose to glabrescent.

Three varieties can be distinguished, although intermediates can be found.

1. Lower surface of leaves glabrous or sparsely pubescent, although nerves may be densely pubescent; hypanthium glabrous or pubescent outside 2
– Lower surface of leaves ± densely pubescent or shortly tomentose all over; hypanthium ± pubescent outside (rarely glabrous) .... **iii)** var. *niembaensis*
2. Hypanthium glabrous outside .... **i)** var. *angolensis*
– Hypanthium ± pubescent outside .... **ii)** var. *lasiocalyx*

i) Var. **angolensis**

Berlinia angolensis Benth. in Trans. Linn. Soc. **25**: 310 (1865). —Oliver in F.T.A. **2**: 296 (1871). —Baker, Legum. Trop. Afr.: 687 (1930).
Westia angolensis (Benth.) Macbride in Contrib. Gray Herb. **59**: 21 (1919).

Lamina glabrous or sparsely pubescent beneath, but nerves pubescent; hypanthium glabrous outside.

Zambia. N: Mbala Dist., Ndundu to Mbala (Abercorn), fl. 23.iv.1966, *Richards* 21457 (K). **Malawi.** N: Mzimba Dist., SW of Luwawa Dam on road towards A1, fr. 10.iv.1990, *Brummitt & Seyani* 18689 (K, MAL, MO, NY, P, WAG).
Also in S Sudan, Tanzania, Congo and Angola. In miombo woodland and wooded grassland; 950–1800 m.

ii) Var. **lasiocalyx** Hoyle & Brenan in Kew Bull. **4**: 78 (1949). —Brenan, Check-list For. Trees Shrubs Tang. Terr.: 103 (1949); in Kew Bull. **17**: 222 (1963); in F.T.E.A. Legum.-Caesalp.: 141, fig.28 (1967). —Troupin in Bull. Jard. Bot. État **20**: 306 (1950). —Hauman in F.C.B. **3**: 380 (1952). Type: Malawi, Nyika Plateau, *Whyte* s.n. (K holotype). FIGURE 3.2.**19**/1–8.

Berlinia densiflora Baker in Bull. Misc. Inform., Kew **1897**: 269 (1897). —Baker, Legum. Trop. Afr.: 687 (1930). Type as above.

Berlinia stolzii Harms in Bot. Jahrb. Syst. **53**: 465 (1915). —Baker, Legum. Trop. Afr.: 687 (1930). Type: Tanzania, Rungwe Dist., Mulinda Forest, *Stolz* 1957 (EA, K).

Isoberlinia densiflora (Baker) Milne-Redh. in Bull. Misc. Inform., Kew **1937**: 415 (1937).

Lamina glabrous or sparsely pubescent beneath; hypanthium pubescent outside.

Zambia. N: 18 km NW of Mbala (Abercorn), fr. 19.vii.1930, *Hutchinson & Gillett* 3937 (K, PRE). C: Kasanka Nat. Park, 11 km E of Musande Tent Camp and 4 km W of pontoon over Kasanka R., fl. 19.xi.1993, *Harder, Merello, Nkhoma & Mupeta* 1979 (K, MO). **Malawi.** N: Rumphi Dist., Katumbi's Village, fl. & imm.fr. 1.v.1952, *White* 2543 (FHO, K).

Also in Tanzania, Congo and Angola. In miombo woodland and wooded grassland; 1000–2100 m.

iii) Var. **niembaensis** (De Wild.) Brenan in Kew Bull. **17**: 222 (1963); in F.T.E.A., Legum.-Caesalp.: 141 (1967). Type: Congo, Niemba Camp, *Delevoy* 289 (BR holotype).

Berlinia niembaensis De Wild., Pl. Bequaert. **3**: 145 (1925). —Baker, Legum. Trop. Afr.: 689 (1930).

Isoberlinia niembaensis (De Wild.) P.A. Duvign. in Bull. Séances Inst. Roy. Colon. Belg. **21**: 434 (1950). —Troupin in Bull. Jard. Bot. État **20**: 308 (1950). —Hauman in F.C.B. **3**: 382 (1952).

Lamina ± densely pubescent or shortly tomentose beneath; hypanthium pubescent or (rarely) glabrous outside.

Zambia. B: Zambezi (Balovale), Kabompo, fl. 10.xi.1952, *Gilges* 235 (K). N: Mbala, near Middle Lunza (Lunzua) Falls by power station, fr. 12.i.1975, *Brummitt & Polhill* 13746 (K). W: Mwinilunga Dist., just N of Mwinilunga, fl. 28.x.1937, *Milne-Redhead* 2995 (K). C: Lusaka Dist., Great East Road, c. 104 km (mile 65) Lusaka to Chipata (Fort Jameson), fr. 16.iv.1952, *White* 2691 (FHO, K). E: 6 km (4.5 miles) Lundazi to Mzimba, fr. 18.x.1958, *Robson & Angus* 149 (K). **Malawi.** N: Mzimba Dist., edge of Champira Forest, 3 km (2 miles) NE of Katete, fl. 26.xii.1975, *Pawek* 10557 (K, MAL, MO, SRGH). C: Kasungu Dist., Kasungu Nat. Park, fl. 25.xi.1972, *Pawek* 6018 (K, MAL, MO).

Also in Tanzania, Congo and Angola. In miombo woodland and wooded grassland; 600–1650 m.

The leaflets of this variety tend to be smaller and more rounded at the apex than those of the other varieties.

2. **Isoberlinia tomentosa** (Harms) Craib & Stapf in Bull. Misc. Inform., Kew **1912**: 93 (1912). —Brenan, Check-list Trees Shrubs Tang. Terr.: 104 (1949); in Kew Bull. **17**: 224 (1963); in F.T.E.A. Legum.-Caesalp.: 142, fig.28/9, 10 (1967). —Troupin in Bull. Jard. Bot. État **20**: 307 (1950). —Hauman in F.C.B. **3**: 381 (1952). —Léonard in Mém. Acad. Roy. Sci. Belg. **30**(2): 175, 176 (1957). Type: Tanzania, Rungwe Dist., Umuamba, L. Likaba, *Goetze* 1315 (B† holotype). FIGURE 3.2.**19**/9, 10.

Berlinia tomentosa Harms in Bot. Jahrb. Syst. **30**: 321 (1901). —Baker, Legum. Trop. Afr.: 689 (1930).

Westia tomentosa (Harms) Macbr. in Contrib. Gray Herb. **59**: 21 (1919).

Tree 3–12 m tall. Leaves: stipules connate, intrapetiolar, often with foliaceous ovate-cordate auricles to ± 2.5–4 × 2–2.5 cm; petiole with rachis 15–27 cm long; leaflets 3–4(5) pairs, (7)10–25 × (4)4.8–12.7 cm (larger, to 30 × 16 cm, on coppice and juvenile shoots — see *Milne-Redhead* 795), ovate-elliptic, elliptic, or sometimes oblong-elliptic, ± densely pubescent or tomentose at

Fig. 3.2.**19**. ISOBERLINIA ANGOLENSIS var. LASIOCALYX. 1, flowering branchlet (× $^2/_3$); 2, flower bud, with bract (× 2); 3, bracteoles and calyx (× 1); 4, flower, with bracteoles removed (× 2); 5, upper petal (× 2), 1–5 from *Stolz* 1957; 6, pod (× $^2/_3$), from *Greenway* 3626; 7, upper suture of pod cut transversely (× $^2/_3$); 8, seed (× $^2/_3$), 7 & 8 from *Procter* 704. ISOBERLINIA TOMENTOSA. 9, flower bud, with bract (× 2); 10, upper petal (× 2), 9 & 10 from *Richards* 1796. Drawn by Lura Ripley. From F.T.E.A.

least on midrib and lateral nerves beneath and sometimes all over; primary lateral nerves (9)12–14 on each side of the midrib in mature leaflets. Inflorescences lax; ultimate branches 5–25 cm long, dark brown-tomentose; bracts 4–8 mm long; pedicels of open flowers 4–13 mm long; bracteoles brown, 11–20 × 10–12 mm. Sepals 5–6. Petals white, the upper one 12–15 × 6–10 mm, obovate or obovate-elliptic, rounded and not bilobed at apex, the other 4–5 smaller, 10–13 × 3.5–5 mm. Stamens 10(14). Pods 15–30 × 5.4–8.5 cm, densely rusty-brown-tomentose, somewhat glabrescent. Seeds 2.5–3.2 × 1.8–3 cm.

Zambia. N: Kitchen Garden Shamba, Chilongowelo, fl. 16.v.1952, *Richards* 1796 (K). W: Solwezi Dist., R. Nyalisonga, st. 29.vii.1930, *Milne-Redhead* 795 (K). **Malawi.** N: Mzimba Dist., 11 km (7 miles) WSW of Chikangawa, fl. 10.xi.1970, *E. Phillips* 4213 (K).

Guinea Republic east to Sudan; Congo, Tanzania. In miombo woodland; 900–1600 m.

20. JULBERNARDIA Pellegr.

Julbernardia Pellegr. in Boissiera **7**: 297 (1943). —Léonard in Mém. Acad. Roy. Sci. Belg. **30**(2): 188 (1957).

Pseudoberlinia P.A. Duvign. in Bull. Séances Inst. Roy. Colon. Belge **21**: 431 (1950).

Trees, often evergreen, unarmed. Leaves paripinnate; leaflets 2–many, opposite, asymmetric at base; translucent dots present or not; petiolules usually twisted with secondary nerves from the base on the proximal side of the main nerve, the venation prominent on both surfaces; stipules intrapetiolar, connate below, bilobed, persistent or caducous. Inflorescence a terminal panicle; bracteoles 2, valvate, enclosing the flower buds, persistent, keeled on the back. Hypanthium absent. Sepals 5, well developed, ± equal. Petals 5, one large and four small (in our area). Stamens 10, all fertile; nine filaments shortly connate, the tenth free. Ovary very shortly stipitate, up to 5-ovulate; style elongate; stigma capitate. Pods flattened, woody, elastically dehiscent into two valves; upper suture usually with a lateral flange-like ridge or wing, 1–5-seeded. Seeds flattened, without areoles, on short funicles.

Leaflets usually in 4–6 pairs; sepals oblong, not or only slightly wider above than below, not contiguous, 1.5 mm wide; largest petal ovate, 6.5–9 mm long; bracts usually half as long as the bracteoles or more . . . . **1.** *globiflora*

Leaflets usually in 3–4 pairs; sepals broadly obovate-spathulate, contiguous or slightly imbricate, 3.5–4.5 mm wide; largest petal c. 6 mm long; bracts much shorter than the bracteoles . . . . **2.** *paniculata*

1. **Julbernardia globiflora** (Benth.) Troupin in Bull. Jard. Bot. État **20**: 314 (1950). —Brenan in Mem. New York Bot. Gard. **8**: 427 (1954); in F.T.E.A. Legum.-Caesalp.: 147, fig.30 (1967). —Léonard in Mém. Acad. Roy. Sci. Belg. **30**(2): 192, 195 (1957). —White, F.F.N.R.: 125 (1962). —Gomes e Sousa, Dendrol. Moçamb. Estudo Geral **1**: 266, fig.64 (1966). —Ross in F.S.A. **16**(2): 43 (1977) as *globifera*. —M. Coates Palgrave, Trees Sthn. Africa: 332 (2002). Type: Malawi, Shire Highlands, *Buchanan* 138 (K holotype). FIGURE 3.2.**20**/1–11.

Brachystegia globiflora Benth. in Hooker's Icon. Pl. **14**: 43 (1881).

Berlinia eminii Taub. in Engler, Pflanzenw. Ost-Afrikas **C**: 199 (1895). —Harms in Engler, Pflanzenw. Afrikas **3**(1): 471, fig.255 (1915). —Eyles in Trans. Roy. Soc. S. Afr. **5**: 366 (1916). Types: Tanzania, Tschaja (?L. Chaya, Dodoma), Karagwe, *Stuhlmann* 498, and Tabora Dist., Igonda, *Boehm* 157a (both B† syntypes).

Berlinia globiflora (Benth.) Harms in Engler, Pflanzenw. Afrikas **3**(1): 472 (1915). —Baker, Legum. Trop. Afr.: 689 (1930).

Isoberlinia globiflora (Benth.) Greenway in Bull. Misc. Inform. Kew **1928**: 203 (1928). —Brenan, Check-list Trees Shrubs Tang. Terr.: 104 (1949). —O. Palgrave, Trees Central Africa: 107 (1957). —Burtt Davy & Hoyle in Topham, Check List For. Trees Shrubs Nyasaland Prot.: 37 (1958).

Pseudoberlinia globiflora (Benth.) P.A. Duvign. in Bull. Séances Inst. Roy. Col. Belge **21**: 434 (1950). —Hauman in F.C.B. **3**: 405 (1952).

Tree 5–15 m tall, with flat or rounded spreading crown, but sometimes shrubby or flowering from coppice shoots 0.3–2 m high; bark rough, grey. Young branches pubescent, glabrescent. Leaves: stipules intrapetiolar, 3–5 mm long, connate at base, bicuspidate, caducous; petiole with rachis (3)5–20 cm long; leaflets in (2)4–6(8) pairs, (1)2–8.5(11.5) × (0.6)1–3.3(5.6) cm, oblong-elliptic, ovate-oblong or obovate-oblong, asymmetrically cuneate at base, obtuse to emarginate at apex, margins white pubescent. Panicles up to 30 cm long and wide, brown pubescent; bracts 2–10 mm long, usually at least half as long as the bracteoles; bracteoles 7–10 × 6–9 mm. Sepals 2.5–4.5 × 1.5 mm, oblong, non-contiguous. Petals white, the larger one 6.5–9 × 4–5 mm, ovate, shortly clawed, others 3–8 × 0.5–3 mm, narrowly obovate to spathulate or linear. Pods 4–9 × 2–3.2 cm, obovate-oblong to oblong, brown tomentose. Seeds dark brown, 1.3–1.6 × 1–1.5 cm.

Botswana. N: Nata–Kazungula road at 18°44.3'S, 25°36.2'E, c. 286 km (179 miles) N of Nata and c. 400 km (250 miles) N of Sibuyu Forest Res. boundary, fl. & imm.fr. 11.iv.1983, *P.A. Smith* 4230 (K, SRGH). **Zambia.** N: Mbala Dist., near Mpulungu, fr. 23.v.1936, *B.D. Burtt* 5899 (K). C: Chongwe, fl. 26.xi.1965, *Lawton* 1333 (K). E: Chipata (Fort Jameson) township, fl. & imm.fr. 25.iv.1952, *White* 2465 (FHO, K). S: Livingstone Forest Res., fl. 8.i.1956, *Gilges* 524 (K, SRGH). **Zimbabwe.** N: Kariba Nat. Park, 40 km from Makuti on Kariba road, fl. 19.ii.1981, *Philcox, Leppard & Dini* 8759 (K). W: Hwange Dist., Zambezi Camp, above Victoria Falls, fl. 25.i.1979, *Mshasha* 171 (K, SRGH). C: Domboshawa Mt, fl. & imm.fr. 5.ii.1987, *Bayliss* 10672 (K, MO). E: Mutare Dist., Mutare (Umtali) Commonage, fl. 6.ii. 1949, *Chase* 1172 (K, LISC, SRGH). S: Kopjes by Lundi R., fr. 30.vi.1930, *Hutchinson & Gillett* 3295 (K, PRE). **Malawi.** N: Mzimba Dist., near Mbelwa Institute, 90 km (62 miles) SW of Mzuzu, fl. 30.iii.1969, *Pawek* 1898 (K). C: Nkhotakota Dist., Chia area, fr. 1.ix.1946, *Brass* 17475 (K, NY). S: Zomba Plateau road, fr. 5.iv.1984, *Banda & Salubeni* 2143 (K, MAL). **Mozambique.** N: Corrane Region, near road to Nampula, fl. 21.ii.1953, *Gomes e Sousa* 4036 (K). Z: Namagoa, 200 km inland from Quelimane, fl. iii-iv.1945, *Faulkner* PRE series 58 (K). T: Tete, between Angónia and Casula, st. 27.iii.1941, *Torre* 3352 (K, LISC). GI: 20 km S of Inhambane, fl. xii.1935, *Gomes e Sousa* 1692 (K, LISC). MS: Chimoio, crossing of Chimoio (Vila Pery) and Tete roads, fl. 31.iii.1948, *Barbosa* 1299 (K, LISC).

Also in Tanzania and Congo. Common and often dominant in deciduous woodland, especially in the drier areas and on poor soils of plateaux and slopes; up to 1950 m.

Conservation notes: Widespread and abundant; Lower Risk, Least Concern.

Ross in F.S.A. (1977) consistently refers to this taxon as *J. globifera.* This is an error.

2. **Julbernardia paniculata** (Benth.) Troupin in Bull. Jard. Bot. État **20**: 316 (1950); Torre & Hillcoat in C.F.A. **2**: 211 (1956). —Léonard in Mém. Acad. Roy. Sci. Belg. **30**(2): 192, 195, 196 (1957). —Brenan in Mem. New York Bot. Gard. **8**: 427 (1954); in F.T.E.A., Legum.-Caesalp.: 147, fig.30 (1967). —White, F.F.N.R.: 125 (1962). —M. Coates Palgrave, Trees Sthn. Africa: 333 (2002). Type: Angola, Huila, *Welwitsch* 581 (LISU lectotype; BM, K isotype). FIGURE 3.2.**20**/12.

Berlinia paniculata Benth. in Trans. Linn. Soc. **25**: 311 (1865). —Oliver in F.T.A. **2**: 295 (1871). —Harms in Engler, Pflanzenw. Afrikas **3**(1): 470 (1915). —Eyles in Trans. Roy. Soc. S. Afr. **5**: 366 (1916). —Baker, Legum. Trop. Afr.: 687 (1930).

Isoberlinia paniculata (Benth.) Greenway in Bull. Misc. Inform., Kew **1928**: 203 (1928). —Brenan, Check-list Trees Shrubs Tang. Terr.: 104 (1949). —Burtt Davy & Hoyle in Topham, Check List For. Trees Shrubs Nyasaland Prot.: 38 (1958).

Pseudoberlinia paniculata (Benth.) P.A. Duvign. in Bull. Séances Inst. Roy. Col. Belge **21**: 434 (1950). —Hauman in F.C.B. **3**: 403 (1952).

Fig. 3.2.**20**. JULBERNARDIA GLOBIFLORA. 1, flowering branchlet (× ²/₃); 2, tip of leaflet showing pubescent margin (× 4); 3, base of leaflet (× 2); 4, flower bud with bract (× 4); 5, flower (× 3); 6, flower, with petals and stamens removed, showing bracteoles, calyx and ovary (× 3); 7, sepal (× 3); 8, larger petal (× 3); 9, lateral petal (× 3); 10, stamens (× 3), 1–10 from *Welch* 290; 11, pod dehisced (× ²/₃), from *Soil Conserv. Dept.* O.F.C. 2. JULBERNARDIA PANICULATA. 12, sepal (× 3), from *Duff* 137/33. Drawn by Lura Ripley. From F.T.E.A.

Berlinia baumii Harms in Warburg, Kunene-Sambezi Exped. Baum: 249 (1903). —Baker in J. Bot. **66**: 143 (1928); Legum. Trop. Afr.: 689 (1930).

Tree 2–10(22) m tall, flat-topped; bark grey, smooth but flaking into rectangular scales exposing rusty brown new bark. Leaves: stipules intrapetiolar, 5–7 mm long, connate below, bicuspidate, early caducous; petiole with rachis (3)8.5–22(28) cm long; leaflets in (2)3–4 pairs, (2)6.5–13(17) × (1.3)2.5–6.3(10) cm, oblong-elliptic to ovate-oblong or rarely obovate-oblong, apex obtuse to rounded, base asymmetric, one side cuneate, the other rounded, pubescent beneath and sometimes above, rarely glabrous, margins pubescent. Panicles to 15–25 × 15–30 cm, brown tomentellous; bracts 2–3 mm long, much shorter than the bracteoles; pedicels of open flowers 2–5 mm long; bracteoles 9–12 × 8–10 mm. Sepals 4–5 × 3.5–4.5 mm, broadly obovate, contiguous to slightly imbricate. Petals white, the larger one 6 × 5 mm, ± circular or triangular, shortly clawed, the others 5.5–7 mm long, spathulate, claw 3–4 mm long, lamina 2–3 × 1.5–2.5 mm. Pods obovate-oblong to oblong, 5–10.5 × 2–3.6 cm, brown tomentose. Seeds 1.4–1.7 × 1.3–1.6 cm.

Zambia. B: Sesheke Dist., NW of Sisisi Forest, Masese, fr. 11.viii.1947, *Brenan & Keay* 7683 (FHO, K). N: Mbala Dist., on road to Mpulungu, fl. 10.vi.1936, *B.D. Burtt* 5889 (K). W: Mufulira, fl. 16.v.1948, *Cruse* 345 (K). C: Mumbwa, between Lubale (Lutale) R. and Tepula Game Camp, fl. 15.v.1963, *van Rensburg* KBS 2152 (K). E: Chipata (Fort Jameson), Katete boma, fl. 23.iv.1952, *White* 2448 (FHO, K). S: Mazabuka, fl., 1932, *Stevenson* AC7 (FHO, K). **Malawi.** N: Viphya Plateau, junction of M1 and M12, between Mzimba and Chikangawa, fl. 25.iv.1986, *R.M. & D. Polhill* 5250 (K). C: Dzalanyama Forest, near nDebvu Village, fl. 10.ii.1954, *Adlard* 76 (FHO, K). S: Limbe, fl., 1930, *Topham* 821 (FHO, K). **Mozambique.** T: Angónia, Vila Mouzinho, fr. 25.viii.1941, *Torre* 3325 (K, LISC).

Also in Congo and Angola. Widespread in deciduous woodland where it may form extensive stands. Mainly in areas between 875 and 1125 mm of rain each year, but also on Kalahari sands; c. 1000–1600 m.

Conservation notes: Widespread species; not threatened.

21. BRACHYSTEGIA Benth.*

Brachystegia Benth. in Bentham & Hooker, Gen. Pl. 1: 582 (1865). —Burtt Davy & Hutchinson in Bull. Misc. Inform., Kew **1923**: 129–163.

Trees, shrubs or suffrutices producing annual shoots from a woody rootstock, 0.25–45 m tall, sometimes low-branching; trunk sometimes buttressed to 1.5 m high; bark deeply to shallowly fissured longitudinally and coarsely reticulate, or smooth, flaking in irregular to rectangular, thick or thin scales. Leaves paripinnate, 15–350 mm long; petioles 2–70 mm long including a 2–8 mm long pulvinus; rachis deeply or shallowly canaliculate above; leaflets in 2–72 pairs, sessile or with petiolules 3–7 mm long. Stipules intrapetiolar, free, shortly connate at base or partially to fully fused; persistent to early-caducous; lamina 2–45 × 0.2–10 mm; auricles 1–35 × 0.5–15 mm if present, lateral or basal, persistent or caducous independently of the stipules. Axillary dormant buds ovoid to globose or much flattened. Inflorescences of terminal and/or axillary panicles or racemes up to 150 mm long; bracts 2–6 × 1–5 mm. Flowers 3–15 × 2–10 mm; bracteoles 4–20 × 2–12 mm; perianth comprising greatly reduced sepals and rudimentary petals, or petals lacking altogether. Hypanthium present, shortly cupular or cylindrical at the base, turbinate above, up to 1.5 mm from base to point of insertion of sepals. Sepals 0–8, 1–8 × 0.5–3 mm, imbricate, valvate or widely spaced. Petals rudimentary, 0–6, 0.3–7 × 0.2–1.5, filiform. Stamens 10–18(20), connate at base for 0.5–7 mm, clearly exserted above the bracteoles, anthers dorsifixed. Ovary 2–10 × 1.5–3 mm, stipitate, stipe 1.5–5 mm long. Pods 45–300 × 15–80

* By A. Chikuni, with additional notes from J.M. Lock, M.G. Bingham, J. Burrows, B. Schrire, P.P. Smith and J.R. Timberlake

mm, oblong to obovate, ventral flanges suberect, spreading, revolute or absent, epicarp smooth, endocarp dull to reddish brown, spongy around the seeds. Seeds up to 11 per pod, 12–30 × 10–25 mm, much flattened.

28 species found across tropical Africa, with the centre of diversity in the northern part of the Flora Zambesiaca area and adjacent parts of Congo. An ecologically important group, many species of which tend to dominate woodlands or even form extensive pure stands. Species show ectomycorrhizal symbiosis.

Previous taxonomic treatments of this difficult genus (Hoyle in Flora du Congo Belge **3**: 446–482 (1952); in Conspectus Flora Angolensis **2**: 221–234 (1956); in F.T.E.A., Legum.-Caesalp.: 167–170 (1967); White & Hoyle in F.F.N.R.: 101–118 (1962); Aubréville in Flore du Gabon **15**: 277–282 (1968); in Flore du Cameroun **9**: 258–265 (1970)) have been regionally rather than broadly based, and somewhat inconsistent in their delimitation of species, particularly in proposing numerous putative hybrids to account for extremes in variation. Recent publications also differ greatly in the number of species recognized, e.g. Hoyle (1967) lists 30 species, Lock (1989) lists 36, and Lebrun & Stork (1992) list 29. Hybridization has been seen as widespread in the genus and as being responsible for much of the observed variation (Hoyle had proposed up to 23 hybrid taxa, see White & Hoyle in F.F.N.R. (1962) and Lebrun & Stork, Enum. Plant. Fleur. Afr. Trop. **2**: 40–41 (1992)), but a recent study (A. Chikuni, A taxonomic study of *Brachystegia* Benth., D.Phil. thesis, University of Oxford (1998)) using morphometric techniques and RAPD markers, has provided little evidence for it. In the present treatment, based on this study, varieties and putative hybrids have not been recognized as separate taxa.

Stipules and leaflet junctions can be useful in separating species. A selection are shown in FIGURE 3.2.**21**.

In addition to 21 readily distinguished species (13 in the Flora area) there are three problematic groupings, which in our area comprise: *B. longifolia/B. boehmii*; *B. spiciformis/B. bakeriana*; and *B. tamarindoides*. The latter species is the only one in the genus that has been subdivided into subspecies.

Key to species

1. Suffrutices; stems annual from a woody rootstock, 0.25–2 m tall 2
– Shrubs or trees; stems persistent above ground, 2–40 m tall 3
2. Inflorescences paniculate; leaflets in 6–19 pairs, each 5–18(20) mm wide; middle leaflet pairs largest; distal leaflet pairs 1–1.5 times larger than the proximal ones; leaflet apices acuminate, acute or obtuse; petioles 10–25 mm long . . . **1.** *russelliae*
– Inflorescences simply racemose; leaves in 17–36 pairs, each 1.5–8 mm wide; proximal leaflet pairs largest; proximal leaflet pairs 1.5–2.5 times larger than the distal ones; leaflet apices rounded, retuse; petioles 5–10 mm long . . **2.** *michelmorei*
3. Leaflets in 10–72 pairs; middle leaflets largest 4
– Leaflets in 2–10 pairs; either the distal or middle leaflets largest 10
4. Leaflet midveins excentric or marginal, the proximal side of the lamina at least 2.5 times broader than the distal 5
– Leaflet midveins central or subcentral, the proximal side of the lamina 1–2 times broader than the distal 6
5. Leaflets in 20–45 pairs, inserted up to 5 mm apart along the rachis, each 1–4 mm wide, narrowly oblong to oblong-falcate; apex acute or rounded; leaves 60–120 mm long; petioles 2–4 mm long . . . . **3.** *taxifolia*
– Leaflets in 12–30 pairs, inserted at least 5 mm apart along the rachis, each 5–15 mm wide, narrowly triangular, ovate or oblong; apex retuse to emarginate; leaves 100–350 mm long; petioles 3–8 mm long . . . . **4.** *wangermeeana*

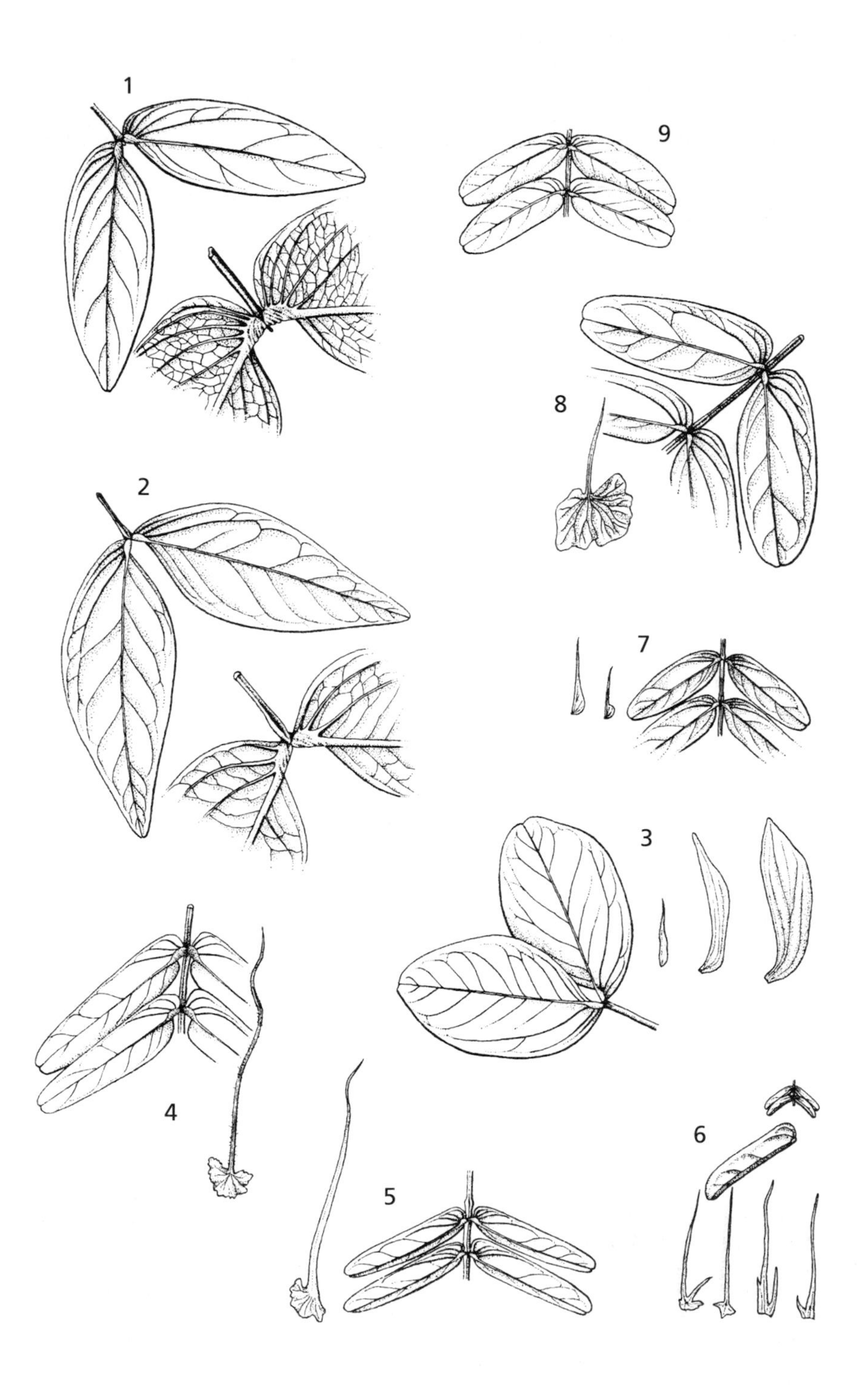
1
9
8
2
7
3
4
6
5

6. Dormant axillary buds flat and sharply keeled, with stipular remnants 7
– Dormant axillary buds ovoid or globose, without stipular remnants 9
7. Petioles (8)10–50 mm long, 3–5 times longer than the pulvinus; leaflets 6–18 pairs, narrowly triangular to ovate, apex acute to round . . . . **5.** *longifolia*
– Petioles 2–10(14) mm long, 1.5–3 times longer than the pulvinus; leaflets 10–30 pairs, oblong-elliptic to lanceolate, apex acute, round or retuse 8
8. Leaflets of the middle pair 15–30 × 5–10 mm, apex acute to obtuse; leaflets 10–22 pairs, surfaces markedly discolorous . . . . **6.** *gossweileri*
– Leaflets of the middle pair 30–65 × 7–18 mm, apex round, emarginate or retuse; leaflets 15–30 pairs, surfaces not markedly discolorous . . . . **7.** *boehmii*
9. Young branchlets and leaves puberulous with antrose short hairs; bark rough, reticulately fissured; stipules free, linear to lanceolate, not auriculate; leaflet apex obliquely retuse . . . . **8.** *puberula*
– Young branchlets and leaves glabrous or pubescent with erect hairs; bark smooth, thinly flaking; stipules shortly connate at the base, filiform, linear to narrowly spathulate, auriculate; leaflet apex not obliquely retuse . . . . **9.** *tamarindoides*
10. Flowers in simple racemes, with the flowers arranged in a close spiral (can be detected from scars up to last fruiting stage); staminal tube 2–7 mm long; up to 8 pairs of leaflets 11
– Flowers in panicles; stamens variously connate at the base to 0.5–2 mm (except in *B. stipulata*); 2–many pairs of leaflets 12
11. Petioles 10–50 mm long, interval between the first leaflet pairs up to 1.5 times petiole length; leaves (30)50–200 mm long; trees; bark deeply to reticulately fissured; pods predominantly dark brown; widespread . . . . **10.** *spiciformis*
– Petioles 3–10 mm long, interval between the first leaflet pairs at least 1.5 times petiole length; leaves 15–40(70) mm long; shrubs; bark smooth; pods predominantly yellowish brown, restricted to Kalahari sands . . . . . **11.** *bakeriana*
12. Dormant axillary buds much flattened and sharply keeled, with stipular remnants 13
– Dormant axillary buds ovoid or globose, without stipular remnants 16
13. Stamens 12–18; filaments more than 20 mm long; bracteoles 11–20 × 5–10 mm; leaflet apex round to retuse; stipules persistent . . . . **12.** *stipulata*
– Stamens 10; filaments up to 20 mm long; bracteoles 2–10 × 2–5 mm; apex acuminate to emarginate; stipules caducous or persistent 14
14. Leaflets in 6–18 pairs, at least twice as long as wide, narrowly triangular, ovate, oblong or lanceolate, apex acute or round . . . . **5.** *longifolia*
– Leaflets in 3–8 pairs; apex rounded, retuse or emarginate 15
15. Leaflets in 3–6 pairs, 1.5–2 times longer than wide, rectangular, oblong-lanceolate, apex obliquely retuse, round or emarginate . . . . **13.** *allenii*
– Leaflets in 4–8 pairs, 2–3 times longer than wide, narrowly to broadly ovate, apex rounded, retuse or emarginate 16

Fig. 3.2.**21**. BRACHYSTEGIA. Leaflets and stipules from a range of species, showing diagnostic differences. 1, *B. floribunda*, distal pair of leaflets (× ½) and base (× 1); 2, *B. bussei*, distal pair of leaflets (× ½) and base (× 1); 3, *B. manga*, distal pair of leaflets (× ½) and three forms of stipule (× 1); 4, *B. boehmii*, two middle pairs of leaflets (× ½) and stipule (× 1); 5, *B. wangermeeana*, two middle pairs of leaflets (× ½) and stipule (× 1); 6, *B. taxifolia*, two middle pairs of leaflets (× ½), single leaflet (× 2) and four forms of stipule (× 1); 7, *B. utilis*, two middle pairs of leaflets (× ½) and two stipules (× 1); 8, *B. stipulata*, two middle pairs of leaflets (× ½) and stipule (× 1); 9, *B. puberula*, two middle pairs of leaflets (× ½). Drawn by J. Chandler. Adapted from F.T.E.A.; based partly on F. White, Forest Flora of Northern Rhodesia.

16. Leaflets increasing in size distally; coastal Mozambique only ....**17.** *oblonga*
– Leaflets next to distal pair longest .... **9.** *tamarindoides* subsp. *torrei*
17. Rachis shallowly canaliculate above, with the margins neither raised nor winged; stipels absent 18
– Rachis deeply canaliculate; margins raised and winged; stipels present 19
18. Leaflets 2–3 times longer than wide, ovate to elliptic, often narrowly so, sometimes somewhat falcate; apex acuminate to acute; leaflets with (3)4–6 secondary veins diverging from the base; mature foliage bluish green; flowers often on old growth .... **14.** *floribunda*
– Leaflets up to twice as long as wide, ovate to elliptic, usually broadly so, even sometimes suborbicular; apex rounded to emarginate; leaflets with 2–3 secondary veins diverging from the base; mature foliage grey to silvery green; flowers borne on new growth .... **15.** *manga*
19. Leaflets 2–4 pairs, increasing in size distally, elliptic or lanceolate, sometimes falcate; base obliquely cuneate with 2–3 secondary veins diverging fanwise from the proximal leaflet base; stipules not auriculate ....**16.** *bussei*
– Leaflets more than 4 pairs; proximal or next-to-distal leaflets largest 20
20. Leaflets minutely appressed-pubescent, in 8–21 pairs; petioles 8–20 mm long .... **8.** *puberula*
– Leaflets glabrous, or spreading-hairy, in 6–12 pairs; petioles 2–15 mm long .... **18.** *utilis*

1. **Brachystegia russelliae** I.M. Johnst. in Contrib. Gray Herb. **73**: 32 (1924). —Baker, Legum. Trop. Afr.: 718 (1930). —Hoyle in C.F.A. **2**: 234 (1956). —White & Hoyle in White, F.F.N.R.: 116 (1962). Type: Angola, *Curtis* 428 (GH holotype, FHO fragm.).

Stems to 2 m high from a woody rootstock, glabrous or pubescent. Leaves up to 25 cm long; petiole 1–2.5 cm long, stout; rachis deeply canaliculate with raised winged margins, glabrous or pubescent; leaflets in 6–19 pairs, contiguous or overlapping, the middle leaflet pair often the largest, distal pairs 1–1.5 times larger than the proximal pairs, 1.5–5 × 0.5–1.8(2) cm, ovate, narrowly triangular, narrowly ovate or elliptic, often falcate, acute to acuminate at the apex, obliquely rounded, truncate or cordate at the base, glabrous or pubescent; midveins central or subcentral, with 3–4 secondary veins arising from the proximal leaflet base. Stipules persistent, 10–30 × 0.5–4 mm, shortly connate at the base, filiform or linear-lanceolate, often falcate; auricles 2–10 × 1–2 mm, reniform, semicircular or oblong, sometimes reduced to a basal tooth. Dormant axillary buds much flattened. Inflorescences of terminal and axillary panicles up to 6 cm long; bracts up to 3 × 3 mm, ovate to deltate, rounded to acute at the apex, caducous; bracteoles 4–8 × 4–6 mm, elliptic to obovate. Flowers 3–6 × 2–4 mm; pedicels 1–4 mm long. Sepals 5, 2–4 × 1–3 mm, elliptic or obovate to lanceolate, imbricate, densely ciliate. Petals 1–6, up to 0.5–1.5 × 0.5 mm, linear to spathulate. Stamens 10(12), united to 1.5 mm at the base, filaments 8–13 mm long. Ovary 3–5 × 1–1.5 mm, densely pubescent with rusty hairs, 3–5-ovulate, stipe to 2 mm long; style 7–10 mm long. Pods 5–10 × 2–4 cm, oblong, glabrous, reddish brown to pale brown with narrow adaxial flanges up to 2 mm wide. Seeds (perhaps immature) 14–17 × 9–12 mm.

Zambia. N: Kalungwishi valley, fl. 19.viii.1933, *Michelmore* 541 (K).

Also in Angola. Locally dominant in open woodland with *Cryptosepalum*; 1400–1800 m.

Conservation notes: Local within the Flora area and probably Vulnerable, but widespread in Angola; Lower Risk, Least Concern globally.

Within our area, *B. russelliae* is only known from the Kalungwishi Valley in Zambia, far removed from its main range in Angola. The leaves of *Michelmore* 541 are pubescent on the midrib beneath (glabrous in Angolan specimens), and have up to

10 leaflet pairs, the middle ones being the largest. The inflorescence axes, bracts and bracteoles of *Michelmore* 541 are virtually glabrous, but are densely pubescent in Angolan material. It would be premature to recognise these differences formally. The specimen is assigned to *B. russelliae* because of the few pairs of leaflets with acute apices. It was collected in the same general area as the type of *B. michelmorei*, and there is a need to establish if these species occur sympatrically.

2. **Brachystegia michelmorei** Hoyle & Brummitt in Kew Bull. **54**: 156 (1999). Type: Zambia, Mporokoso Dist., Kalungwishi Basin, Kaseli Valley, 19.viii.1933, *Michelmore* 539 (K holotype, NDO).

Brachystegia sp. nr. *russelliae* sensu White & Hoyle in White, F.F.N.R.: 117 (1962).

Brachystegia astlei Hoyle in Kew Bull. **54**: 155 (1999). Type: Zambia, NE of Kawambwa, 6.5 km from source of Mibamba stream, x.1965, *Claydon* in *Astle* 3980 (FHO holotype, LISC, NDO, SRGH).

Stems 10–45 cm tall from a woody rootstock. Leaves 3–15 cm long; rachis deeply canaliculate with raised winged margins; leaflets in 17–36 pairs, contiguous to overlapping, progressively decreasing in size distally, the proximal leaflet pairs 1.5–2.5 times larger than the distal ones, 10–30 × 1.5–8 mm, linear-falcate, narrowly triangular or oblong, obtuse or retuse at the apex, obliquely rounded to truncate or subcordate at the base, glabrous or pubescent; midveins subcentral, with 3–5 secondary veins diverging from the proximal leaflet base. Stipules caducous, 10–20 × 2–4 mm, shortly connate at the base, linear, falcate, sometimes subfoliaceous, auricles 1–5 × 0.5–3 mm, reniform, digitate, pubescent. Dormant axillary buds much flattened. Inflorescences of terminal simple racemes up to 6 cm long; bracteoles 6–8 × 3–6 mm, orbicular to obovate. Flowers 3–7 × 3–5 mm; pedicels 1–3 mm long. Sepals 5, 2–2.5 × 0.5–2 mm, narrowly oblong, lanceolate, obovate to elliptic, imbricate, margins densely ciliate. Petals 1–5, up to 6 × 1.5 mm, linear to spathulate. Stamens 10–12, united to 1–2 mm at the base; filaments 8–13 mm long. Ovary 3–4 × 1–1.5 mm, densely pubescent, up to 5-ovulate, stipe 1.5–2.5 mm long; style 8–10 mm long. Pods up to 6 × 3 cm; ventral flanges suberect. Seeds not seen.

Zambia. N: Mporokoso Dist., upper Kalungwishi valley, fl. 19.viii.1933, *Michelmore* 539 (K, NDO); Kawambwa Dist., NE of Kawambwa, c. 6.5 km (4 miles) from source of Mibamba stream, fl. x.1965, *Claydon* in *Astle* 3980 (K, FHO); near Mupamba steam, fr.1.vii.1961, *Astle* 797 (FHO, K, SRGH).

Known only from 4 collections in the upper Kalungwishi Valley of N Zambia. Undershrub in open woodland in association with *B. longifolia* and *Cryptosepalum*; 1200–1700 m.

Conservation notes: A narrow endemic with an area of occupancy estimated at less than 500 km^2. Although this upper catchment area is sparsely populated and not at present threatened, the limited range must make a rating of Vulnerable appropriate.

Brachystegia astlei and *B. michelmorei* were published from Hoyle's original notes (Hoyle & Brummitt 1999). They are both known from populations in the upper Kalungwishi valley in N Zambia; *B. astlei* was based on *Claydon* in *Astle* 3980 collected near Mibamba stream, whilst *B. michelmorei* was based on *Michelmore* 536, collected from near the source of the Mibamba stream. The leaflet and floral characters of these two species overlap, with leaflet size appearing to be the only distinguishing character. Leaflet size and number are highly variable in *Brachystegia*, and leaflet width alone cannot be used as a basis for species delimitation hence *B. astlei* has been placed in synonymy under *B. michelmorei*. The name *B. michelmorei* is chosen here since it is based on the earlier collection (*Michelmore* 536, 17.viii.1933). It is distinguished from *B. russelliae* by its leaflets that tend to decrease in size distally and which have a retuse apex. In *B. russelliae* the leaflet apices are predominantly acute.

3. **Brachystegia taxifolia** Harms in Bot. Jahrb. Syst. **33**: 155 (1902); in R.E.Fries, Wiss. Ergebn. Schwed. Rhod.-Kongo-Exped. 1911–1912, **1**: 69 (1914); in Engler, Pflanzenw. Afrikas **3**(1): 480 (1915). —Burtt Davy & Hutchinson in Bull. Misc. Inform., Kew **1923**: 153 (1923). —De Wild., Contrib. Fl. Katanga, Suppl. **2**: 61 (1929). —Baker, Legum. Trop. Afr.: 717 (1930). —Jackson in J. S. Afr. Bot. **6**: 39 (1940). —Brenan & Hoyle in Brenan, Check-list For. Trees Shrubs Tang. Terr.: 90 (1949). —Hoyle in F.C.B. **3**: 480 (1952). —White & Hoyle in White, F.F.N.R.: 116–118, fig.22N (1962). —Hoyle in F.T.E.A., Legum.-Caesalp.: 195 (1967). —White, Dowsett-Lemaire & Chapman, For. Fl. Malawi: 306 (2001). Type from Tanzania, SW Iringa Dist., Ngominyi–Bueni, *von Prittwitz & Gaffron* 54 (B† holotype, K fragm.). FIGURE 3.2.**21**/6.

Brachystegia mimosifolia Hutch. & Burtt Davy in Bull. Misc. Inform., Kew **1923**: 153 (1923). —De Wild., Contrib. Fl. Katanga, Suppl. **2**: 49 (1929). —Baker, Legum. Trop. Afr.: 717 (1930). Type: Zambia, Kabwe (Broken Hill), ix.1909, *Rogers* 8552 (K holotype).

Brachystegia subfalcato-foliolata De Wild., Contrib. Fl. Katanga, Suppl. **2**: 59 (1929). —Baker, Legum. Trop. Afr.: 718 (1930). —Hoyle in F.C.B. **3**: 482 (1952). Type: Congo, Katanga, 1911, *Hock* s.n. (BR holotype).

Tree 1.5–12 m tall; bark reticulate to deeply fissured longitudinally, grey. Leaves 6–12 cm long; petioles 2–4 mm long; rachis deeply canaliculate with raised and winged margins; stipels present, rusty-pubescent; leaflets in 20–45 overlapping pairs, the middle leaflet pairs the largest, 5–20 × 1–4 mm, narrowly oblong-falcate, acute to round at the apex, rounded to cordate at the base, glabrous or pubescent; midveins excentric, with 2–3 secondary veins diverging fanwise from the proximal leaflet base. Stipules persistent, up to 15 × 1.5 mm, shortly connate at the base, linear, pubescent; auricles up to 7 × 5 mm, reniform, tooth-like to digitate, pubescent. Dormant axillary buds much flattened. Inflorescences of compact, terminal and/or axillary panicles up to 5 cm long, densely puberulous; bracts 2–4 × 2–3 mm, ovate to deltate, caducous, pubescent; bracteoles 6–10 × 3.5–6 mm, orbicular, pubescent. Flowers 4–6 × 2.5–3 mm; pedicels up to 2.5 mm long. Sepals 5; 2–4 × 1–3 mm, oblong to obovate, imbricate, margins densely ciliate. Petals 0–3, filiform, linear. Stamens 10, united to 1 mm at base, filaments 9–15 mm long. Ovary 3–5 × 1–2 mm, up to 8-ovulate, stipe to 2.5 mm long. Pods 4–13 × 2.5–4 cm, round to obtuse at the apex, flanges suberect or revolute. Seeds up to 15 × 15 mm, circular to oblong.

Zambia. N: Mbala, Kawimbe, fl. 29.ix.1959, *Richards* 11480 (FHO, K). W: Chondwe Forest Res., imm.fl. 11.ix.1938, *Greenway & Miller* 5660 (K). C: Mumbwa Dist., 22.ix.1930, *Stevenson* 108/30 (FHO). **Malawi.** N: Rumphi, edge of Nyika Plateau, fr. 29.v.1952, *White* 2813 (FHO, K).

Also in Congo and Tanzania. Appears evergreen as old and new leaves occur together; generally has a layered canopy. Locally dominant especially on wetland margins, hill slopes, river banks, rocky plateau areas; common in sandy and poor stony leached soils, especially at high altitude and in high rainfall areas. Also grows as an undershrub in mixed open woodlands; 1000–1800 m.

Conservation notes: A widespread taxon; Lower Risk, Least Concern.

4. **Brachystegia wangermeeana** De Wild. in Repert. Spec. Nov. Regni Veg. **11**: 513 (1913). —Harms in Engler, Pflanzenw. Afrikas **3**(1): 480 (1915). —Burtt Davy & Hutchinson in Bull. Misc. Inform., Kew **1923**: 152, fig.1 (1923). —De Wild., Contrib. Fl. Katanga, Suppl. **2**: 65 (1929). —Baker, Legum. Trop. Afr.: 718 (1930). —Jackson in J. S. Afr. Bot. **6**: 40 (1940). —Brenan & Hoyle in Brenan, Check-list For. Trees Shrubs Tang. Terr.: 90 (1949). —Hoyle in F.C.B. **3**: 478, fig.40A (1952); in C.F.A. **2**: 233 (1956). —White & Hoyle in White, F.F.N.R.: 116, fig.22M (1962). —Hoyle in F.T.E.A., Legum.-Caesalp.: 192 (1967). Type from Congo, Katanga, Lubumbashi (Elizabethville), 1911, *Hock* s.n. (BR holotype; K fragm.). FIGURE 3.2.**21**/5.

Brachystegia sapinii De Wild., Contrib. Fl. Katanga, Suppl. **2**: 56 (1929). Type: Congo, vi.1908, *Sapin* s.n. (BR holotype, K photo).

Brachystegia wildemanniana R.E. Fries, Wiss. Ergebn. Schwed. Rhod.-Kongo-Exped. 1911–12, **1**: 68 (1914). —Burtt Davy & Hutchinson in Bull. Misc. Inform., Kew **1923**: 153 (1923). —De Wild., Contrib. Fl. Katanga, Suppl. **2**: 65 (1929). —Baker, Legum. Trop. Afr.: 719 (1930). Type: Zambia, Malolo–Katwe, between Lakes Bangweulu and Tanganyika, *R.E. Fries* 1207 (UPS holotype, K photo).

Brachystegia kalongoensis De Wild. in Contrib. Fl. Katanga, Suppl. **2**: 40 (1929). —Hoyle in F.C.B. **3**: 481 (1952). Type: Congo, Kalongo–Kiambi, 9.i.1922, *Delevoy* 591 (BR holotype).

Tree up to 15 m tall; bark deeply fissured longitudinally, dark grey. Young branchlets brown-pubescent. Leaves 10–3.5 cm long; petioles 3–8 mm long; rachis deeply canaliculate with raised and winged margins; leaflets in 12–30 pairs, overlapping or contiguous, the middle leaflet pairs the largest, 15–50 × 5–15 mm, narrowly triangular, ovate or oblong, usually falcate, obliquely retuse to emarginate at the apex, obliquely rounded to truncate at the base, glabrous or pubescent; midveins submarginal to excentric, with 4–5 secondary veins diverging from the proximal leaflet base. Stipules caducous, 15–45 × 1–3 mm, shortly connate at the base, filiform, liner or lanceolate, pubescent; auricles 2–10 × 1–5 mm, reniform or semi-circular, palmately nerved, pubescent or glabrous. Dormant axillary buds much flattened. Inflorescences of terminal and axillary panicles up to 6 cm long; bracts 1.5–2.5 × 1–2 mm, ovate to deltate, caducous, pubescent; bracteoles 5–8 × 3–5 mm, pubescent. Flowers 4–5 × 2–3 mm; pedicels up to 2 mm long. Sepals 4–5, 2–2.5 × 1–2 mm, oblong, lanceolate, imbricate, margins densely ciliate. Petals 0–1, filiform, linear. Stamens 10, united to 1–2 mm at the base; filaments 10–12 mm long. Ovary to 2.5 × 2 mm, up to 9-ovulate, the stipe to 1.5 mm long. Pods 6.5–15 × 3–4 cm, apex round to obtuse, ventral flanges suberect to revolute. Seeds 15–20 × 12–15 mm, circular to oblong.

Zambia. N: Kawambwa, c. 19 km (12 miles) along Mansa (Fort Rosebery) road, fl. 30.x.1952, *Angus* 672 (FHO, K). W: Kasempa, by road to Kasempa c. 120 km (75 miles) S of Solwezi, 30.ix.1947, *Brenan & Greenway* 7993 (K).

Also in Congo, Tanzania and Angola. Deciduous woodland with *Julbernardia* and *B. floribunda*, common in flat high plateau areas, on sandy brown soils and ridge slopes; up to 1800 m.

Conservation notes: A fairly widespread species; Lower Risk, Least Concern.

Brachystegia wangermeeana can be confused with some specimens of *B. boehmii*, especially those with subcentral midveins. However, *B. boehmii* differs in its stout petiole, half of which is the pulvinus, and in its persistent stipules and large auricles (10–20 × 5–10 mm). In *B. wangermeeana* the stipules are caducous with smaller auricles (2–10 × 1–5 mm). Flowering individuals of *B. wangermeeana* can be recognized in the field by their conspicuous pale bracts. This treatment of *B. wangermeeana* differs from Hoyle's (1952) in treating *B. kalongoensis* as conspecific with *B. wangermeeana*. The types of the two taxa are indistinguishable.

5. **Brachystegia longifolia** Benth. in Hooker's Icon. Pl. 14: fig.1359 (1881). —R.E. Fries, Wiss. Ergebn. Schwed. Rhod.-Kongo-Exped. **1**: 67 (1914). —Burtt Davy & Hutchinson in Bull. Misc. Inform., Kew **1923**: 154 (1923). —Topham in Bull. Misc. Inform., Kew **1930**: 357 (1930). —Baker, Legum. Trop. Afr.: 723 (1930). —Jackson in J. S. Afr. Bot. **6**: 38 (1940). —Brenan & Hoyle in Brenan, Check-list For. Trees Shrubs Tang. Terr.: 93 (1949). —Hoyle in F.C.B. **3**: 471, fig.40 (1952). —White & Hoyle in White, F.F.N.R.: 111, 118, fig.22H (1962). —Hoyle in F.T.E.A., Legum.-Caesalp.: 188–191, fig.41 (1967) as *B. × longifolia*. Type: Malawi, Shire Highlands, *Buchanan* 22 (K holotype, E).

Brachystegia goetzei Harms in Bot. Jahrb. Syst. **30**: 318, fig.13 (1901). —Burtt Davy & Hutchinson in Bull. Misc. Inform., Kew **1923**: 150 (1923). —Baker, Legum. Trop. Afr.: 721 (1930). Type: Tanzania, Mbeya Dist., Unyiha, near Pisaki, xi.1899. *Goetze* 1423 (B† holotype, E, K photo).

Brachystegia holtzii Harms in Bot. Jahrb. Syst. **33**: 154 (1902). —Burtt Davy & Hutchinson in Bull. Misc. Inform., Kew **1923**: 150 (1923). —Baker, Legum. Trop. Afr.: 724 (1930). Type: Tanzania, Uzaramo Dist., Mogo Forest Reserve, xi.1901, *Holtz* 31 (B† holotype, EA).

Brachystegia glaberrima R.E. Fr. in Wiss. Ergebn. Schwed. Rhod.-Kongo-Exped. **1**: 66 (1914). —Burtt Davy & Hutchinson in Bull. Misc. Inform., Kew **1923**: 157 (1923). —De Wild., Contrib. Fl. Katanga, Suppl. **2**: 35 (1929). —Baker, Legum. Trop. Afr.: 723 (1930). —White & Hoyle in White, F.F.N.R.: 110, 117, fig.22G (1962). —Hoyle in F.T.E.A., Legum.-Caesalp.: 186–188, fig.40 (1967). Lectotype, chosen by Hoyle (1967): Zambia, Mporokoso, between Lakes Bangweulu and Tanganyika, *R.E. Fries* 1177 (UPS lectotype).

Brachystegia apertifolia Hutch. & Burtt Davy in Bull. Misc. Inform., Kew **1923**: 154 (1923). —De Wild., Contrib. Fl. Katanga, Suppl. **2**: 16 (1929). —Baker, Legum. Trop. Afr.: 726 (1930). —Topham in Bull. Misc. Inform., Kew **1930**: 358 (1930). —Hoyle in F.C.B. **3**: 481 (1952). Type: Malawi, Thondwe, 21.x.1905, *Cameron* 134 (K holotype).

Brachystegia bournei Greenway in Bull. Misc. Inform., Kew **1928**: 200 (1928). —Baker, Legum. Trop. Afr.: 724 (1930). Type: Zambia, Nchanga, 16.viii.1927, *Bourne* 78 (K holotype).

Brachystegia luishiensis De Wild. in Ann. Soc. Scient. Brux. **1928**: 76 (1928). —De Wild., Contrib. Fl. Katanga, Suppl. **2**: 45 (1929). —Baker, Legum. Trop. Afr.: 720 (1930). Type: Congo, Luishia, 29.vi.1922, *Delevoy* 973 (BR holotype, FHO fragm.).

Brachystegia falcato-appendiculata De Wild., Contrib. Fl. Katanga, Suppl. **2**: 22 (1929). —Baker, Legum. Trop. Afr.: 724 (1930). Type: Congo, Musambo, Tshilungo, 7.viii.1922, *Delevoy* 965 (BR holotype, FHO fragm.).

Brachystegia homblei De Wild., Contrib. Fl. Katanga, Suppl. **2**: 39 (1929). —Baker, Legum. Trop. Afr.: 725 (1930). Type: Congo, Kapiri Valley, ii.1913, *Homblé* 1242 (BR holotype).

Brachystegia longifoliolata De Wild. in Contrib. Fl. Katanga, Suppl. **2**: 44 (1929). Type: Congo, 19.xii.1921, *Delevoy* 470 (BR holotype, FHO).

Brachystegia longifolia var. *parviflora* Topham in Bull. Misc. Inform., Kew **1930**: 358 (1930). Type: Malawi, Dedza Mt., *Burtt Davy* 21451 (FHO holotype).

Tree 4–30 m tall; bark deeply or shallowly fissured longitudinally, often coarsely reticulate, flaking in thick irregular scales, dark grey. Leaves 7–25 cm long; petiole (0.8)1–5 cm long; rachis deeply canaliculate with raised and winged margins; leaflets in 6–18 pairs, widely spaced or overlapping; middle leaflet or the one next to the proximal the largest; proximal leaflet pair 1–5 × 1–2 cm, the middle 2.5–8 × 1–3 cm, each 2–3 times longer than broad, narrowly triangular, ovate to oblong, sometimes falcate, acute or rounded at the apex, obliquely rounded, cordate or truncate at the base, glabrous or pubescent; midvein central or subcentral, with 3–5 secondary veins diverging from the proximal leaflet base. Stipules caducous, 15–35 × 1–5 mm, shortly connate at the base, filiform, linear, lanceolate-falcate to subfoliaceous; auricles 1–17 × 0.5–9 mm, reniform to subcircular, caducous independently of the stipules. Dormant axillary buds much flattened. Inflorescences in terminal and/or axillary panicles to 10 cm long, glabrous or pubescent with dark or rusty hairs; bracts 2.5–4 × 1.5–3 mm, ovate to oblong, acute at the apex. Flowers 4–6 × 3–4 mm; pedicels up to 2.5 mm long; bracteoles 6–8 × 3–4 mm, obovate or orbicular. Sepals 5, 2–4 × 0.5–5 mm, oblong, obovate, lanceolate or imbricate, the proximal and lateral sepals often fused. Petals 0–4, up to 2.5–3 × 0.5 mm, linear to spathulate. Stamens 10, united to 1 mm at the base, filaments 10–20 mm long. Ovary 4–6 × 1.5–2 mm, up to 8-ovulate; stipe to 2 mm long, densely pubescent; style up to 12 mm long. Pods 7–20 × 2.5–5 cm, oblong to obovate, rounded to obtuse at the apex with a 4–8 mm long apical beak, light to dark brown, with spreading or suberect ventral flanges up to 7 mm wide. Seeds 2–6, 12–15 × 10–15 mm, oblong to ovoid.

Zambia. N: Mbala, Saise Valley Nature Res., 16.x.1938, *Miller* 238 (FHO); W. Miengwe, c. 50 km (30 miles) S of Ndola, fl. 3.xi.1955, *Fanshawe* 2545 (K, NDO). C: Mkushi Dist., Munchiwemba (Munshiwenta), fl. 27.ix.1941, *Stohr* 609 (BOL, K). S: near Chief Siachitema's Court, 23 km (15 miles) NW of Tara, fr. 23.xi.1963, *Bainbridge* 933 (K). **Malawi.** N: Mzimba Dist., Mzuzu, Marymount, fr. 28.ix.1975, *Pawek* 10197 (K, MAL, MO, SRGH). C: Dedza Dist., Chongoni Forest, fl. 7.x.1966, *Banda* 878 (K, SRGH). S: Chigumula Forest, *Topham* 646 (FHO). **Mozambique.** N:

Niassa, Litunde, road to Marrupa, 12 km after R. Licuir, fl. 7.x.1983, *Groenendijk* 712 (K, WAG). Z: between Gúricò and Alto Molócuè, st. 4.x.1941, *Torre* 3586 (FHO, K, LISC). T: between Furancungo and Vila Mouzinho, fl. 19.x.1943, *Torre* 6030 (FHO, K, LISC).

Also in Congo, Tanzania and Angola. Locally dominant on hill slopes, in deciduous woodlands, flooded areas with sandy clay soils, on reddish leached soils, high plateau areas with sandy brown soils and (in Zambia) on Kalahari sands; 200–1700 m.

Conservation notes: Widespread; Lower Risk, Least Concern.

The leaflets of *Brachystegia longifolia* are extremely variable in number, shape, size and indumentum. Hoyle (1967) considered that *B. longifolia* combined characters of *B. boehmii* and *B. glaberrima*. However, the characters used to distinguish *B. longifolia* from *B. glaberrima* are unreliable, and *B. glaberrima* is here treated as synonymous with *B. longifolia*. Baker (1930) also considered *B. glaberrima* conspecific with *B. longifolia*.

B. longifolia can be confused with *B. boehmii*, but *B. longifolia* has longer petioles (10–50 mm) that are 3–5 times longer than the pulvinus, and generally fewer and larger leaflets in 6–18 pairs. These are narrowly triangular to ovate, with an acute to round apex.

Hoyle (1952) disagreed with the placement of *B. luishiensis* in the synonymy of *B. longifolia*, but the type of *B. luishiensis* is indistinguishable from *B. longifolia*. As here circumscribed, including *B. glaberrima* and *B. luishiensis*, *B. longifolia* is distinguished from *B. boehmii* by its short pulvini that occupy less than half the petiole length, and by its predominantly narrowly ovate leaflets. The inflorescences of *B. longifolia* are lax and rarely have secondary branches, while those of *B. boehmii* are compact and always heavily branched. *Brachystegia apertifolia*, which has been considered to represent the putative hybrid *B. longifolia* × *B. boehmii*, is here treated as a synonym of *B. longifolia*; the types of the two taxa are indistinguishable.

6. **Brachystegia gossweileri** Hutch. & Burtt Davy* in Burtt Davy & Hutchinson, Bull. Misc. Inform., Kew **1923**: 152 (1923). —De Wild., Contrib. Fl. Katanga, Suppl. **2**: 35 (1929). —Baker, Legum. Trop. Afr.: 718 (1930). —Hoyle in C.F.A. **2**: 232 (1956). —White & Hoyle in White, F.F.N.R.: 112 (1962). Type: Angola, Benguela, ix.1905, *Gossweiler* 2025 (K holotype).

Tree up to 12 m tall; bark deeply or shallowly fissured longitudinally, flaking in thick irregular scales, grey to dark grey. Young branchlets brown-pubescent, occasionally glabrous. Leaves 7–18.5 cm long; petiole 2–8 mm long; pulvinus 1.5–3 mm long; rachis deeply canaliculate with raised and winged margins, glabrous or pubescent; leaflets in 10–22 pairs, widely spaced or overlapping, middle leaflets or those next to the proximal largest, proximal leaflet pair (0.7)1–2.5 × 0.3–0.6 cm, the middle ones 1.5–3 × 0.5–1 cm, the distal ones 1.5–2.5 x 0.5–0.8 cm, linear-lanceolate, oblong-elliptic or narrowly triangular, acute or obtuse at the apex, truncate or subcordate at the base, glabrous or pubescent, discolorous, glaucous-grey; midvein central or subcentral with 3–4 secondary veins diverging from the base. Stipules caducous, 10–25 × 1–6 mm, shortly connate at the base, linear, lanceolate-falcate to spathulate; auricles 5–10 × 3–5 mm, reniform, caducous independently of the stipules, pubescent. Dormant axillary buds much flattened. Inflorescences in terminal and/or axillary panicles, to 7 cm long; bracts 2–3 × 2–4 mm, ovate, acute at the apex. Flowers 4–5 × 3–4 mm; pedicels to 2 mm long; bracteoles 4–6.5 × 3–4.5 mm, obovate or orbicular. Sepals 5, 2–2.5 × 1–2 mm, ovate to oblong, imbricate, margins densely ciliate. Petals up to 2, filiform to linear. Stamens 10,

* Burtt Davy & Hutchinson in Bull. Misc. Inf., Kew **1923**: 129–163 (1923) consistently cite species first described in that publication as being "Hutch. & Burtt Davy". The incorrect inversion of authorities has been widespread in many publications since (e.g. Hoyle in White F.F.N.R., 1962).

united to 0.5–1 mm at the base; filaments 6–10 mm long. Ovary 3–5 × 1–2 mm, oblong, densely pubescent; stipe up to 3 mm long. Pods 9–13 × 3–4 cm, oblong to obovate, rounded to obtuse at the apex, with a 4–8 mm long apical beak, with spreading or suberect ventral flanges. Seeds not seen.

Zambia. W: Chingola, fl. 12.x.1955, *Fanshawe* 2502 (K). C: Mumbwa, fl. vii.1912, *Macaulay* 840 (K). S: Siamambo Forest Res., near Choma, fr. 9.vi.1952, *White* 2940 (FHO).

Also in Congo and Angola. In deciduous woodlands, on margins of wetlands or shallow lateritic soils, often associated with *B. boehmii, B. floribunda, B. spiciformis* and *B. taxifolia*; up to 1700 m.

Conservation notes: Probably Lower Risk, Least Concern, but has a rather more restricted range than many species in the genus.

B. gossweileri can be confused with *B. tamarindoides* when young. Mature trees of *B. gossweileri* are separable on their habit (a small tree) and rough bark.

7. **Brachystegia boehmii** Taub. in Engler, Pflanzenw. Ost-Afrikas **C**: 197 (1895). —Burtt Davy & Hutchinson in Bull. Misc. Inform., Kew **1923**: 151 (1923). —De Wild., Contrib. Fl. Katanga, Suppl. **2**: 17 (1929). —Baker, Legum. Trop. Afr.: 721 (1930). —Jackson in J. S. Afr. Bot.. **6**: 37 (1940). —Miller, Check-list For. Trees Shrubs Bech. Prot.: 23 (1948). —Brenan & Hoyle in Brenan, Check-list For. Trees Shrubs Tang. Terr.: 93 (1949). —Hoyle in F.C.B. **3**: 474, fig.33 (1952). —O. Palgrave, Trees Central Africa: 77 (1957). —White & Hoyle in White, F.F.N.R.: 111, 117, fig.24 (1962). —Gomes e Sousa, Dendrol. Moçamb. Estudo Geral **1**: 270, fig.66 (1966). —Hoyle in F.T.E.A., Legum.-Caesalp.: 191–192 (1967). Type: Tanzania, Tabora Dist., Igonda, *Boehm* 159a (B† holotype, K fragm.). FIGURES 3.2.**21**/4 & 3.2.**22**.

Brachystegia flagristipulata Taub. in Engl., Pflanzenw. Ost-Afrikas **C**: 198 (1895). —Burtt Davy & Hutchinson in Bull. Misc. Inform., Kew **1923**: 152 (1923). —Baker, Legum. Trop. Afr.: 722 (1930). Type: Tanzania, Uzaramo Dist., *Stuhlmann* 6400 (B† holotype, K fragm.).

Brachystegia woodiana Harms in Bot. Jahrb. Syst. **30**: 82 (1901). —Burtt Davy & Hutchinson in Bull. Misc. Inform., Kew **1923**: 151 (1923). —De Wild., Contrib. Fl. Katanga, Suppl. **2**: 65 (1929). —Baker, Legum. Trop. Afr.: 720 (1930). Type: Malawi, *Buchanan* in *Herb. Wood* 6960 (BR holotype, NU).

Brachystegia katangensis De Wild. in Ann. Mus. Congo Belge, sér. 4, Bot. 2-4: 204 (1903). —Burtt Davy & Hutchinson in Bull. Misc. Inform., Kew **1923**: 152 (1923). —De Wild. in Contrib. Fl. Katanga, Suppl. **2**: 41 (1929). —Baker, Legum. Trop. Afr.: 719 (1930). Type: Congo, Katanga, x.1899, *Verdick* 117 (BR holotype).

Brachystegia filiformis Hutch. & Burtt Davy in Bull. Misc. Inform., Kew **1923**: 150, fig.2 (1923). —Baker, Legum. Trop. Afr.: 722 (1930). —Topham in Bull. Misc. Inform., Kew **1930**: 356 (1930). Type: Zambia, Kabwe (Broken Hill), xi.1904, *Rogers* 8605 (K holotype, BM isotype).

Brachystegia ferruginea De Wild. in Ann. Soc. Sci. Brux. **73** (1928); Contrib. Fl. Katanga, Suppl. **2**: 23 (1929). —Baker, Legum. Trop. Afr.: 719 (1930). Type: Congo, 14.ix.1921, *Delevoy* 23 (BR holotype).

Brachystegia malengensis De Wild., Contrib. Fl. Katanga, Suppl. **2**: 47 (1929). —Baker, Legum. Trop. Afr.: 722 (1930). Type: Congo, Malenge, 22.ix.1921, *Delevoy* 295 (BR holotype).

Brachystegia hopkinsii Suesseng. in Suessenguth & Merxmüller, Contrib. Fl. Marandellas Dist.: 86 (1951). Type: Zimbabwe, Marondera (Marandellas), 26.ix.1947, *Dehn* 452, 453 (BM holotype).

Brachystegia boehmii var. *katangensis* (De Wild.) Hoyle in F.C.B. **3**: 477 (1952).

Tree up to 15 m tall; bark deeply or shallowly fissured longitudinally and often coarsely reticulate, flaking in thick irregular scales, dark grey; young branchlets glabrous or brown-pubescent. Leaves 8–35 cm long; petioles 3–12 mm long including a 2–7 mm long pulvinus;

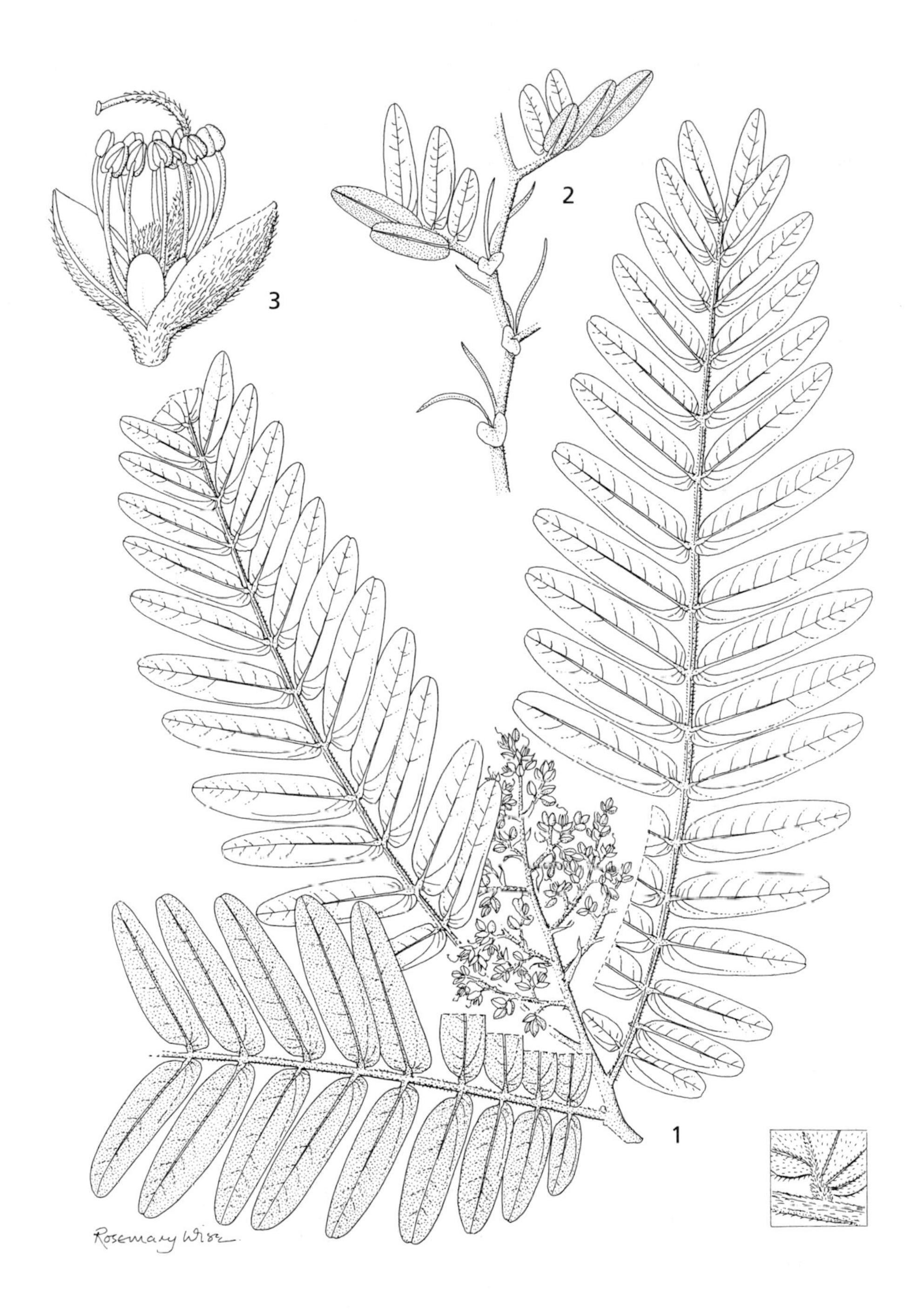

Fig. 3.2.**22**. BRACHYSTEGIA BOEHMII. 1, flowering branch with both leaflet sides (× ½); 2, young branchlet with stipules (× ½); 3, flower (× 3); 1 & 3 from *Proctor* 522, 2 from *Lindeman* 369. Drawn by Rosemary Wise. From A. Chikuni, D.Phil. thesis, University of Oxford.

rachis deeply canaliculate with raised and winged margins, glabrous or pubescent; leaflets in 15–30 overlapping or widely spaced pairs, middle leaflets or those next to the proximal largest; proximal leaflet pairs 1.5–4.5 × 0.5–1.5 cm, the middle ones 3–6.5 × (0.5)0.7–1.8 cm, the distal ones 2–5 × 0.5–1.5 cm, oblong or oblong-elliptic, rounded, obtuse, retuse or emarginate at the apex, obliquely round or cordate at the base, glabrous or pubescent; midvein central or subcentral, with 3–5 secondary veins diverging from the proximal leaflet base. Stipules persistent, 15–55 × 1–3 mm, filiform, linear, lanceolate or spathulate; auricles 10–20 × 5–10 mm, reniform to subcircular, caducous independently of the stipules, lamina palmately nerved, pubescent. Dormant axillary buds much flattened. Inflorescences in much-branched terminal and/or axillary panicles up to 10 cm long; bracts 2–3.5 × 2–3 mm, ovate to oblong, acute at the apex. Flowers 5–8 × 3–4 mm; pedicels to 3 mm long; bracteoles 6–8 × 3–5 mm, obovate or orbicular, densely pubescent. Sepals 4–5, the proximal median sometimes fused with one or both of the laterals, 3–4 × 1.5–3 mm, oblong, orbicular or oblong-lanceolate, imbricate, entire to densely ciliate on the margins. Petals 0–3, 3–5 × 0.5–1 mm, filiform to spathulate. Stamens 10, united to 1 mm at the base; filaments 10–12 mm long. Ovary 3–6 × 2–3 mm, to 8-ovulate, densely pubescent; stipe to 4 mm long. Pods 7–25 × 3–5.5 cm, oblong to obovate, round to obtuse at the apex, dark brown, with spreading, suberect or revolute ventral flanges. Seeds up to 6, 15–25 × 12–16 mm, oblong to ovoid.

Botswana. N: Sitengu Pan, fr. viii.1937, *Miller* 145 (FHO, K). **Zambia.** N: Mbala, Kayimbi, fr. iv.1937, *Trapnell* 1757 (FHO, K). W: Ndola Dist., E side of Botondo R., near Mufulira, fl. 28.ix.1947, *Brenan & Greenway* 7970 (K). C: Lusaka, Woodlands, Mutanda Road, fl. 22.xi.1995, *Bingham* 10691b (K). S: Livingstone Dist., Kalembara road, fl. 20.x.1956, *Gilges* 659 (K). **Zimbabwe.** N: Mutoko Dist., c. 1.5 km (1 mile) from Mutoko towards Murewa, fl. 23.x.1965, *Corby* 1418 (K, SRGH). W: 200 km (130 miles) N of Bulawayo on Victoria Falls road, fl. 26.x.1970, *Vahrmeyer* 2217 (K, PRE). E: Mutare, fr., 13.xi.1930, *R.E. Fries* 2968 (K, LUND). S: Bikita Dist., Bikita to Save (Sabi), fl. 21.x.1930, *Fries, Norlindh & Weimarck* 2160 (K). **Malawi.** N: Chitipa Dist., N end of Nyika Plateau, S slopes of Kawozya, fr. 10.viii.1972, *Brummitt & Synge* in *WC* 195 (K). C: Kasungu Nat. Park, fl. 1.xii.1970, *Hall-Martin* 1015 (K). S: Mulanje Mt. slopes, fr. 8.v.1957, *Chapman* 369 (FHO, K). **Mozambique.** N: Cabo Delgado, Lúrio, fr. 19.x.1948, *Andrada* 1419 (FHO, K, LISC). T: Estima–Songo road, Marueira, imm.fr. 4.ii.1972, *Macedo* 4777 (K, LISC, LMA). MS: Manica e Sofala, Chibuli–Bandula, fl. n.d., *Honey* 777 (FHO).

Also in Congo, Tanzania and Angola. Locally dominant in deciduous woodland on hill slopes; also common on reddish rocky soils; 900–1600 m.

Conservation notes: Widespread species; Lower Risk, Least Concern.

Brachystegia boehmii can be confused with *B. gossweileri* because of the leaflet shape and shape of the leaflet apex. However, *B. boehmii* is distinct in its persistent stipules and stout petioles, whilst *B. gossweileri* has discolorous leaflet surfaces. It can be separated from *B. longifolia* as *B. boehmii* has short stout petioles up to 10 mm long, and 15–30 pairs of leaflets which are oblong-elliptic to lanceolate with a round to emarginate apex.

8. **Brachystegia puberula** Hutch. & Burtt Davy in Bull. Misc. Inform., Kew **1923**: 156 (1923) (see note under *B. gossweileri*). —Baker, Legum. Trop. Afr.: 721 (1930). —Hoyle in C.F.A. **2**: 231, figs.23, 32 (1956). —White & Hoyle in White, F.F.N.R.: 114, fig.1M (1962). —Hoyle in F.T.E.A., Legum.-Caesalp.: 174 (1967). Type: Angola, Bié, between R. Cusaba & R. Cunene, ii.1907, *Gossweiler* 2867 (BM holotype, COI, K, LISC). FIGURE 3.2.**21**/9.

Tree up to 12 m tall; bark reticulately fissured. Young branchlets densely puberulous with antrorse hairs. Leaves 10–21 cm long; petioles (6)8–20 mm long including a 2–4 mm long pulvinus; rachis deeply canaliculate with raised winged margins, pubescent; leaflets in 8–21 pairs, overlapping or contiguous, 2–3 times longer than broad, middle leaflets or those next to the proximal largest, proximal leaflet pairs 1–4 × 0.5–1.5 cm, the middle ones 2.5–6.5 × (0.5)0.8–2.5 cm, the distal ones 2–5.5 × 0.8–1.6 cm, oblong, lanceolate or elliptic, rounded or obliquely obtuse at the apex (so that the distal side of the lamina appears to be longer than the proximal side), obliquely round or cordate at the base, puberulous with antrorse hairs; midvein central or subcentral, with 3–4(5) secondary veins from the proximal leaflet base. Stipules caducous, 10–22 × 1.5–3 mm, free, linear, lanceolate, sometimes falcate, puberulous; auricles absent. Dormant axillary buds ovoid or globose. Inflorescences in terminal and/or axillary panicles up to 7 cm long; pubescent. Flowers 4–5 × 3–4 mm; pedicels up to 2–3 mm long; bracteoles 4–6 × 3–4 mm, elliptic or orbicular, densely puberulous. Sepals 5, 2–3 × 1–1.5 mm, elliptic to orbicular, imbricate. Petals absent. Stamens 10, united to 1–2 mm at the base; filaments 7–10 mm long. Ovary 2–3 × 1–1.5 mm; stipe to 1–2 mm long; style to 7–8 mm long. Pods 8–14 × 3–4 cm, oblong to obovate, with spreading ventral flanges. Seeds 15–20 × 12–15 mm, oblong to ovoid.

Zambia. W: Mwinilunga, Upper Lunga basin, fr. ix.1924, *Trapnell* 1571 (FHO, K); Mwinilunga [no locality], *Brachystegia* woodland, not on Kalahari sands, imm.fl. 13.xi.1955, *Holmes* 1327 (K).

Also in Tanzania and Angola. Locally dominant in deciduous woodlands on the margins of Kalahari sands and on rocky hill slopes; 900–1700 m.

Conservation notes: Lower Risk, Least Concern, although somewhat more restricted in range than many species in the genus.

Brachystegia puberula is similar in some leaf characters (e.g. number of leaflet pairs, petiole length, size and shape of leaflets) to *B. longifolia*. However, *B. puberula* differs in its globose dormant buds, obliquely retuse leaflet apices and puberulous leaves and inflorescences.

9. **Brachystegia tamarindoides** Benth. in Trans. Linn. Soc. **25**: 312 (1866). —Burtt Davy & Hutchinson in Bull. Misc. Inform., Kew **1923**: 153 (1923). —De Wild., Contrib. Fl. Katanga, Suppl. **2**: 60 (1929). —Baker, Legum. Trop. Afr.: 717 (1930). —Pardy in Rhod. Agric. J. **50**: 462 (1953). —Hoyle in C.F.A. **2**: 232 (1956). —O. Palgrave, Trees Central Africa: 77 (1957). Type: Angola, v.1859, *Welwitsch* 585 (K holotype, BR).

Tree 4–35 m tall; bark smooth, thinly flaking, exposing yellow patches beneath, pale silvery grey to dark grey. Young branchlets glabrous or pubescent with erect hairs. Leaves 6–15 cm long; petioles 2–15 mm long including a 2–7 mm long pulvinus; rachis deeply canaliculate above with raised and winged margins, glabrous or pubescent; leaflets in 6–72 pairs, contiguous, overlapping or widely spaced, the middle pairs or those next to the proximal or distal ones the largest, each 0.5–6 × 0.15–2 cm, linear, triangular, ovate, oblong, lanceolate or elliptic, acute, rounded, obtuse, retuse emarginate or apiculate at the apex, obliquely rounded, cordate or truncate at the base, glabrous or pubescent; midvein central or subcentral, 3–4 secondary veins diverging from the proximal leaflet base. Stipules usually caducous, 6–30 × 0.5–3.5 mm, shortly connate at the base, erect, filiform, linear, lanceolate or spathulate, sometimes subfoliaceous; auricles 2–6 × 1–3.5 mm, reniform, digitate, sometimes reduced to a tooth, caducous independently of the stipules. Dormant axillary buds ovoid or globose. Inflorescences in terminal and/or axillary panicles up to 8 cm long; bracts 1.5–3 × 1–2 mm, ovate, pubescent. Flowers 3–6 × 2–4 mm; pedicels 1.5–3 mm long; bracteoles 4–8 × 3–5 mm, obovate or orbicular. Sepals 4–5, the proximal sometimes fused with one or both of the laterals, 2–2.5 × 0.5–2 mm, ovate, oblong, orbicular or oblong-lanceolate, imbricate or valvate, margins entire to densely ciliate. Petals 0–1, 1.5 × 0.3 mm. Stamens 10, united to 0.5–1 mm at the base; filaments 8–10 mm long. Ovary 3–4 × 1–2 mm, up to 9-ovulate, densely pubescent; stipe up to 3 mm long; style to 6–10 mm long. Pods 5–10 × 1.5–3 cm, oblong to obovate, dark brown, with suberect or revolute spreading ventral flanges. Seeds up to 7, 15–20 × 10–15 mm, oblong to ovoid.

Leaflets in 15–72 pairs, inserted 2–5 mm apart, linear to triangular; distal leaflet pairs 1.5–5 mm wide, apex round, obtuse or acute, often apiculate; petioles 2–5(7) mm long, more than half of which is pulvinus .... **i)** subsp. *microphylla*
Leaflets in 6–20 pairs, inserted 5–20 mm apart, narrowly to broadly ovate, oblong, elliptic or lanceolate; distal leaflet pairs 8–20 mm wide, apex round, retuse, subacute, acuminate; petioles 5–15 mm long, of which less than half is pulvinus .... **ii)** subsp. *torrei*

i) subsp. **microphylla** (Harms) Chikuni, comb. & stat. nov. Type: Tanzania, Iringa Dist., Ndegere, ii.1899, *Goetze* 603 (B† holotype, K fragm.).

Brachystegia fischeri Taub. in Engler, Pflanzenw. Ost-Afrikas **C**: 197 (1895). —Burtt Davy & Hutchinson Bull. Misc. Inform., Kew **1923**: 154 (1923). —De Wild., Contrib. Fl. Katanga, Suppl. **2**: 32 (1929). —Baker, Legum. Trop. Afr.: 726 (1930). Type: Tanzania, Usuri, x.1885, *Fischer* 148 (B holotype, K).

Brachystegia microphylla Harms in Bot. Jahrb. Syst. **28**: 397 (1900). —Burtt Davy & Hutchinson in Bull. Misc. Inform., Kew **1923**: 153 (1923). —Baker, Legum. Trop. Afr.: 716 (1930). —Jackson in J. S. Afr. Bot. **6**: 35, 39 (1940). —Hoyle in F.C.B. **3**: 477 (1952). —White & Hoyle in White, F.F.N.R.: 115, 118, fig.22L (1962). —Hoyle in F.T.E.A., Legum.-Caesalp.: 174–177, fig.38 (1967). —White, Dowsett-Lemaire & Chapman, For. Fl. Malawi: 305 (2001). —M. Coates Palgrave, Trees Sthn. Africa: 323 (2002).

Brachystegia pectinata Sim in For. Fl. Port. E. Afr.: 50 (1909). —Burtt Davy & Hutchinson in Bull. Misc. Inform., Kew **1923**: 154 (1923). —De Wild., Contrib. Fl. Katanga, Suppl. **2**: 53 (1929). Type: Mozambique, Quelimane, between Maganja da Coasta and Namacurra, *Sim* 6100 (NU).

Brachystegia reticulata Hutch. & Burtt Davy in Bull. Misc. Inform., Kew **1923**: 154 (1923). —De Wild., Contrib. Fl. Katanga, Suppl. **2**: 54 (1929). —Baker, Legum. Trop. Afr.: 719 (1930). Type: Zimbabwe, Mutare, Odzani River valley, 1915, *Teague* 299 (K holotype).

Brachystegia glaucescens Hutch. & Burtt Davy in Bull. Misc. Inform., Kew **1923**: 152 (1923). —Baker, Legum. Trop. Afr.: 716 (1930). —Hoyle in F.C.B. **3**: 477 (1952). —White & Hoyle in White, F.F.N.R.: 115, 117 (1962). —Drummond, Common Trees C. Watershed Woodl. Zimbabwe: 64 (1981). —M. Coates Palgrave, Trees Sthn. Africa: 322 (2002). Type: Zimbabwe, Mutare, 1915 *Teague* 300 (K holotype).

Brachystegia robynsii De Wild., Contrib. Fl. Katanga, Suppl. **2**: 55 (1929). —Baker, Legum. Trop. Afr.: 716 (1930). Type: Congo, Katanga, *Robyns* 1835 (BR holotype).

Brachystegia letestui De Wild., Contrib. Fl. Katanga, Suppl. **2**: 42 (1929). —Baker, Legum. Trop. Afr.: 721 (1930). Type: Congo, 26.xi.1905, *Le Testu* 893 (BR holotype).

Tree 3–35 m tall; leaves 70–90 mm long; petioles 2–5(7) mm, including a pulvinus 2–3 mm long; leaflets in 15–72 pairs, 2–5 mm apart, middle leaflet pairs or pairs next to the proximal ones the largest.

Zambia. N: Mbala, gorge leading to Saise Valley, fl. 12.x.1936, *B.D.Burtt* 5748 (K). W: SW side of Ndola–Nkana road, c. 2.5 km (1.5 miles) on Nkana side of Mwekera turn, fl. 2.x.1947, *Brenan & Greenway* 8008 (FHO). C: Lusaka, Mt. Makulu Research Station, 20.v.1956, *Angus* 1292 (FHO). **Zimbabwe.** N: Mazoe Dist., 12.ix.1946, *Wild* 1228 (FHO). C: Marondera (Marandellas), Ruzawi School, fl. *Pitt* 79 (FHO). E: Chimanimani, E of Timbiri valley, iv.1969, *Goldsmith* 46/69 (FHO). S: near Lundi R., st. 30.vi.1930, *Hutchinson & Gillett* 3283 (K). **Malawi.** N: Chisenga, foothills of Mafinga Mts., fl. 9.xi.1958, *Robson & Fanshawe* 529 (K, LISC). C: Ntcheu, Sharp Vale, 22.vii.1958, *Jackson* 2240 (FHO). S: Mulanje Mt., Lukulezi Valley, fl. 2.i.1958, *Chapman* 511 (FHO, K). **Mozambique.** N: Pemba (Porto Amélia), from Mahate to Metuge, 1.x.1948, *Andrada* 1391 (FHO, LISC). Z: between Gurué and Namarrói, 3–4 km to Gurué, 16.ix.1949, *Barbosa & Carvalho* 4102 (FHO, LISC). MS: Chimoio, base of Monte de Belas, fr. 1.iv.1948, *Garcia* 822 (K, LISC).

Also in Congo and Tanzania. Locally dominant on rocky hills, escarpments, rocky granite soils and leached reddish soils, in high altitude and rainfall areas; often co-dominant with *B. taxifolia* and *B. spiciformis*; 800–2000 m.

Conservation notes: Widespread species; Lower Risk, Least Concern.

Between 1930 and 1950 *Brachystegia microphylla* and *B. glaucescens* were considered conspecific with *B. tamarindoides* (see Brenan & Hoyle 1949), but were reinstated as distinct species by White & Hoyle (1962). Chikuni (A Taxonomic Study of *Brachystegia*, D.Phil. thesis, Wolfson College, Oxford, 1998) found that *B. microphylla, B. glaucescens* and the newly described *B. torrei* intergrade morphologically with each other, and so a single species, *B. tamarindoides*, is recognized here as comprising three geographically and ecologically isolated subspecies. Subsp. *tamarindoides* is restricted to Angola. Subsp. *microphylla* is widespread and is distinguished by a combination of a short petiole, half of which is accounted for by the pulvinus length, apiculate leaflet apices and a large number of leaflets (up to 72 leaflet pairs). As leaves of the holotype of *B. glaucescens* have up to 20 leaflet pairs, 15 × 5 mm in size, *B. glaucescens* falls within the broad concept of subsp. *microphylla* and it is here considered conspecific. Subsp. *torrei* is distinguished by having less than 10 widely-spaced leaflet pairs which tend to increase in size distally.

In the Flora area the two taxa present are somewhat geographically and ecologically separated, with subsp. *torrei* occurring below 500 m altitude and *B. microphylla* being found on the central plateau and higher ground.

Local opinion in the region, however, suggests that there are perhaps two distinct forms of subsp. *microphylla* — *glaucescens* (leaves with 9–16 leaflet pairs, and broader oblong leaflets 13–30 × 4–10 mm), a widespread variant especially across Zimbabwe, and *microphylla* (leaves with 25–55 leaflet pairs, and slender, narrowly triangular, often curved, closely-set leaflets 5–15 ×1–3 mm), which occurs locally on granite outcrops in higher rainfall areas. The leaf differences are shown in Chikuni (1998, p. 121, fig.4.3). Some botanists prefer to keep these closely-related taxa separate as *B. glaucescens* and *B. microphylla.*

ii) subsp. **torrei** (Hoyle) Chikuni, comb. & stat. nov. Type: Mozambique, Manica e Sofala, bewteen Espungabera (Spungabera) & Chibabava, 10.xi.1943, *Torre* 6141 (FHO holotype, K, LISC). FIGURE 3.2.**23**.

Brachystegia sp. nov. 1 sensu K. Coates Palgrave, Trees Sthn. Africa: 274 (1977).

Brachystegia torrei Hoyle in Kew Bull. **54**: 159 (1999). —M. Coates Palgrave, Trees Sthn. Africa: 325 (2002).

Tree 4–18 m tall; leaves 50–120 mm long; petioles 5–15 mm, including a pulvinus 3–5 mm long; leaflets in 6–12 overlapping or widely spaced pairs, 7–20 mm apart, leaflets next to the distal pairs the longest.

Zimbabwe. E: Lower Save Valley, 28.i.1948, *Wild* 2353 (K); Save-Lundi junction, Chuhenje range, fr. 9.vi.1950, *Wild* 3473 (K, SRGH). **Malawi.** C: Chikwawa, 4.vii.1955, *Jackson* 714 (FHO). S: Nsanje (Port Herald), *Topham* 512 (FHO). **Mozambique.** N: Cabo Delgado, Mueda, c. 47 km from Mueda to Nantulo, 4.i.1964, *Torre & Paiva* 9852 (FHO, LISC). T: 26 km from the Tete–Changara crossroads from Chioco, fl. 4.i.1966, *Torre & Correia* 14003 (FHO, LISC). MS: between Entroncamento and Tete, vi.1949, *Barbosa & Carvalho* 3241 (FHO). GI: between Inharrine and Inhambane, fr. 1.viii.1944, *Torre* 6819 (K, LISC).

Confined to the Flora area. Deciduous woodlands, on rocky hill slopes; 50–200 m.

Conservation notes: Lower Risk, Least Concern.

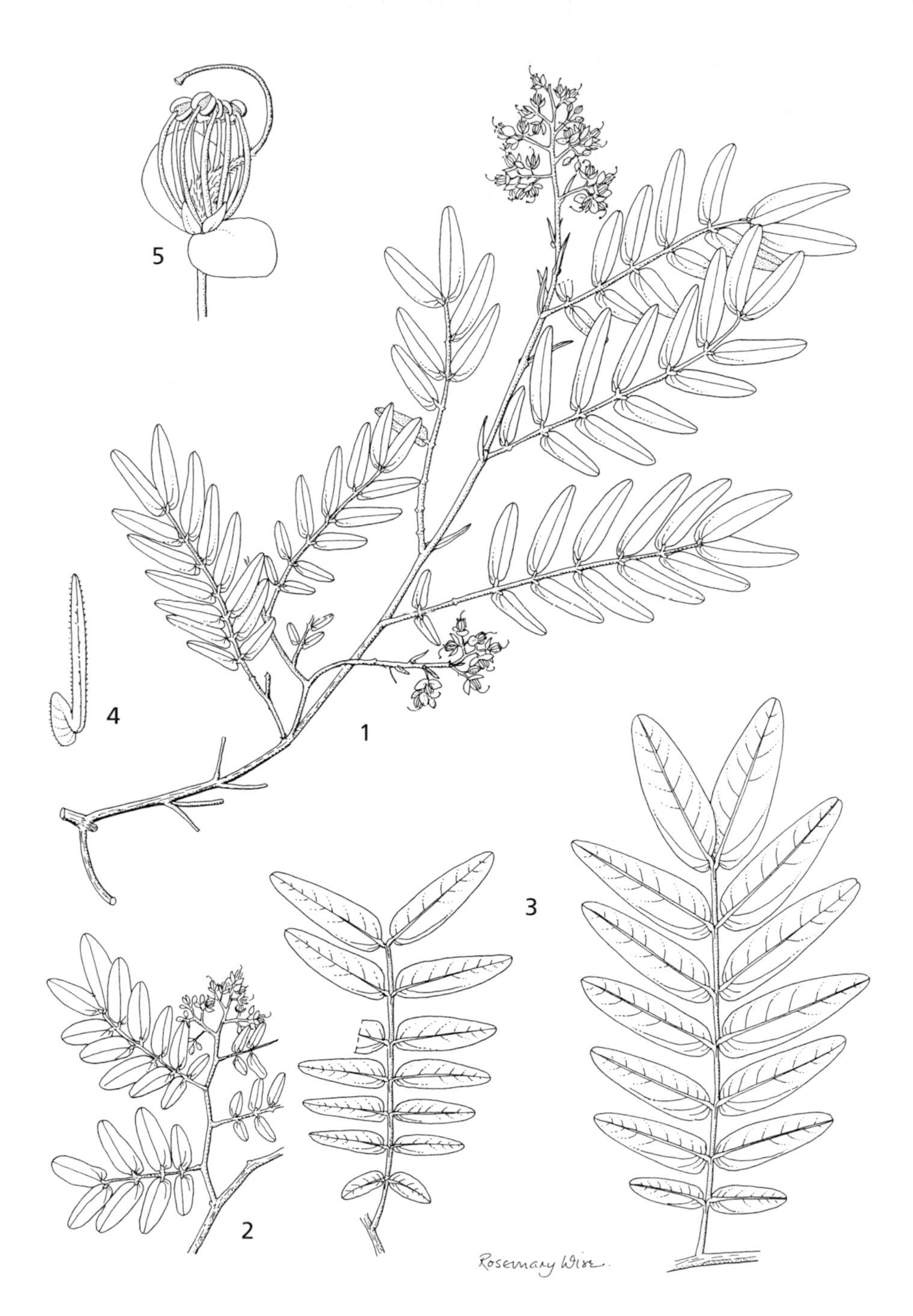

Fig. 3.2.**23**. BRACHYSTEGIA TAMARINDOIDES subsp. TORREI. 1, flowering branchlet with stipules (× $^1/_3$) from *Torre* 6141; 2, flowering branchlet (× $^1/_2$) from *Chikuni* 450; 3, leaves (× $^2/_3$) from *Torre & Correia* 17743 & 13914; 4, auriculate stipule (× 3); 5, flower with deflexed bracteoles (× 3), 4–5 from *Torre* 6141. Drawn by Rosemary Wise. From A. Chikuni, D.Phil. thesis, University of Oxford.

10. **Brachystegia spiciformis** Benth. in Trans. Linn. Soc. **25**: 312 (1866) as *spicaeformis*. —Sim, For. Fl. Port. E. Afr.: 49 (1909). —Burtt Davy & Hutchinson in Bull. Misc. Inform., Kew **1923**: 159 (1923). —Baker, Legum. Trop. Afr.: 727 (1930). —Jackson in J. S. Afr. Bot. **6**: 39 (1940). —Hoyle in F.C.B. **3**: 452, figs.38, 32 (1952); in C.F.A. **3**: 225 (1956); in Mem. New York Bot. Gard. **8**: 427 (1956). —O. Palgrave, Trees Central Africa: 81 (1957). —White & Hoyle in White, F.F.N.R.: 107, 117, 118, fig.23 (1962). —Gomes e Sousa, Dendrol. Moçamb. Estudo Geral **1**: 268, fig.65 (1966). —Hoyle in F.T.E.A., Legum.-Caesalp.: 167, figs.36, 37 (1967). —White, Dowsett-Lemaire & Chapman, For. Fl. Malawi: 305 (2001). —Hurter & van Wyk in Bothalia **31**: 43 (2001). —M. Coates Palgrave, Tres Sthn. Africa: 324 (2002). Type: Angola, Huìla, Mumpulla–Nene, *Welwitsch* 578 (LISU holotype, BM). FIGURE 3.2.**24**.

Brachystegia appendiculata Benth. in Trans. Linn. Soc. **25**: 313 (1866). —Oliver in F.T.A. **2**: 305 (1871). —Burtt Davy & Hutchinson in Bull. Misc. Inform., Kew **1923**: 162 (1923). —De Wild., Contrib. Fl. Katanga, Suppl. **2**: 17 (1929). Type: Malawi, Zomba, Magamero, E of Lake Chirwa, fl. x.1861, *Meller* s.n. (K holotype).

Brachystegia spiciformis var. *glandiflora* Benth. in Trans. Linn. Soc. **25**: 313 (1866). Type: Angola, Huila, *Welwitsch* 579 (FHO fragm.).

Brachystegia spiciformis var. *parviflora* Benth. in Trans. Linn. Soc. **25**: 313 (1866). Type from Tanzania.

Brachystegia itoliensis Taub. in Engler, Plfanzenw. Ost-Afrikas **C**: 197 (1895). —Burtt Davy & Hutchinson in Bull. Misc. Inform., Kew **1923**: 158 (1923). —Baker, Legum. Trop. Afr.: 728 (1930). —De Wild., Contrib. Fl. Katanga Suppl. **2**: 40 (1929). Type: Tanzania, Bukoba Dist., Itolio, 13.xi.1890, *Stuhlmann* 925 (B† holotype).

Brachystegia mpalensis Micheli in Compt. Rend. Soc. Bot. Belg. **36**: 73 (1897). —Burtt Davy & Hutchinson in Bull. Misc. Inform., Kew **1923**: 158 (1923). —De Wild., Contrib. Fl. Katanga, Suppl. **2**: 52 (1929). Type: Congo, Moba (Baudouinville) Dist., Mpala, *Deschamps* 27 (BR lectotype).

Brachystegia randii Baker f. in J. Bot. **37**: 433 (1899). —Burtt Davy & Hutchinson in Bull. Misc. Inform., Kew **1923**: 160 (1923). —Baker, Legum. Trop. Afr.: 727 (1930). Lectotype, chosen by Hoyle (F.T.E.A. 1967): Zimbabwe, Harare (Salisbury), *Rand* 610 (BM lectotype).

Brachystegia bragaei Harms in Bot. Jahrb. Syst. **30**: 82 (1901). —Burtt Davy & Hutchinson in Bull. Misc. Inform., Kew **1923**: 161, (1923). —De Wild., Contrib. Fl. Katanga, Suppl. **2**: 17 (1929). Type: Mozambique, Beira, *Braga* 152 (B† holotype, K).

Brachystegia euryphylla Harms in Bot. Jahrb. Syst. **30**: 82 (1901). —Burtt Davy & Hutchinson in Bull. Misc. Inform., Kew **1923**: 162 (1923). —De Wild., Contrib. Fl. Katanga, Suppl. **2**: 22 (1929). —Baker, Legum. Trop. Afr.: 730 (1930). Type: Tanzania, Morogoro Dist., E Ukami, 8.x.1894, *Stuhlmann* 8666 (B holotype).

Brachystegia hockii De Wild. in Repert. Spec. Nov. Regni Veg. **11**: 512 (1913). —Burtt Davy & Hutchinson in Bull. Misc. Inform., Kew **1923**: 159 (1923). —De Wild., Contrib. Fl. Katanga, Suppl. **2**: 35 (1929). Type: Congo, Haut Katanga, *Hock* s.n. (BR holotype).

Brachystegia hockii var. *papyracea* De Wild. in Ann. Mus. Congo Belge **4**: 46 (1913). Type: Congo, *Bequeart* 170 (K photo).

Brachystegia trijuga R.E. Fr., Wiss. Ergebn. Schwed. Rhod.-Kongo-Exped.: 65 (1914). —De Wild., Contrib. Fl. Katanga, Suppl. **2**: 62 (1929). Type: Zambia, *R.E. Fries* 586 (UPS holotype, K photo).

Brachystegia lujae De Wild. in Bull. Jard. Bot. État. **7**: 252 (1920). —Burtt Davy & Hutchinson in Bull. Misc. Inform., Kew **1923**: 157 (1923). —De Wild., Contrib. Fl. Katanga, Suppl. **2**: 47 (1929). —Baker in Legum. Trop. Afr.: 729 (1930). —Hoyle in F.C.B. **3**: 450 (1952). Type: Congo, Kasai, viii.1902, *Luja* 51 (BR holotype).

Brachystegia edulis Hutch. & Burtt Davy in Bull. Misc. Inform., Kew **1923**: 162 (1923). —De Wild., Contrib. Fl. Katanga, Suppl. **2**: 22 (1929). Type: Zambia, Batoka Highlands, 1860, *Kirk* s.n. (K holotype).

Brachystegia taubertiana Hutch. & Burtt Davy in Bull. Misc. Inform., Kew **1923**: 150 (1923). —De Wild., Contrib. Fl. Katanga, Suppl. **2**: 61 (1929). Type: Tanzania, Tanga Dist., Usambara, Doda, iv.1894, *Holst* 3023 (K holotype).

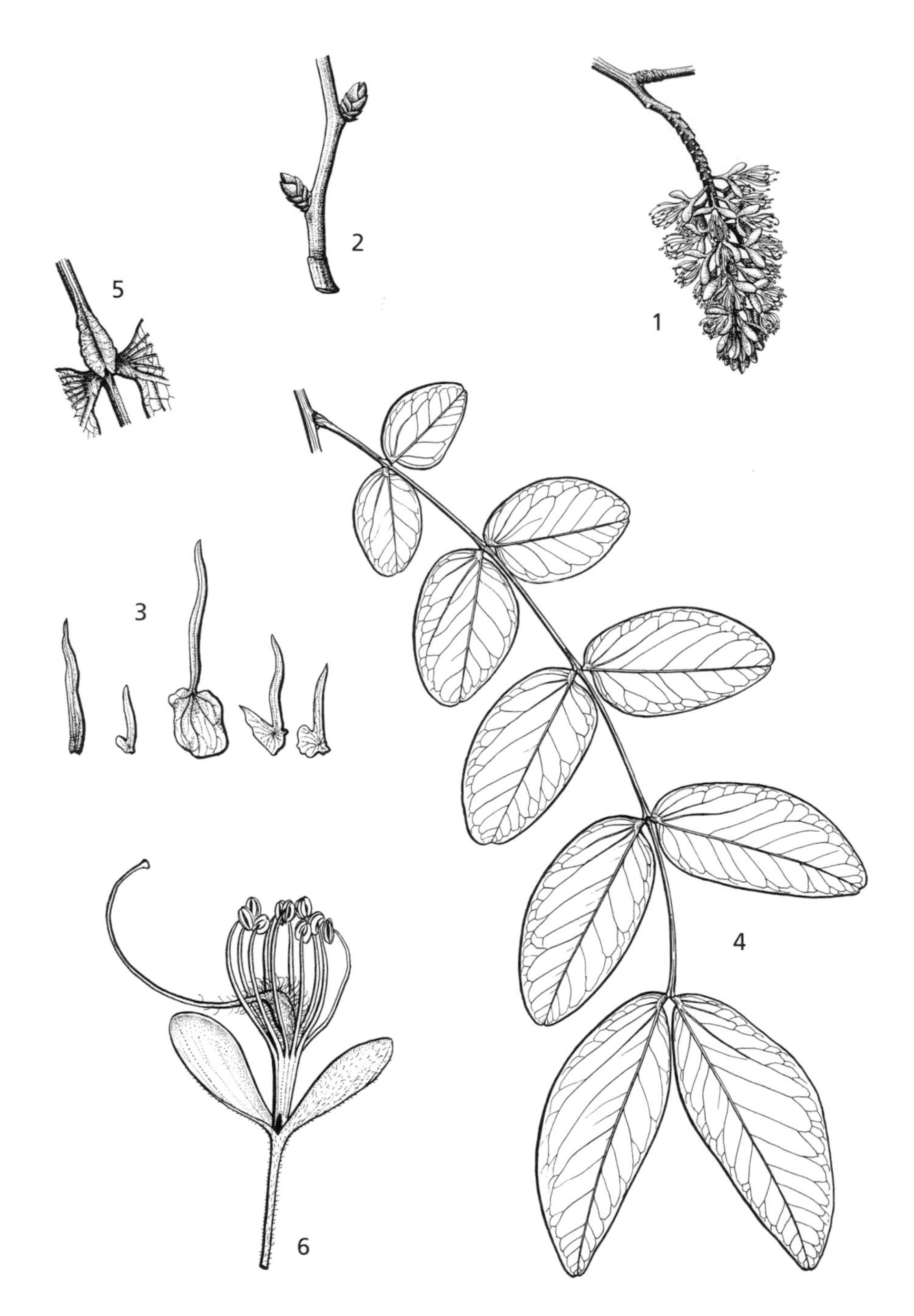

Fig. 3.2.**24**. BRACHYSTEGIA SPICIFORMIS. 1, raceme showing scars of fallen flowers (× ½); 2, main shoot buds on leafless mature branchlet (× ½); 3, five forms of stipule (× 1); 4, leaf; 5, part of upper surface of leaf-rachis showing stipels and local expansions (× 2); 6, flower with bracteoles (× 2); 4–6 from *Harley* 9150. Adapted from F.T.E.A.

Brachystegia venosa Hutch. & Burtt Davy in Bull. Misc. Inform., Kew **1923**: 158 (1923). — De Wild., Contrib. Fl. Katanga, Suppl. **2**: 65 (1929). Type: Zambia, near Victoria Falls, 15.x.1906, *Allen* 165 (K holotype, PRE).

Brachystegia hockii var. *piloso-pedicellata* De Wild., Contrib. Fl. Katanga, Suppl. **2**: 37 (1929). —Baker, Legum. Trop. Afr.: 727 (1930). Type: Angola, *Wild* s.n. (K photo).

Brachystegia spiciformis var. *kwangensis* Hoyle in Bull. Jard. Bot. État. **21**: 361 (1951); in F.C.B. **3**: 454 (1952). Type: Congo, Kwango, viii.1944, *Germain* 2526 (BR holotype).

Brachystegia spiciformis var. *latifoliolata* (De Wild.) Hoyle in F.C.B. **3**: 453, fig.32 (1952). Type: Congo, Katanga, *Quarré* 508 (K photo).

Brachystegia spiciformis var. *mpalensis* (Michel.) Hoyle in F.C.B. **3**: 456 (1952). Type: Congo, *Descamps* 27 (K photo).

Brachystegia spiciformis var. *schmitzii* Hoyle in Bull. Jard. Bot. État. **21**: 101 (1951). —Hoyle in C.F.A. **2**: 229 (1956); in F.C.B. **3**: 456 (1952). Type: Congo, Katanga, *Schmitz* 982 (BR holotype, K).

Tree up to 30 m tall; bark deeply to shallowly fissured longitudinally, often coarsely reticulate or smooth, flaking in thick irregular scales, non-scaly if smooth, dark grey. Branchlets glabrous or pubescent. Leaves (3)5–20 cm long; petioles 1–5 cm long; rachis deeply canaliculate above with raised and winged margins; leaflets in 2–8 contiguous or widely spaced pairs, increasing in size distally; proximal leaflet pairs 1.5–5 × 0.5– 2.5 cm, the others 3–20 × 1.5–8 cm, narrowly to broadly ovate, oblong, elliptic or lanceolate, acute, acuminate, rounded, obtuse, retuse or emarginate at the apex, obliquely round, cordate or truncate at the base, midvein central or subcentral, with 3–5 secondary veins from the proximal leaflet base. Stipules 10–40 × 1–4 mm, shortly connate at the base, erect or reflexed, filiform, linear or lanceolate, early caducous; auricles usually present, 5–25 × 1–6.5 mm, reniform to semi-circular, caducous independently of the stipules. Dormant axillary buds globose or ovoid. Inflorescences of terminal and/or axillary racemes up to 8 cm long; bracts 4–5 × 1–2 mm, caducous, ovate to oblong. Flowers 5–8 × 3–5 mm; pedicels 2–8 mm long; bracteoles 5–10 × 3–6 mm, obovate or orbicular. Sepals 0–4, 0.5–3 × 0.5–1.5 mm, linear, linear-lanceolate or rudimentary. Petals absent. Stamens 10, united to 2–7 mm at the base; filaments 10–20 mm long. Ovary 2–5 × 1–2 mm, densely pubescent, up to 7-ovulate, the stipe to 3–6 mm long; style up to 15 mm long. Pods dark brown, 8–15 × 3–5 cm, oblong to obovate; ventral flanges spreading, suberect or revolute. Seeds up to 6, 12–25 × 10–20 mm, oblong to ovoid.

Zambia. B: Zambezi (Balovale), fl. 15.viii.1952, *Gilges* 203 (K). N: Mbala Dist., Kambole Estate, above Lusamo Waterfall, Katanga R., fl. 14.ix.1980, *Richards* 13252 (K). W: Ndola, fl. 14.ix.1938, *Greenway & Miller* 5681 (K). C: Kabwe Dist., Luwonde Forest, fr. 30.vi.1998, *P.P. Smith* 1763 (K). E: hill to W of Chipata (Fort Jameson), fl. 9.x.1958, *Robson & Angus* 34 (K). S: c. 60 km (41 miles) N of Livingstone, fl. 16.ix.1935, *Galpin* 15100 (K, PRE). **Zimbabwe.** N: Mafungabusi Plateau, 26.vi.1947, *Keay* in FHI. 21370 (FHO). W: Matopos Nat. Park, 6 km S of Circular Drive towards Mtsheleli Dam, fr. 9.i.1972, *Bullock* in *SRGH* 217539 (K, LISC, SRGH). C: NNW of Harare, Dombashawa, 1.vi.1927, *Eyles* 7503 (FHO). E: Nyanga, xi.1946, *Miller* 369 (FHO). S: Wedza Mt., 18.iii.1931, *Pardy* 4763 (FHO). **Malawi.** N: Rumphi, edge of Nyika Plateau, fr. 9.v.1952, *White* 2812 (FHO, K). C: Dzalanyama Reserve, 25.ix.1927, *Topham* 43 (FHO). S: Mangochi, Idulusi, Chowe Escarpment, fl. 8. x. 1985, *Salubeni, Balaka & Kwatha* 4290 (K, MAL). **Mozambique.** N: Niassa, Litunde, fl. 3.x.1983, *Groenendijk, Costa & Cruz* 673 (K, LMU). Z: Maganja da Costa, 3.3 km from Namacurra, 27.ix.1949, *Barbosa & Carvalho* 3830 (FHO, LISC). MS: Chimoio, 31.iii.1948, *Barbosa* 1301 (FHO, LISC). GI: near Jangamo, 21 km S of Inhambane, fr. 6.v.1971, *Edwards & Vahrmeyer* 4222 (K, PRE). M: Marracuene, fl. 15.ix.1989, *Groenendijk* 2169 (K, LMU).

Also in Kenya, Tanzania, Burundi, Congo, Angola and South Africa. In a variety of habitats from coastal to upland, generally in deciduous woodlands and open forest, on hill slopes and river banks; 50–2000 m.

Conservation notes: Very widespread; Lower Risk, Least Concern.

A very variable species. Specimens from wetter areas tend to have more leaflet pairs (up to 7) while those from drier areas have up to 3 pairs. Hoyle originally recognised a number of varieties but later abandoned these and recognised a single variable species. *Brachystegia lujae*, kept separate by Hoyle, is here treated as synonymous with *B. spiciformis*.

A number of putative hybrids of *B. spiciformis* × *B. tamarindoides* subsp. *microphylla* (previously termed *B. glaucescens*), with intermediate characters and both presumed parents present nearby, have been identified by botanists in Zimbabwe. These are here grouped under *B. spiciformis* as their characters generally fall within the wide range given for that species, although they may key out as *B. tamarindoides* subsp. *torrei* based on number of leaflets. The putative hybrid has a spreading form and rounded branch stumps on the trunk, similar to *B. tamarindoides* subsp. *microphylla*.

11. **Brachystegia bakeriana** Hutch. & Burtt Davy in Bull. Misc. Inform., Kew **1923**: 159 (1923) (see note under *B. gossweileri*). —De Wild., Contrib. Fl. Katanga, Suppl. **2**: 17 (1929). —Baker, Legum. Trop. Afr.: 729 (1930). —Hoyle in C.F.A. **2**: 224 (1956). —White & Hoyle in White, F.F.N.R.: 106 (1962). Type: Angola, Benguela, 6.vi.1906, *Gossweiler* 2595 (K holotype).

Brachystegia gairdnerae Hutch. & Burtt Davy in Bull. Misc. Inform., Kew **1923**: 161 (1923). —De Wild., Contrib. Fl. Katanga, Suppl. **2**: 33 (1929). —Baker, Legum. Trop. Afr.: 730 (1930). Type: Zambia, Sesheke, *Gairdner* 220 (K holotype).

Brachystegia obliqua Hutch. & Burtt Davy in Bull. Misc. Inform., Kew **1923**: 160 (1923). —De Wild., Contrib. Fl. Katanga, Suppl. **2**: 55 (1929). Type: Angola, Benguela, vi.1906, *Gossweiler* 2711 (K holotype).

Brachystegia bakeriana var. *pubescens* Baker f., Legum. Trop. Afr.: 729 (1930). Type as for *B. obliqua*.

Shrub to 6(10) m tall, branching low down or at ground level; bark smooth, non-scaly, reticulately fissured at the base, greyish white. Young branchlets glabrous, or pubescent with brown hairs. Leaves 1.5–4(7) cm long; petioles 3–8(10) mm long; rachis deeply canaliculate above with raised and winged margins, glabrous or pubescent; leaflets in 2–4(6) contiguous or overlapping pairs, increasing is size distally; proximal leaflet pairs 1–4 × 0.8–2 cm, the rest 2–6 × 1–2.5 cm, oblong, elliptic, oval or obovate, rounded, obliquely retuse or emarginate at the apex, obliquely cuneate, rounded or cuneate at the base, midvein central or subcentral, with 3–4 veins diverging from the proximal leaflet base. Stipules early-caducous, up to 15 × 2 mm, shortly connate at the base, erect, filiform, linear or lanceolate; auricles reniform to semi-circular, caducous independently of the stipules. Dormant axillary buds globose or ovoid. Inflorescences terminal and/or axillary racemes, up to 5 cm long, pubescent with dark or brown hairs; bracts 2–3 × 1.5–2 mm, caducous, ovate to oblong, acute at the apex. Flowers up to 9 × 5 mm; pedicels 2–4 mm long; bracteoles 5–10 × 3–5 mm, obovate or orbicular. Sepals 0–4, 0.5–3 × 0.5–1.5 mm, linear, linear-lanceolate or rudimentary, ciliate on the margins. Petals absent. Stamens 10, united to 6 mm at the base; filaments 7–15 mm long. Ovary 2–5 × 1–2 mm, up to 7-ovulate, densely pubescent, stipe to 3 mm long. Pods yellowish brown or light brown, 5–9 × 2–3 cm, oblong to obovate, with spreading ventral flanges. Seeds up to 4, 12–20 × 10–15 mm, oblong to ovoid.

Zambia. B: Kataba, Sesheke north, fl. 1.ix.1960, *Fanshawe* 5803 (FHO, K, LISC); Kataba, fr. 12.xii.1960, *Fanshawe* 5968 (K).

Also in Angola. Locally dominant on white Kalahari sands, forming dense thickets. Also common on margins of wetlands, generally forming pure stands but occasionally intermixed with *B. spiciformis*; 900–1300 m.

Conservation notes: Probably Lower Risk, Least Concern.

Brachystegia bakeriana resembles *B. spiciformis* in the minute sepals, the racemose inflorescences and the large conspicuous stipels. It differs in its short petioles and leaves, smooth non-scaly bark, and in its fruits that are predominantly yellowish brown. It is generally a shrub and is characteristic of Kalahari sands.

12. **Brachystegia stipulata** De Wild. in Ann. Mus. Congo Bot., sér. 4, **1**: 44 (1902). —R.E. Fries, Wiss. Ergebn. Schwed. Rhod.-Kongo-Exped. **1**: 67 (1914). —Harms in Engler, Pflanzenw. Afrikas **3**(1): 480 (1915). —Burtt Davy & Hutchinson in Bull. Misc. Inform., Kew **1923**: 150 (1923). —De Wild., Contrib. Fl. Katanga, Suppl. **2**: 59 (1929). —Baker, Legum. Trop. Afr.: 724 (1930). —Jackson in J. S. Afr. Bot. **6**: 39 (1940). —Brenan & Hoyle in Brenan, Check-list For. Trees Shrubs Tang. Terr.: 92 (1949). —Hoyle in F.C.B. **3**: 457 (1952). —White & Hoyle in White, F.F.N.R.: 106, fig.22F (1962). —Hoyle in F.T.E.A., Legum.-Caesalp.: 181 (1967). Type: Congo, Katanga, Lukafu, *Verdick* 18 (BR holotype). FIGURE 3.2.**21**/8.

Brachystegia bequaertii De Wild. in Repert. Spec. Nov. Regni Veg. **11**: 512 (1913). —Burtt Davy & Hutchinson in Bull. Misc. Inform., Kew **1923**: 150 (1923). —De Wild., Contrib. Fl. Katanga: 69 (1921); Contrib. Fl. Katanga, Suppl. **2**: 17 (1929). —Hoyle in F.C.B. **3**: 460 (1952). Type: Congo, Katanga, 24.ix.1911, *Bequeart* 211 (BR holotype, K fragm.).

Brachystegia velutina De Wild. in Repert. Spec. Nov. Regni Veg. **11**: 512 (1913). —Harms in Engler, Pflanzenw. Afrikas **3**(1): 479 (1915). —Burtt Davy & Hutchinson in Bull. Misc. Inform., Kew **1923**: 150 (1923). —De Wild., Contrib. Fl. Katanga, Suppl. **2**: 62 (1929). —Baker, Legum. Trop. Afr.: 722 (1930). —Topham in Bull. Misc. Inform., Kew **1930**: 359 (1930). Type: Congo, Katanga, Lubumbashi (Elizabethville), 1911, *Hock* s.n. (BR holotype).

Brachystegia lufirensis De Wild. in Repert. Spec. Nov. Regni Veg. **11**: 512 (1913). —Burtt Davy & Hutchinson in Bull. Misc. Inform., Kew **1923**: 149 (1923). —De Wild., Contrib. Fl. Katanga, Suppl. **2**: 45 (1929). —Baker, Legum. Trop. Afr.: 730 (1930). Type: Congo, Lufira R. valley, 6.x.1911, *Hock* s.n. (BR holotype).

Brachystegia thomasii De Wild. in Ann. Soc. Sci. Brux. **48**: 77 (1928). —De Wild., Contrib. Fl. Katanga, Suppl. **2**: 61 (1929). —Baker, Legum. Trop. Afr.: 730 (1930). Type: Congo, 14.ix.1923, *Thomas* 1339 (BR holotype).

Brachystegia velutina var. *quarrei* De Wild., Contrib. Fl. Katanga, Suppl. **2**: 63 (1929). —Baker, Legum. Trop. Afr.: 723 (1930). Type: Congo, 10.ii.1927, *Quarre* 123 (BR holotype).

Brachystegia kassneri Baker f. in J. Bot. **67**: 195 (1929). —Baker, Legum. Trop. Afr.: 730 (1930). Type: Congo, Matumbi, *Kassner* 2506 (BR holotype).

Brachystegia stipulata var. *velutina* (De Wild.) Hoyle in F.C.B. **3**: 459 (1952). —Jackson in Soc. Malawi J. **21**: 14 (1968).

Small to medium tree to 8 m tall; bark deeply or reticulately fissured, flaking in thick scales. Young branchlets densely pubescent. Leaves 6–20 cm long; petiole 10–25 mm long, rachis deeply canaliculate above with raised and winged margins; leaflets in 5–8(10) overlapping or contiguous pairs, the middle leaflet pairs the largest, the proximal ones 2–7 × 1–3.5 cm, the middle ones 4–8 × 1.5 cm, the distal ones 3–6 × 1.2–2.5 cm, narrowly ovate, oblong-lanceolate to narrowly elliptic, acute, emarginate to retuse at the apex, obliquely cordate at the base, midvein subcentral or central, 3–4 secondary veins diverging from the base. Stipules persistent, 10–20 × 1–5 mm, shortly connate at the base, linear to falcate; auricles 8–35 × 5–15 mm, persistent independently of stipules, reniform to semicircular, palmately nerved. Dormant axillary buds much flattened. Inflorescences in terminal and/or axillary panicles up to 4 cm long; bracts 3–3.5 × 2–2.5 mm, ovate to oblong, acute at the apex. Flowers 7–15 × 4–6 mm; pedicels 2–15 mm long; bracteoles 11–20 × 5–10 mm, orbicular. Sepals 5–8, 4–8 × 1–3 mm, obovate, lanceolate, oblong, or linear, densely pubescent on the margins. Petals 1–4, 3–7 × 0.4–1.5 mm, filiform, linear or spathulate. Stamens 12–18(20), united to 6 mm at the base; filaments 25–45 mm long. Ovary 8–10 × 2–3 mm, up to 12-ovulate, stipe to 4 mm long; style to 25 mm long. Pods 9–23 × 3.5–5 cm, oblong, with spreading or suberect ventral flanges up to 10 mm wide. Seeds up to 11, 20–25 × 15–20 mm, round to oblong.

Zambia. N: c. 5 km (3 miles) S of Chinsali, fr. 19.x.1968, *Lawton* 1544 (FHO, K). W: Ndola, between Mwekera Rest House and Mufulira, fl. 28.ix.1947, *Brenan & Greenway* 7979 (FHO, K). E: Lundazi to Chama, Tigone Dam, fr. 18.x.1958, *Robson & Angus* 154A (FHO, K). C: Serenje Dist., 1.5 km (1 mile) from Kundalila Falls towards Kanona, fr. 29.iii.1984, *Brummitt, Chisumpa & Nshingo* 16984 (K). E: Petauke, 48 km

(30 miles) E of Petauke Boma, fr. 21.iv.1952, *White* 2432 (FHO, K). **Malawi.** C: Chitedze Agric. Research Station, 21.i.1954, *Jackson* 1216 (FHO). **Mozambique.** T: Mualadze (Vila Gamito), fl. 19.x.1943, *Torre* 6059 (FHO, LISC). MS: between Fingoè and Fronteira, fr. 28.vi.1949, *Barbosa & Carvalho* 3366 (FHO, LISC).

Also in Tanzania and Congo. Locally dominant on lateritic ridges, rocky hill slopes, open woodland and seasonal swamps; common on sandy clay and sandy loam soils; occurs in association with *B. floribunda* and *B. spiciformis*; 800–1500 m.

Conservation notes: Probably Lower Risk, Least Concern.

13. **Brachystegia allenii** Hutch. & Burtt Davy in Bull. Misc. Inform., Kew **1923**: 156 (1923) (see note under *B. gossweileri*). —De Wild., Contrib. Fl. Katanga, Suppl. **2**: 16 (1929). —Baker, Legum. Trop. Afr.: 731 (1930). —Jackson in J. S. Afr. Bot. **6**: 37 (1940). —Brenan & Hoyle in Brenan, Check-list For. Trees Shrubs Tang. Terr.: 92 (1949). —Hoyle in F.C.B. **3**: 467 (1952). —White & Hoyle in White, F.F.N.R.: 108 (1962). —Hoyle in F.T.E.A., Legum.-Caesalp.: 182, fig.39 (1967). —Corby in Kirkia **9**: 313 (1974). Type: Mozambique, Cabo Delgado (Niassa), mouth of Messalo (Msalu) R., 25.xi.1911, *Allen* 93 (K holotype). FIGURE 3.2.**25**.

Brachystegia giorgii De Wild. in Ann. Soc. Scien. Brux. **48**: 74 (1928); Contrib. Fl. Katanga, Suppl. **2**: 33 (1929). Type: Congo, Moba (Baudounville), ix.1922, *De Giorgi* 48 (FHO fragm.).

Brachystegia pruinosa De Wild., Contrib. Fl. Katanga, Suppl. **2**: 53 (1929). —Baker, Legum. Trop. Afr.: 730 (1930). Type: Congo, Katanga, Kiambi, NW L. Mweru, 4.xi.1922, *Delevoy* 537 (BR holotype).

Brachystegia schliebenii Harms in Notizbl. Bot. Gart. Berlin. **12**: 508 (1935). Type: Tanzania, Lukuledi valley, 23.x.1934, *Schlieben* 5520 (B† holotype).

Tree to 15 m tall; bark deeply or shallowly fissured longitudinally, often coarsely reticulate, flaking in thick rectangular scales, dark grey. Young branchlets pubescent with brown or cream hairs. Leaves 6–15 cm long; petioles 10–30 mm long; rachis deeply canaliculate above with raised and winged margins; leaflets glaucous in 3–6 overlapping to widely spaced pairs, broadly oblong or rectangular, 1.5–2 times as long as broad, proximal leaflet pair 2–5 × 1.5–3.5 cm, the rest 3–7 × 2–4 cm, obliquely rounded, obtuse or emarginate at the apex, obliquely cordate at the base, glabrous, midvein central, 3–5 veins from the proximal leaflet base. Stipules persistent, 10–26 × 1.5–3 mm, shortly connate at the base, erect, linear or falcate; auricles 7–22 × 5–15 mm, reniform to semicircular, palmately nerved, caducous independently of the stipules. Dormant axillary buds much flattened. Inflorescences in terminal and/or axillary panicles to 10 cm long; bracts 2–3 × 1.5–2 mm, ovate. Flowers up to 3–6 × 3–4 mm; pedicels up to 2.5 mm long; bracteoles 5–8 × 3–4 mm, obovate or orbicular. Sepals 5, 2–4 × 1–2 mm, oblong, orbicular or oblong-lanceolate, imbricate, densely cilate on the margins. Petals 0–3, up to 3 × 1.5 mm, filiform or linear. Stamens 10, united to 2 mm at the base filaments 8–12 mm long. Ovary 3–5 × 1.5–2.5 mm, densely pubescent, oblong, stipe to 3.5 mm long. Pods dark brown, 9–16 × 3–4.5 cm, oblong to obovate, with suberect ventral flanges. Seeds up to 7, 15–20 × 10–15 mm, oblong or circular.

Zambia. N: Mpulungu, fl. 3.xii.1936, *Gamwell* 246 (K). C: S Luangwa Nat. Park, S of Kateti R, 4.v.1966, *B.L.Mitchell* 2814. E: Tigone Dam, 2.5 km (1.5 miles) from Lundazi to Chama, fl. 17.x.1958, *Robson & Angus* 143 (K, LISC). **Zimbabwe.** N: Guruve (Sipolilo), foot of Rukowakuona Mts., fl. 19.x.1962, *Wild & Barbosa* 5911 (K, LISC). **Malawi.** N: foothills of Misuku-Yembe Hills near Ngerenge, Karonga, fr. 23.viii.1956, *Jackson* 2038 (FHO, K). **Mozambique.** N: Mogincual, between Liupe and Quixaxe, 18.x.1948, *Barbosa* 2476 (FHO, LISC). T: between Furancuango and Mualadze (Vila Gamito), fl. 20.x.1943, *Torre* 6069 (BR, FHO, K, LISC, LMU, LUN, SRGH).

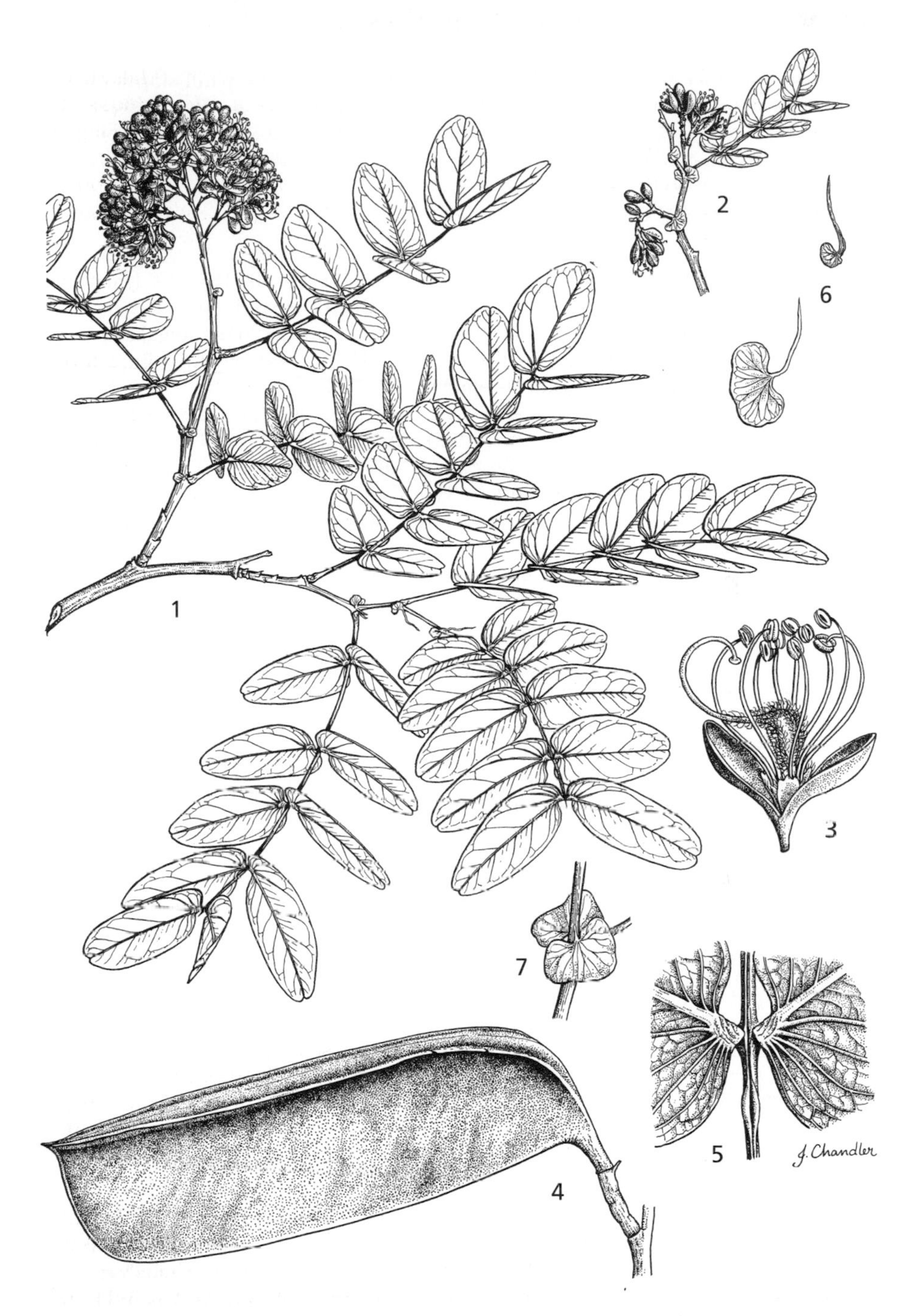

Fig. 3.2.**25**. BRACHYSTEGIA ALLENII. 1, flowering branchlet (× 1/2); 2, part of inflorescence with reduced leaf and auriculate stipules; 3, flower (× 2); 4, pod (× 1/2); 5, base of leaflets of middle pair (× 2); 6, two forms of stipules (× 1); 7, auriculate stipule (× 1). 1, 3, 5 from *Allen* 93, 2 from *R.G. Miller* 244, 4 from *Hoyle* 1080, 6 from *Allen* 93 & *Lindeman* 793, 7 from *Gamwell* 246. Drawn by J. Chandler. Adapted from F.T.E.A.

Also in Tanzania and Congo. Locally abundant on steep stony hill slopes in dry deciduous woodland and on dambo margins. In N Zimbabwe *B. allenii* is found in rugged areas on the Zambezi escarpment, while in N Mozambique, especially to the east, its ecology is very different and it appears to be confined to dambo margins. On sandy clay or stony soils; 500–1400 m.

Conservation notes: Lower Risk, Least Concern.

14. **Brachystegia floribunda** Benth. in Hooker's Icon. Pl. **14**: 43 (1881). —Harms in Engler, Pflanzenw. Afrikas **3**(1): 479 (1915). —Burtt Davy & Hutchinson in Bull. Misc. Inform., Kew **1923**: 157 (1923). —De Wild., Contrib. Fl. Katanga, Suppl. **2**: 33 (1929). —Baker, Legum. Trop. Afr.: 732 (1930). —Topham in Bull. Misc. Inform., Kew **1930**: 363 (1930). —Jackson in J. S. Afr. Bot. **6**: 38 (1940). —Brenan & Hoyle in Brenan, Check-list For. Trees Shrubs Tang. Terr.: 91 (1949). —Hoyle in F.C.B. **3**: 465, fig.39b (1952); in C.F.A. **2**: 229 (1956). —White & Hoyle in White, F.F.N.R.: 107, fig.22A (1962). —Hoyle in F.T.E.A., Legum.-Caesalp.: 179 (1967). —Brummitt in Wye Coll. Malawi Proj. Rep.: 62 (1973). Lectotype (chosen by Hoyle 1967): Malawi, Shire Highlands, Zomba, *Buchanan* 10 (K lectotype). FIGURE 3.2.**21**/1.

Brachystegia polyantha Harms in Bot. Jahrb. Syst. **30**: 319 (1901); in Engler, Pflanzenw. Afrikas **3**(1): 480 (1915). —Burtt Davy & Hutchinson in Bull. Misc. Inform., Kew **1923**: 117 (1923). —De Wild., Contrib. Fl. Katanga, Suppl. **2**: 53 (1929). —Baker, Legum. Trop. Afr.: 731 (1930). Type: Tanzania, Rungwe Dist., Kiwira valley, *Goetze* 1478 (B† holotype, E).

Brachystegia nchangensis Greenway in Bull. Misc. Inform., Kew **1928**: 201 (1928). —Baker, Legum. Trop. Afr.: 728 (1930). Type: Zambia, Nchanga, 14.viii.1927, *Bourne* 68 (K holotype).

Tree to 12 m tall; bark shallowly fissured longitudinally or coarsely reticulate, flaking in thick rectangular scales, dark grey. Young branchlets pubescent or glabrous. Leaves 7–18 cm long; petioles 20–70 mm long; rachis shallowly canaliculate above, margins neither raised not winged; leaflets in 2–5 widely spaced pairs, increasing in size distally, 2–3 times longer than broad, glabrous, proximal leaflet pairs 4–10 × 1.5–4 cm, distal ones 6–12.5 × 2.5–5 cm, ovate, oblong, lanceolate, elliptic or falcate, acuminate, acute or retuse at the apex, obliquely cuneate or truncate at the base, midvein central, (3)4–5(6) secondary veins from the proximal leaflet base. Stipules early caducous, 10–30 × 1–8 mm, lanceolate, subfoliaceous, free; auricles absent. Dormant axillary buds ovoid or globose. Inflorescences in terminal and/or axillary panicles to 8 cm long, often on the old wood; bracts up to 3 × 3 mm, ovate. Flowers up to 4–5 × 3–3.5 mm; pedicels up to 2.5 mm long; bracteoles 5–6 × 3–4 mm, obovate or circular. Sepals 5, 2–3 × 1–2 mm, ovate to triangular, imbricate, sparsely ciliate on the margins. Petals 1–2, up to 2 × 1 mm, filiform or linear. Stamens 10, united to 1–2 mm at the base; filaments 8–10 mm long. Ovary 2.5–3 × 1–2 mm, densely pubescent, up to 7-ovulate. Pods 8–14 × 2.5–4.5 mm, oblong to obovate, deep brown, with suberect ventral flanges. Seeds up to 8, 10–20 × 10–15 mm, oblong or circular.

Zambia. N: Mbala (Abercorn), Ndundu road, fl. & fr. 23.ix.1965, *Richards* 20499 (K). W: Solwezi, by Kasempa road, c. 15 km (10 miles) S of Solwezi, fl. 30.ix.1947, *Brenan & Greenway* 7991 (K). C: W of Chinama, fr. 10.viii.1938, *Greenway & Trapnell* 5581 (K). **Malawi.** N: Chitipa Dist., N end of Nyika Plateau, lower part of Monwe R., fr. 20.viii.1972, *Synge* 279 (K). C: Dedza, Mphunzi Hill, fl., *Topham* 123 (FHO). S: Mulanje, Likabula, fr. 11.iii.1957, *Chapman* 344 (FHO, K). **Mozambique.** T: Angónia, fl. 28.ix.1942, *Mendonça* 478 (FHO, LISC); Furancungo, Ulongué (Vila Coutinho), 15.vii.1949, *Barbosa & Carvalho* 3636 (FHO, LISC).

Also in Tanzania, Congo and Angola. Locally abundant or dominant, often in pure stands, over extensive areas of red clay soils where rainfall exceeds 1100 mm/year. In deciduous woodland, in plateau areas, steep escarpments and on hill slopes; 1400–1600 m.

Conservation notes: Lower Risk, Least Concern.

B. floribunda can be difficult to distinguish from *B. spiciformis*, but it generally has only 2 leaflet pairs, the leaves are slightly glaucous, never glossy, and panicles are borne on the previous year's growth. The leaf rachis of *B. floribunda* is not or only shallowly channelled and the margins are not winged, while the rachis in *B. spiciformis* is deeply channelled and winged. In addition the stamens are ± free or with only a 1 mm tube, while *B. spiciformis* has a distinct staminal tube.

15. **Brachystegia manga** De Wild., Contrib. Fl. Katanga, Suppl. **2**: 48 (1929). —Baker, Legum. Trop. Afr.: 728 (1930). —Hoyle in F.C.B. **3**: 464 (1952). —White & Hoyle in White, F.F.N.R.: 110, fig.22 (1962). —Hoyle in F.T.E.A., Legum.-Caesalp.: 180 (1967). —Brummitt in Wye College Malawi Proj. Rep.: 62 (1973). Type: Congo, Katanga, Kapiri valley, ii.1912, *Homblé* 1245 (BR lectotype). FIGURE 3.2.**21**/3.

Brachystegia burttii C.H.N. Jackson in J. S. Afr. Bot. **6**: 36 (1940) nom. nud. —Brenan & Hoyle in Brenan, Check-list For. Trees Shrubs Tang. Terr.: 90 (1949).

Tree to 20 m tall; bark shallowly fissured longitudinally or coarsely reticulate, flaking in thick rectangular scales, light to dark grey. Leaves 7–20 cm long; petioles (15)20–40 mm long; rachis shallowly canaliculate above, margins neither raised nor winged; leaflets in 3–5 widely spaced pairs, increasing in size distally, 1–2 times as long as broad, 3–11 × 1.5–6 cm, quadrate, lanceolate or narrowly ovate, acute, rounded, obtuse or emarginate at the apex, obliquely round or cuneate at the base, with 2–3 secondary veins from the proximal leaflet base. Stipules caducous, free, erect, 10–30 × 3–10 mm, lanceolate, elliptic, falcate or subfoliaceous, midveins subcentral, acute at the apex; auricles absent. Dormant axillary buds globose. Inflorescences in terminal and/or axillary panicles 5 cm long. Flowers up to 3–5 × 1.5–2.5 mm; pedicels up to 2 mm long; bracteoles 3–5 × 2.5–4 mm, obovate or orbicular. Sepals 5, up to 3 × 1.5 mm, ovate or oblong, imbricate or valvate, sparsely ciliate on the margins. Petals 1–2, 2 × 1 mm, filiform or linear. Stamens 10, united to 2 mm at base; filaments 6–10 mm long. Ovary 2.5–3 × 1.5 mm, densely pubescent, up to 7-ovulate, stipe to 2 mm long. Pods 7–12 × 2.5–4 cm, oblong to obovate, with suberect or revolute ventral fringes. Seeds 15–20 × 10–15 mm, oblong to circular.

Zambia. N: Mpika Dist., Mulobezi Nat. Park (Lavushi Manda Game Res.), 16 km W of Great North Rd., 1.x.1969, *Lawton* 1575 (K). C: 80 km (50 miles) E of Lusaka, 22.viii.1929, *Burtt Davy* 20816 (FHO). E. 8 km (5 miles) from Lundazi to Mzimba, fr. 29.iv.1952, *White* 2502 (K, FHO). **Malawi.** N: Karonga Dist., between Mwenzo and Chitipa (Fort Hill), st. 23.vi.1936, *B.D. Burtt* 5816 (FHO); C: Lilongwe, between Kazira and Sundwi villages on Bua R. between Mchinji (Fort Manning) and Lilongwe, fr. 30.vii.1936, *B.D. Burtt* 5729 (FHO). S: 16 km (10 miles) from L. Chilwa (Chirwa) to Zomba, 20.ix.1929, *Burtt Davy* 21910 (FHO). **Mozambique.** N: Mandimba, Luampula Region, fl. 24.x.1948, *Andrada* 1444 (FHO, LISC). Z: Gurué, 3 km from Lioma, fl. 10.xi.1967, *Torre & Correia* 16049 (K, LISC). T: between Chicoa and Fingoé, fr. 26.vi.1949, *Barbosa & Carvalho* 3300 (FHO, LISC).

Also in Tanzania and Congo. Deciduous woodland on stony hill slopes, sometimes co-dominant with *B. longifolia* and *B. allenii*; 45–1500 m.

Conservation notes: Lower Risk, Least Concern.

16. **Brachystegia bussei** Harms in Bot. Jahrb. Syst. **33**: 155 (1902). —Harms in Engler, Pflanzenw. Ost-Afr. **C**: 480 (1915). —Burtt Davy & Hutchinson in Bull. Misc. Inform., Kew **1923**: 157 (1923). —De Wild., Contrib. Fl. Katanga, Suppl. **2**: 18 (1929). —Baker, Legum. Trop. Afr.: 732 (1930). —Jackson in J. S. Afr. Bot. **6**: 38 (1940). —Brenan & Hoyle in Brenan, Check-list For. Trees Shrubs Tang. Terr.: 91 (1949). —Hoyle in F.C.B. **3**: 463, fig.39 (1952). —White & Hoyle in White, F.F.N.R.: 108, fig.22 (1962). —Hoyle in F.T.E.A., Legum.-Caesalp.: 170

(1967). Type: Tanzania, Songea Dist., Ungoni, xii.1900, *Busse* 729 (B† holotype; EA). FIGURE 3.2.**21**/2.

Tree to 20 m tall; bark smooth, flaking in large thin irregular scales. Young branchlets puberulous or glabrous. Leaves 5–10 cm long; petioles (15)20–45 mm long; rachis deeply canaliculate above, with raised and winged margins; leaflets in 2–4 widely spaced pairs, 2–3 times longer than broad, increasing in size distally, glabrous or sparsely pubescent, the proximal leaflet pairs 3.5–8 × 1.5–4 cm, the middle ones 5–8.5 × 2–3.5 cm, the distal ones 6–10 × 2.5–4.5 cm, ovate, elliptic or lanceolate, often falcate, acuminate, acute, obtuse or emarginate at the apex, obliquely cuneate at the base, midvein central or subcentral, 2–3 secondary veins from the proximal leaflet base. Stipules caducous, free, erect, 8–30 × 1.8–12 mm, linear, lanceolate or falcate; auricles absent. Dormant axillary buds globose. Inflorescences in terminal and/or axillary panicles to 6 cm long; bracts up to 3 × 3 mm, ovate. Flowers up to 5 × 4 mm, pubescent; pedicels up to 2 mm long; bracteoles 4–7 × 3–5 mm, obovate or circular, pubescent. Sepals 5, 2–3 × 1–1.5 mm, ovate or oblong, imbricate or valvate, ciliate on the margins. Petals 1–4, up to 2 × 1 mm, filiform or linear. Stamens 10, shortly united at the base, filaments 5–10 mm long. Ovary 2–3 × 1.5 mm, densely pubescent, up to 7-ovulate; style up to 8 mm long. Pods 6–15 × 2–4 cm, oblong to obovate, rounded at the apex, ventral fringes suberect or revolute. Seeds not seen.

Zambia. N: Isoka Dist., 50–65 km (30–40 miles) from Tunduru by road to Mbala (Abercorn), fl. 22.x.1947, *Brenan & Greenway* 8189 (FHO, K). C: 48 km (30 miles) from Mkushi, 28.viii.1930, *Stevenson* 122/30 (FHO). E: Petauke Dist., c. 48 km (30 miles) from Luangwa R. to Petauke, st. 18.iv.1952, *White* 2410 (FHO, K). **Malawi.** N: Kaningina Mts, near Luwazi R., n.d., *Topham* 810 (FHO). C: escarpment between Dowa and Nkhota Kota (Kota Kota), st. 21.vii.1936, *B.D. Burtt* 5721 (K). S: Mangochi, Tambala Estate, 15.xii.1941, *Hornby* 2188 (FHO, K). S: Zomba Dist., fl., *Clements* 580 (FHO). **Mozambique.** N: Ribáuè, fr. 12.ii.1960, *Guerreiro & Carvalho* 2 (FHO). Z: between Mulevala and Nampero, 10 km from Mulevala, st. 31.v.1949, *Barbosa & Carvalho* 2935 (K). T: Serra da Pandalanja, fr. 15.v.1948, *Mendonça* 4255 (FHO, LISC).

Also in Tanzania and Congo. Dominant in deciduous woodland, on mountain slopes on reddish soils; also on flat areas with sandy-clay soils; 250–1500 m.

Conservation notes: Lower Risk, Least Concern.

Brachystegia bussei is not very variable, although sterile specimens can be confused with some collections of *B. spiciformis.* In the absence of bark and floral characters, *B. bussei* can be distinguished from *B. spiciformis* by the size and position of the stipels; in *B. spiciformis* these are large and located less than 2 mm below each leaflet pair; in *B. bussei* they are more than 2 mm below each leaflet pair.

17. **Brachystegia oblonga** Sim, For. Fl. Port. E. Afr.: 48 (1909). Lectotype (chosen by Hoyle, unpublished): Mozambique, Maganja da Costa, 1908, *Sim* 21108 (PRE lectotype, not located).

Brachystegia woodiana sensu Burtt Davy & Hutchinson in Bull. Misc. Inform., Kew **1923**: 151 (1923) in part as regards Mozambique material, non Harms in Bot. Jahrb. Syst. **30**: 82 (1901).

Brachystegia obliqua sensu De Wild., Contrib. Fl. Katanga, Suppl. **2**: 53 (1929), non Hutch & Burtt Davy.

Tree 10–15 m tall; bark shallowly fissured or coarsely reticulate. Young branchlets glabrous. Leaves 3–16 cm long; petioles 4–16 mm long; rachis deeply canaliculate above with raised and winged margins; leaflets in 4–8 contiguous or widely spaced pairs, increasing in size distally, 2–3 times longer than broad, glabrous or pubescent, the proximal leaflet pairs 1–2.5 × 0.7–1.5 cm, the rest 1.5–5 × 1–2.5 cm, broadly oblong, elliptic or rarely obovate, rounded, retuse or emarginate at the apex, obliquely rounded or subcordate at the base, midvein central or subcentral. Stipules caducous, shortly connate at the base, erect, 10–16 × 1–4 mm, elliptic, lanceolate or falcate, midvein subcentral; auricles up to 10 × 4 mm, reniform to subcircular.

Dormant axillary buds much flattened. Inflorescences in terminal and/or axillary panicles up to 7 cm long; bracts 2–3 × 2–3 mm, ovate. Flowers up to 6 × 4 mm; pedicels up to 3 mm long; bracteoles 5–7 × 4–5 mm, obovate or orbicular. Sepals 5, 2–3 × 1.5–2 mm, ovate or circular, imbricate, ciliate on the margins. Petals absent. Stamens 10, united for 1 mm at the base; filaments 10–12 mm long. Ovary 2–3 × 1–1.5 mm, densely pubescent, up to 8-ovulate; style up to 10 mm long. Pods 8–14 × 2.5–3.5 cm, oblong to obovate, ventral flanges spreading, up to 5 mm wide. Seeds 5, 10–15 × 10–15 mm, circular to oblong.

Mozambique. N: Moma, 15 km from R. Ligonha on Naburi road, fl. 17.i.1968, *Torre & Correia* 17233 (LISC). Z: Maganja da Costa, Gobene Forest, 50 km from Vila da Maganja, fl. 12.ii.1966, *Torre & Correia* 14536 (LISC); Maganja da Costa, Gobene Forest near Ramaga, 35 km from Vila da Maganja, fl. 10.i.1968, *Torre & Correia* 17039 (LISC).

Endemic to coastal Mozambique. Deciduous woodlands in coastal areas; 20–50 m.

Conservation notes: Endangered A3(cd)B12(ab); narrowly endemic to coastal Mozambique in a region also known for the monotypic caesalpinoid legume *Icuria dunensis* Wieringa and *Scorodophloeus torrei* Lock.

B. oblonga was inadequately described by Sim in 1909 and was then confused with *B. obliqua* (= *B. bakeriana*), so that the name became lost from use.

18. **Brachystegia utilis** Hutch. & Burtt Davy in Bull. Misc. Inf., Kew **1923**: 155 (1923) (see note under *B. gossweileri*). —Greenway in Bull. Misc. Inf., Kew **1928**: 202 (1928). —De Wild., Contrib. Fl. Katanga, Suppl. **2**: 62 (1929). —Topham in Bull. Misc. Inf., Kew **1930**: 359 (1930). —Baker, Legum. Trop. Afr.: 725 (1930). —Jackson in J. S. Afr. Bot. **6**: 39 (1940). —Brenan & Hoyle in Brenan, Check-list For. Trees Shrubs Tang. Terr.: 93 (1949). —Hoyle in F.C.B. **3**: 468, fig.40 (1952); in Mem. New York Bot. Gard. **8**: 428 (1956); in C.F.A. **2**: 231 (1956). —White & Hoyle in White, F.F.N.R.: 114, 117, 118, fig.22 (1962). —Chapman, Veg. Mulanje Mt.: 35 (1962). —Hoyle in F.T.E.A., Legum.-Caesalp.: 172 (1967). Type: Malawi, Mulanje, x.1905, *Purves* 193 (K holotype). FIGURE 3.2.**21**/7.

Brachystegia diloloensis De Wild., Contrib. Fl. Katanga, Suppl. **2**: 18 (1929). —Baker, Legum. Trop. Afr.: 725 (1930). Type: Congo, Dilolo, vi.1908, *Sapin* s.n. (BR holotype).

Brachystegia diloloensis var. *subglabra* De Wild., Contrib. Fl. Katanga, Suppl. **2**: 19 (1929). —Baker, Legum. Trop. Afr.: 725 (1930). Type from Congo.

Brachystegia diloloensis var. *brevifoliolata* De Wild., Contrib. Fl. Katanga, Suppl. **2**: 20 (1929). —Baker, Legum. Trop. Afr.: 725 (1930). Type: Congo, Luishia, 13.vii.1927, *Ritschard* 1464 (BR holotype).

Tree up to 20 m tall; bark shallowly fissured or coarsely reticulate. Young branchlets pubescent. Leaves 5–13 cm long; petioles 2–8(10) mm long; rachis deeply canaliculate above with raised and winged margins; leaflets in 6–10 contiguous pairs, the middle pairs the largest, the proximal pairs 1.5–2.5 × 0.5–1 cm, the middle ones 2.5–6 × 0.8–1.8 cm, the distal ones 2–6 × 0.8–1.2 cm, oblong or lanceolate, acute, rounded, retuse or emarginate at the apex, obliquely rounded at the base, midvein central or subcentral, 2–3(4) veins from the proximal leaflet base. Stipules caducous, free, erect, 2.5–15 × 3–5 mm, subulate; auricles absent. Dormant axillary buds globose. Inflorescences in terminal and/or axillary panicles up to 8 cm long; bracts 2–3 × 1.5–2 mm, ovate. Flowers up to 3–5 × 2–3 mm; bracteoles 4–7 × 3–4 mm, obovate or circular. Sepals 5, 1.5–3 × 1–2.5 mm, linear, obong, rhombic or lanceolate, imbricate or valvate, ciliate on the margins. Petals absent. Stamens 10, united to 0.5 mm at the base, filaments 6–10 mm long. Ovary 2–3 × 1–2 mm, densely pubescent, oblong, stipe up to 2 mm long. Pods 5–12 × 2–3.5 cm, oblong or obovate, ventral flanges spreading, revolute or suberect, up to 4 mm wide. Seeds up to 6, up to 20 × 15 mm, circular to oblong.

Zambia. N: Kasama township, fl. 11.xi.1952, *Angus* 745 (FHO, K). W: Kitwe, 26.x.1968, *Fanshawe* 10407 (FHO, K, NDO). C: Lusaka Dist., c. 77 km (48 miles) on

Great East Road from Lusaka to Katete (Fort Jameson), st. 16.iv.1952, *White* 2686 (FHO, K). **Zimbabwe.** N: Lomagundi, Nyanga's Kraal, Magali Reserve, SW of Chinhoyi (Sinoia), st. vii.1959, *Aylen* in SRGH 96165 (K, SRGH). E: Mutare Dist., c. 14.5 km (9 miles) S of Mutare (Umtali), 20.xi.1956, *Chase* 5863 (K, LISC, SRGH). **Malawi.** N: 6.5 km (4 miles) N of Mzimba, fr. 29.iv.1952, *White* 2318 (FHO). C: Dowa Dist., Kasakaila, fl. 28.x.1941, *Greenway* 6372 (EA, K). S: Mulanje (Mlanje), 1.iii.1957, *Chapman* 342 (FHO). **Mozambique.** N: Massangulo, xi.1933, *Gomes e Sousa* 1606 (FHO). Z: Gurué, 5 km from Lioma, Patapane Mts., fl. & imm.fr. 13.ix.1967, *Torre & Correia* 16083 (COI, K, LISC, LMA). T: Zobuè region, st. 24.viii.1943, *Torre* 5790 (K, LISC, SRGH). MS: Manica, Dombe, between Guza and Coia, 11 km from Guza, fl. 22.viii.1953, *Gomes Pedro* 4379 (K, LISC).

Also in Tanzania and Congo. Locally dominant in deciduous woodlands, on hill slopes and flat sandy areas; 300–1800 m.

Conservation notes: Lower Risk, Least Concern.

22. **ICURIA** Wieringa

Icuria Wieringa in Wageningen Agric. Univ. Papers **99**(4): 241 (1999).

Trees, apparently evergreen, unarmed. Leaves paripinnate, with 1–2 pairs of leaflets; leaflets sessile, with only slightly excentric main nerve, the basal one with 0–1(2) glands on the proximal half close to the base; stipules free, valvate in bud, caducous. Flowers in axillary panicles, almost sessile, ± distichous, some flowers functionally male, others bisexual; bracts small, early caduous; bracteoles pinkish, valvate, shortly fused near base. Hypanthium very short or absent, when present fused with the bracteoles. Sepals reduced, only the adaxial one sometimes present and often fused to the bracteoles. Petals usually 0, rarely a vestigial lateral one present. Stamens 6–8, fused at the base; anthers large, dorsifixed. Ovary fully developed only in some flowers; where fully developed shortly stipitate, (1)2(3)-ovulate, velvety; style excentric, geniculate; stigma peltate. Pods oblong-obovate, broadest near the apex, beaked, compressed, woody, dehiscent, the valves velvety, reticulately nerved and with 1(several) lateral marked nerves from base to apex in upper half, the upper suture narrowly winged. Seeds elliptic in outline, with a protruding base; testa very thin.

A monospecific genus confined to a small area of coastal Mozambique.

Icuria dunensis Wieringa in Wageningen Agric. Univ. Papers **99**(4): 242, fig.12.1 (1999). Type: Mozambique, Nampula, Moma region, in SW of BHP Concession, 68 km SW of Angoche, fl. 19.ix.1998, *Johnson & Avis* 694 (WAG holotype; GRA). FIGURE 3.2.**26**.

Tree to 40 m; bark smooth, pale grey. Young branchlets grey, white pubescent. Leaves paripinnate: stipules 7–16 × 1.5–4 mm, caducous; leaflets coriaceous, in 1–2 pairs, if 2 the apical pair larger, (14)30–98 × (4)8–26 mm, narrowly elliptic to narrowly ovate, asymmetric, often somewhat falcate, rounded at the base, obtuse to rounded at the apex, glabrous. Inflorescences paniculate, 1.5–12 cm long, with 2–10 lateral branches; bracts early caducous; bracteoles 3.5–4.4 × 2.3–3.4 mm, circular to obovate, pubescent. Flowers male or hermaphrodite. Sepals 0–1. Petals 0(1). Stamens 6–8; filaments 6–8 mm long, fused for 0.1–0.3 mm at the base. Ovary (1)2(3)-ovulate; style 5–7 mm long. Pod 6–11 × 3.6–5.1 cm, obovate, with a strong lateral ridged nerve just above the middle, the rest of the surface reticulate at maturity, puberulous to velvety, 1–3-seeded. Seeds 2.5–2.8 × 2–2.3 cm, elliptic in outline.

Mozambique. N: between Beira and Angoche, fr. 5.xii.1997, *Dold* 3362 (GRA, K, WAG); Angoche (António Enes), between village and beach, fr. 15.x.1965, *Gomes e Sousa* 4865 (LISC, WAG).

Fig. 3.2.**26**. ICURIA DUNENSIS. 1, flowering branchlet (× 1), from *Johnson & Avis* 694; 2, undersurface of basal leaflet showing gland (× ²/₃), from *Gomes e Sousa* 4865; 3, stipule from outside (× 2), from *Dold* 3362; 4, flower (× 4); 5, abaxial sepal (× 20), 4 & 5 from *Johnson & Avis* 694; 6, fruiting branchlet (× ²/₃), from *Gomes e Sousa* 4865. Drawn by W. Wessel-Brand. Reproduced with permission from the Biosystematics Group, Wageningen University, Netherlands.

Endemic to N Mozambique; only known from about 200 km of coast between Angoche and Moebase. Forming almost monospecific stands on old sandy well-drained dunes.

Conservation notes: Often abundant in its restricted range; probably Vulnerable.

The bark is used to make canoes. Some state that the wood is of poor quality, while others say that it is a valuable timber.

In Wieringa's (1999) cladogram, *Icuria* is clearly embedded within *Bikinia*, so that its recognition as a separate genus makes *Bikinia* polyphyletic. However, in his discussion (p. 83), he states that the large numbers of characters coded as 'Unknown' makes the analysis unreliable. After his cladistic analysis had been completed, flowers of *Icuria* became available and these, he states (p. 294), confirm placement in the "Macrolobieae" but 'do not fit any known genus'. In his earlier discussion (p. 83) he suggests that a full analysis might place *Icuria* as basal to *Bikinia*.

23. APHANOCALYX Oliv.

Aphanocalyx Oliv. in Hooker's Icon. Pl. **11**: 53, fig.1066 (1870). —Wieringa in Wageningen Agric. Univ. Papers **99**(4): 115 (1999).

Shrubs or trees, evergreen. Leaves paripinnate, with 2–many leaflets; leaflets opposite, sessile, very asymmetric at the base, the distal half of each usually absent so that the midrib lies along the distal margin, usually with a few glands near the base; stipules intrapetiolar, completely fused, with numerous parallel veins. Inflorescence racemose, usually simple; bracts densely parallel-veined, caducous; bracteoles valvate, fused with the hypanthium. Hypanthium up to 5 mm long. Sepals 5, variously reduced. Petals 1–5 but only one fully developed, the rest much reduced or absent. Stamens 9–10, usually 9 united at the base and 1 free; anthers often with dorsal teeth. Ovary hairy, 1–6-ovulate; style elongate; stigma capitate or peltate. Pods oblong-obovate, compressed, dehiscent, with a lateral nerve; upper suture winged or not. Seeds with a thin testa.

A genus of about 14 species, mainly in equatorial forests but a few extending into seasonal tropical gallery forests, occurring in Africa mainly south of the Equator. Formerly restricted to a few species of equatorial forest, the circumscription has recently been extended by Wieringa, following a cladistic analysis of the group, to include many species formerly included in *Monopetalanthus* Harms. The only other member of the group in the Flora area is *Icuria*.

Leaflets in a single pair .... **1.** *richardsiae*
Leaflets usually in 5–13 pairs .... **2.** *trapnellii*

1. **Aphanocalyx richardsiae** (J. Léonard) Wieringa in Wageningen Agric. Univ. Papers **99**(4): 165, fig.10.9 (1999). Type: Zambia, Mbala Dist., Inono Stream, close to Mpulungu road, *Richards* 4483 (K holotype, B, BM, BR, EA, LISC, SRGH).

Monopetalanthus richardsiae J. Léonard in C.F.A. **2**: 204 (1956); in Mém. Acad. Roy. Sci. Belg. **30**(2): 257 (1957). —White, F.F.N.R.: 126 (1962). —Brenan in F.T.E.A. Legum.-Caesalp.: 196, fig.42 (1967).

Monopetalanthus leonardii Devred & Bamps in Bull. Jard. Bot. État **30**: 111–114 (1960). Type from Congo.

Tree 3–12(25) m high. Young branchlets glabrous to shortly pubescent, glabrescent. Leaves: stipules early caducous; petiole 1–6 mm long, glabrescent; leaflets 2, sessile, coriaceous, 1.7–7.5(10) × 0.6–3.2(4.5) cm, semi-elliptic or semi-ovate-elliptic, acutely or sometimes obtusely acuminate at apex, rounded on outer side at base; glabrous or almost so, with 0–6(10) glands along proximal margin of leaflet, prominently veined on both surfaces. Racemes 1–3.5 cm long, brownish-pubescent; bracteoles 4.5–7 mm long, up to 6 mm wide, almost circular. Sepals 0–2, the larger ones, if present, to 3 × 2.3 mm, others minute. Large petal white, 5–7 × 3.5–6.5 mm, with a short claw and a subcircular lamina; other petals absent or minute. Stamens 9–10. Pods asymmetric, 2.5–3.7 × 1.5–2.3 cm, oblong-elliptic or obovate-elliptic, sparsely pilose when young, glabrescent, each valve longitudinally and prominently 1(2)-nerved, 1–2-seeded. Seeds 8–14 × 5–12 mm. Germination epigeal; first pair of leaves opposite, with four leaflets; subsequent leaves alternate, with 2 leaflets.

Zambia. N: Kawambwa, fl. 12.i.1958, *Fanshawe* 4343 (K); Mbala (Abercorn), Upper Lufu Basin, fr. iv.1937, *Trapnell* 1750 (K).

Also in NE Angola, S Congo and W Tanzania. Gallery forests in savanna, and in dry forest; 700–1500 m.

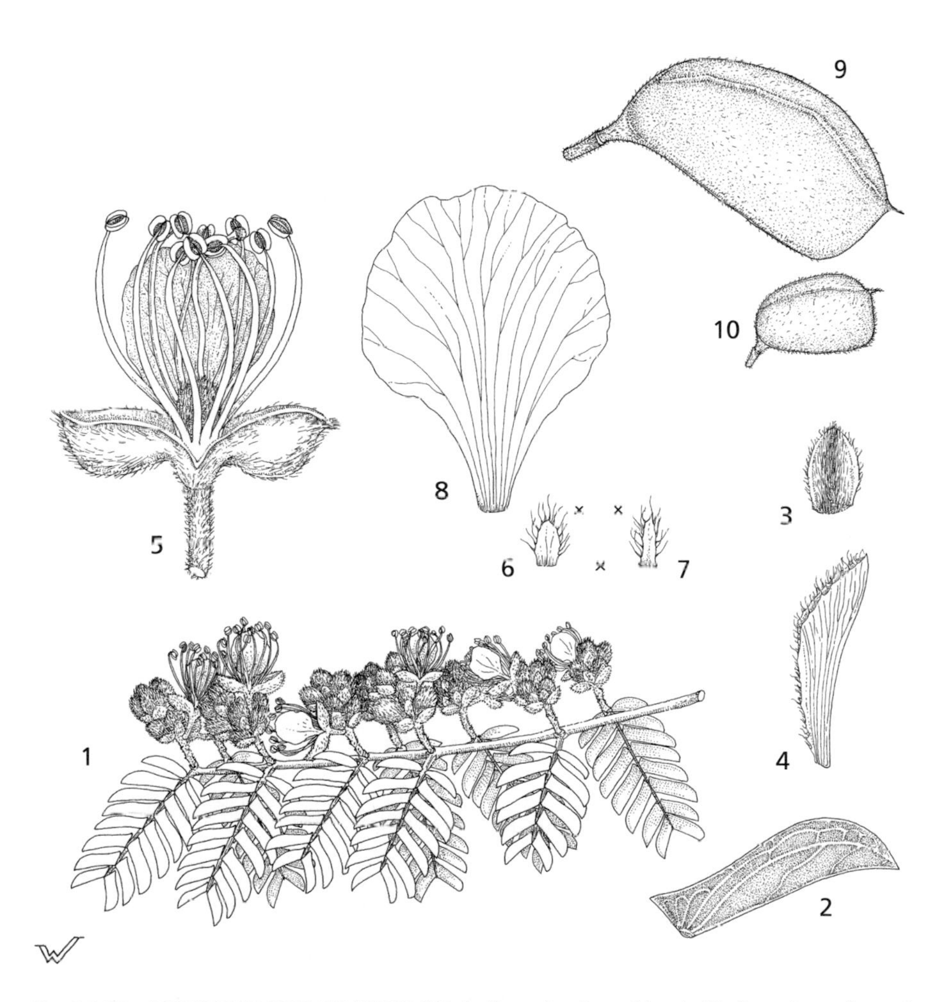

Fig. 3.2.**27**. APHANOCALYX TRAPNELLII. 1, flowering branchlet (× 1); 2, undersurface of leaflet (× 4), 1 & 2 from *Richards* 19612; 3, two fused stipules (× 4), from *Breteler* 12001; 4, bract (× 4); 5, flower (× 4); 6, adaxial sepal (× 6); 7, abaxial sepal (× 6); 8, adaxial petal (× 6), 4–8 from *Richards* 19612; 9, 10, pods (× 1), from *Richards* 10060. Drawn by W. Wessel-Brand. Reproduced with permission from the Biosystematics Group, Wageningen University, Netherlands.

Conservation notes: The fairly extensive distribution, mainly in gallery forest along the southern margins of the Congo Basin forests, suggests that it is not presently endangered. However, these habitats are under pressure for timber, fuel and clearance for cultivation, so a ranking of Vulnerable is probably appropriate.

According to Wieringa (1999), *Aphanocalyx richardsiae* cannot be distinguished when sterile from *A. djumaensis* J. Léonard, known from riverine forests in Gabon and NW Congo. However, he found that *A. richardsiae* has minutely toothed margins to the inflorescence bracts; these are entire in *A. djumaensis*. The two taxa appear to be vicariants on opposite sides of the Congo Basin.

Wieringa (1999) considered that *Richards* 10198 may represent a hybrid between *A. richardsiae* and *A. trapnellii*, as may *Richards* 10199, which is closer to *A. trapnellii*. I have not seen this material.

2. **Aphanocalyx trapnellii** (J. Léonard) Wieringa in Wageningen Agric. Univ. Papers **99**(4): 171 (1999). Type: Zambia, Lufila R., *Greenway & Trapnell* 5580 (K holotype; BR, EA, FHO, WAG). FIGURE 3.2.**27**.

Monopetalanthus trapnellii J. Léonard in Bull. Jard. Bot. État **21**: 133–137, figs.34 & 35 (1951). —White, F.F.N.R.: 126 (1962).

Tree 8–20 m high. Young branchlets pubescent, glabrescent. Leaves: stipules early caducous; petiole 0.5–1.5 mm long; leaflets sessile, coriaceous, 10–26, 1.5–7(13) × 1–3(5) mm, oblong to semi-obovate, apex acute or bluntly so, rounded to sharply angled on outer side at base, glabrous or sparsely villous, with 0–2(3) glands near base. Racemes 0.8–1.7 cm long, brownish-pubescent; bracteoles 4–5.5 × 2.5–4 mm, broadly elliptic. Sepals 0–5, the two larger ones 0.3–1.9 × 0.2–0.8 mm, the others, if present, minute. Large petal white, 5.5–7 × 3–6 mm, with a short claw and an obovate lamina; other petals absent or minute. Stamens 10. Pods 1.2–3.8 × 0.8–1.9 cm, obovate, sparsely long-villous, glabrescent, each valve longitudinally and prominently 1-nerved, 1–2(3)-seeded. Seeds c. 6–11 × 5–9 mm, obovate. Germination epigeal; first pair of leaves opposite, with 18–24 leaflets; subsequent leaves alternate, with more leaflets.

Zambia. N: Mbala (Abercorn), edge of Mshinda R. gorge near Mkoma, 23 km (15 miles) NW of Mbala, fr. 25.ii.1948, *Glover* s.n. (K); Mbala Dist., Kalambo Falls, fl. 9.ii.1965, *Richards* 19612 (B, BR, K, MO).

Confined to N Zambia, although it probably also occurs on the Tanzanian side of the Kalambo River. Gallery forests; 800–1500 m.

Conservation notes: The small number of sites in gallery forest habitat that is often under pressure from settlement and cutting, means that a rating of Vulnerable is probably appropriate.

24. DIALIUM L.

Dialium L., Syst. Nat., ed. 12, **2**: 56 (1767); Mant. Pl.: 3 (1767). —Steyaert in Bull. Soc. Roy. Bot. Belg. **84**: 29–45 (1951).

Trees or rarely shrubs, unarmed, evergreen or deciduous. Leaves imparipinnate (sometimes 3-foliolate), with leaflets opposite or alternate, often with a rather dense vein reticulum; stipules inconspicuous, lateral, caducous. Flowers small, in many-flowered terminal and lateral panicles; bracts and bracteoles small, caducous. Flowers hermaphrodite, usually somewhat zygomorphic, with a well-developed disk. Sepals normally 5, rarely 6 or 7, imbricate. Petals greatly reduced or lacking. Stamens 2–10; anthers basifixed, dehiscing by longitudinal slits. Ovary sessile or shortly stipitate, often oblique, with 2 ovules. Fruit ellipsoid to ± spherical, sometimes laterally compressed, indehiscent; exocarp hard and brittle, the mesocarp pulpy and brown to orange or red when dry, edible. Seeds either 1 or 2, often on the same plant, those in 2-seeded fruits differing markedly in shape from those in 1-seeded fruits, the testa smooth and shiny, without areoles.

A genus of about 35 species, in tropical regions, mostly in the Guineo-Congolian domain of Africa, a few species in South America and Malesia. Species in the Flora Zambesiaca area are all placed by Steyaert (1951) in the larger subgenus *Arouna* (Aubl.) Steyaert, with species 1 in sect. *Geniculata* Steyaert, characterised by the geniculate stamen filaments, and species 2–4 in sect. *Recta* Steyaert with straight filaments.

Evolution within the genus has resulted in reduction in the numbers of petals and stamens, and these numbers appear to be still unstable within some species. For example, *D. schlechteri*, often regarded as having 10 stamens, in fact is found to have any number from 6 to 10. Similarly in the number of leaflets per leaf — *D. angolense* generally has 3 leaflets but occasional leaves are seen with 4 leaflets. Field observations on the constancy of such characters might be interesting.

1. Leaflets 3 .... **1.** *angolense*
– Leaflets 7–15 2
2. Buds 4–6 mm long; petals 5, small; stamens 5; fruits laterally compressed, the style base persisting as a beak or short apiculum .... **2.** *englerianum*
– Buds 2–3.5 mm long; petals 0(1); stamens 2 or 6–10; fruits spheroidal, the style base not persisting 3
3. Leaflets rounded or rarely slightly acuminate at apex; petiolule 1–2 mm long; buds 3–3.5 mm long; stamens 6–10 .... **3.** *schlechteri*
– Leaflets acuminate at apex; petiolule 2.5–4 mm long; buds 2–2.5 mm long; stamens 2 .... **4.** *holtzii*

1. **Dialium angolense** Oliv. in F.T.A. 2: 283 (1871). —Hiern, Cat. Afr. Pl. Welw. **1**: 294 (1896). —De Wild. in Ann. Mus. Congo Belge, sér. 4, Bot. **1**: 47 (1902); Contrib. Fl. Katanga: 73 (1921). —Harms in Engler, Pflanzenw. Afrikas **3**(1): 490, 495 (1915). —Baker, Legum. Trop. Afr.: 648 (1930). —Steyaert in Bull. Soc. Roy. Bot. Belg. **84**: 42 (1951); in F.C.B. **3**: 533 (1952). —Torre & Hillcoat in C.F.A. **2**: 188, fig.40 (1956). —White, F.F.N.R.: 122 (1962). Type from Angola.

Small to medium-sized tree up to 15(20) m high with a broad crown, or sometimes a low spreading shrub 4 m high or less; bark smooth, grey. Young branches pubescent. Leaves: petiole and rachis together 1.5–3(–5) cm long, pubescent; leaflets 3 or rarely with a fourth inserted laterally shortly below the terminal one on isolated leaves, the lower pair opposite or subopposite and (2)2.5–4.5(6.5) × (1.2)1.5–3(4) cm, all elliptic or broadly elliptic or somewhat ovate, cuneate to rounded at the base, rounded or emarginate to obtuse or somewhat acuminate at the apex, coriaceous, glabrous above, glabrous or finely pubescent (particularly on the midrib) beneath; petiolule 1.5–3(4) mm long, pubescent. Flowers 7.5–9.5 mm across, in panicles up to 12 × 10 cm; inflorescence axes brown pubescent or tomentose; pedicels 2.5–4 mm. Buds up to 3–4 mm long; sepals 5, 3–3.5 mm long, grey or grey-brown pubescent outside, white to deep cream inside. Petals 1, c. 2.5 mm long, ovate, inserted between the stamens and overlapping them in bud. Stamens 2, inserted opposite two adjacent sepals; filaments strongly geniculate about their middle; anthers 1.7–2 mm long. Ovary c. 1.5 mm long, brown or blackish-brown tomentose; style 2–2.5 mm long, glabrous. Fruits 1.7–2 × 1.5–1.7 cm, laterally compressed, oblique, with the style persisting as a short apiculum at the apex, brown or blackish-brown tomentellous. Seeds 1 per fruit (or perhaps 2 but not yet recorded in this species), brown, shiny, 8–10 × 6.5–7.5 × 3.5 mm, ± lenticular.

Zambia. B: Zambezi Dist., sandbank in Zambezi R. at Chavuma, fr. 12.x.1952, *Holmes* 950 (FHO, K). N: shore of L. Bangweulu near Samfya, fr. 22.viii.1952, *Angus* 279 (BR, FHO, K). W: Kitwe, fl. 23.xii.1955, *Fanshawe* 2678 (BR, K, LISC). S: Mazabuka, fl. 1931, *Stevenson* 262/31 (K, LISC).

Also in Angola and Congo. In fringing evergreen forest, in evergreen thicket on lake shores, on sand hills, in rocky places, and in woodland on sandy river banks; between 900 and 1275 m.

Conservation notes: Localised species; probably Lower Risk, Near Threatened.

Specimens seen from Angola, including the type collection, have the leaflets conspicuously acuminate, whereas in the F.Z. area the leaflets are usually broadly rounded at the apex. However, acuminate leaflets are occasionally seen also in Zambia, and it seems unwise to suggest any formal taxonomic split.

2. **Dialium englerianum** Henriq. in Bol. Soc. Brot. **16**: 48 (1899). —Baker, Legum. Trop. Afr.: 650 (1930). —Steyaert in Bull. Soc. Roy. Bot. Belg. **84**: 40 (1951); in F.C.B. **3**: 542 (1952). —Torre & Hillcoat in C.F.A. **2**: 187, fig.39 (1956). —White, F.F.N.R.: 122, fig.21L (1962). —Breitenbach, Indig. Trees Sthn. Africa **3**: 344 (1965). —Schreiber in Merxmüller, Prodr. Fl. SW Afrika, fam. 59: 13 (1967). —Palmer & Pitman, Trees Sthn. Africa 2: 875 (1973). —Corby in Kirkia **9**: 314 (1974). —Drummond in Kirkia **10**: 243 (1977). —Ross in F.S.A. **16**(2): 67 (1977). —M. Coates Palgrave, Trees Sthn. Africa: 338 (2002). Type: Angola, Lunda Dist., between Machinge & Ma-Lunda, *S. Marques* 176 (COI holotype, LISU).

Dialium guineense sensu Eyles in Trans. Roy. Soc. S. Afr. **5**: 367 (1916) non Willd.

Dialium simii E.P. Phillips in Bull. Misc. Inform., Kew **1922**: 194 (1922). —Baker in J. Bot. **66**: 138 (1928); Legum. Trop. Afr.: 650 (1930) as *simsii*. —Miller in J. S. Afr. Bot. **18**: 30 (1952). Types: Zimbabwe, Victoria Falls, near railway station, *Sim* 19004 (K syntype, PRE); same locality, *Rogers* 5307 (K syntype).

Dialium lacourtianum Vermoesen, Man. Essenc. For. Congo Belge: 97 (1923); Pl. Bequaert. **3**: 237 (1925). —Baker, Legum. Trop. Afr.: 650 (1930). Type from Congo.

Tree up to 23 m high with a rounded spreading crown; bark grey, smooth but flaking. Young branches brown puberulous to pubescent. Leaflets usually 9, occasionally 7 or 11, opposite or subopposite except sometimes the uppermost pair, coriaceous, 2.8–7(8.5) × 2–3.7 cm, broadly ovate to occasionally lanceolate or ± elliptic, rounded and only slightly asymmetrical at the base, obtusely acuminate at the apex, glabrous above, minutely appressed-pubescent beneath when young but usually glabrescent; petiolule 1–3.5 mm long, pubescent. Flowers 0.8–1.2 cm across (less if sepals are reflexed), in panicles up to 35 × 20 cm; inflorescence axes appressed brown-pubescent; pedicels 2.5–6 mm. Buds up to 4–6 mm long; sepals normally 5 (recorded as 6–7 in F.F.N.R.), 4–5 mm long, brown pubescent outside, cream or brownish inside, often reflexed in full flower. Petals normally 5, insignificant, whitish, 1–1.5 mm long, elliptic or lanceolate with a short claw. Stamens normally 5, rarely 6, sometimes with up to 5 staminodes; anthers 2–2.5 mm long; filaments 1.7–2.2 mm long, straight. Ovary c. 1.5 mm long, brown tomentose; style c. 2.5 mm, glabrous except at the base. Fruits 2.2–3.7 cm long, 1.4–2.5 cm broad, somewhat laterally compressed, obliquely positioned on the pedicel, the style base persisting at the apex as a beak or short apiculum, brown puberulous or tomentellous. Seeds 1 per fruit (or 2 in Flora du Congo Belge, but not confirmed in material seen from the Flora area), reddish-brown to blackish, shiny, 8–13 × 7–8.5 × 3.5–5 mm, ± lenticular,.

Caprivi Strip. About 32 km (20 miles) W of Zambezi, fr. 8.vii.1947, *Brenan & Keay* 7648 (K). **Botswana.** N: Chobe Dist., S of Deka, fl. vi.1952, *Miller* B 1332 (K, PRE). **Zambia.** B: Lusu, fl. 6.vii.1962, *Fanshawe* 6911 (BR, K). W: Mwinilunga Dist., Kanudu's, on Congo-Zambezi watershed, fr. ix.1934, *Trapnell* 1588 (BR, K). S: 32 km (20 miles) W of Namwala Boma, fl. 25.vii.1952, *White* 2987 (BR, FHO, K). **Zimbabwe.** W: Hwange (Wankie), fl. viii.1921, *Henkel* in *Eyles* 3195 (K, PRE).

Also in Namibia, Angola and Congo, and possibly in Gabon (as *D. quinquepetalam* Pellegr. — see Steyaert 1951), recorded perhaps in error by Breitenbach (1965) from "north-western Transvaal". On Kalahari sands in woodlands of *Baikiaea, Cryptosepalum, Burkea, Erythrophleum*; c. 950–1450 m.

Conservation notes: Lower Risk, Least Concern.

3. **Dialium schlechteri** Harms in Bot. Jahrb. Syst. **26**: 276 (1899). —Baker, Legum. Trop. Afr.: 650 (1930). —Gomes e Sousa, Dendrol. Moçamb. Estudo Geral **1**: 258, fig.58 (1966). —Steyaert in Bull. Soc. Roy. Bot. Belg. **84**: 37 (1951) excl. *Busse* 2587. —Breitenbach, Indig. Trees Sthn. Africa **3**: 343 (1965). —Ross, Fl. Natal: 195 (1973); in F.S.A. **16**(2): 65, fig.15 (1977). —M. Coates Palgrave, Trees Sthn. Africa: 339 (2002). Type: Mozambique, Maputo (Lourenço Marques), *Schlechter* 11603 (B† holotype, BR, K). FIGURE 3.2.**28**.

Andruli arborea Sim, For. Pl. Port. E. Afr.: 23 (1909). Type: Mozambique, without locality, *Sim* 6141 (not seen).

Tree 5–10 m high; bark smooth, ± grey. Young branches minutely puberulous. Leaves: petiole and rachis together (3)5–9(14) cm long, puberulous; leaflets (7)9–11(13), alternate or the lower opposite, or occasionally all opposite except the terminal one, 1–3.5(4.3) × 0.7–2 cm, the lower ones broadly ovate and rounded at the base, the upper ones elliptic and cuneate at the base, asymmetric at the base, broadly rounded or rarely slightly acuminate at the apex, glabrous or with a few hairs on the midrib beneath; petiolule 1–2 mm long, puberulous. Flowers 6–8 mm across, in panicles up to 25 × 20 cm; inflorescence axes puberulous to appressed-pubescent; pedicels 1–3 mm. Buds and sepals up to 3–3.5 mm long, densely covered in appressed brown hairs outside; sepals white or cream or yellowish-green inside. Petals 0, or rarely two sepal-like structures present in addition to the five sepals in some flowers. Stamens 6–10; anthers 1–1.5 mm long; filaments 1.5–2.5 mm long. Ovary c. 1 mm long, brown tomentose; style 2–3 mm, glabrous or with a few hairs towards the base. Fruits 1.2–2 × 1–1.5 cm, slightly elongate, smoothly rounded at apex with style base not showing, densely dark brown-puberulous to -tomentellous. Seeds 1 or 2, brown, shiny, 8–9 × 3.5–5 mm and ± lenticular when only one developing, c. 8 × 5 × 4 mm and elongate with ± triangular cross-section when two per fruit.

Mozambique. GI: Cheline, Maxixe–Nova Mambone road, fl. 5.x.1963, *Leach & Bayliss* 11839 (K, LISC, SRGH). M: Matutuine Dist., between Catuane and Santaca, c. 25 km from Catuane, fr. 22.ii.1961, *Myre & Rosa* (LMA).

Also in northern KwaZulu-Natal. In lowland dry woodland and dry forest on sand or alluvial soils associated with *Acacia, Albizia, Afzelia, Terminalia, Balanites, Euphorbia*; below 200 m.

Conservation notes: Probably Lower Risk, Least Concern.

This species has hitherto been described as having 10 stamens, but dissection of young flowers of a number of collections now reveals that there may be any number from 6 to 10, often (but not always) apparently a constant number in any one plant. One collection, *Barbosa & Lemos* 7962, has sometimes had 7 perianth segments, poorly differentiated between sepals and petals.

Some collections from near Inhambane (*Gomes e Sousa* 1650 and 1675) have rather large coriaceous leaflets with a slightly acuminate apex, rather approaching those of *D. holtzii*, but in all other respects they agree well with *D. schlechteri*.

4. **Dialium holtzii** Harms in Bot. Jahrb. Syst. **49**: 427 (1913); in Engler, Pflanzenw. Afrikas **3**(1): 491 (1915). —Baker, Legum. Trop. Afr.: 647 (1930). —Brenan, Check-list For. Trees Shrubs Tang. Terr.: 102 (1949); in F.T.E.A. Legum.-Caesalp.: 104 (1967). —Steyaert in Bull. Soc. Roy. Bot. Belg. **84**: 37 (1951). —Torre in Mendonça, Contrib. Conhec. Fl. Moçamb. **2**: 75 (1954). Types: Tanzania, Livule R., near Mhondo, *Holtz* 1182 (B† syntype) and Bagamoyo Dist., Chakenge, *Holtz* 1117 (B† syntype).

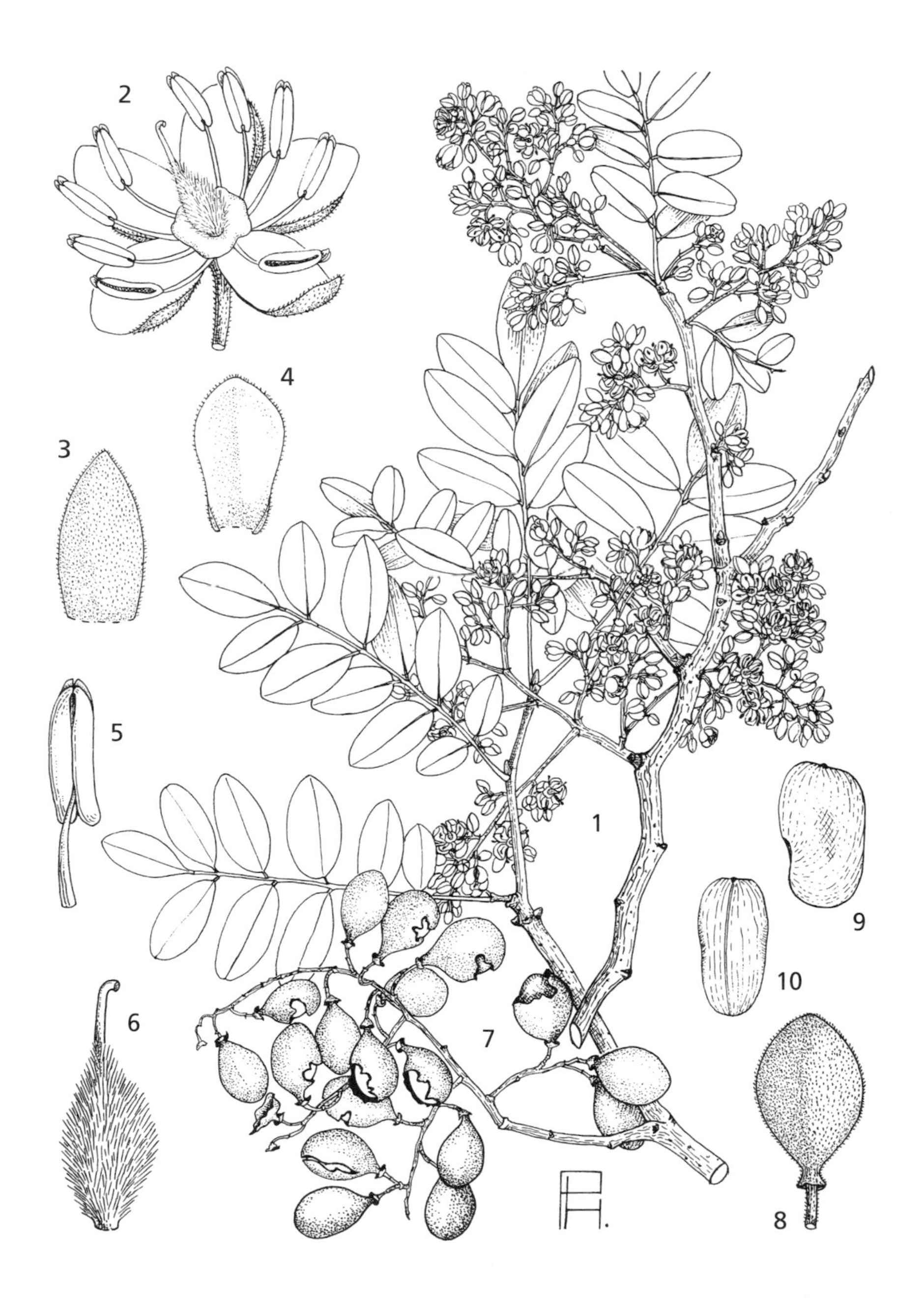

Fig. 3.2.**28**. DIALIUM SCHLECHTERI. 1, flowering branchlet with young leaves (× $^1/_2$); 2, flower (× 5); 3, 4, petal, outer and inner sides (× 5); 5, stamen (× 10); 6, gynoecium (× 10), 1–6 from *De Winter & Vahrmeijer* 8623; 7, fruiting branchlet with mature leaves (× $^1/_2$); 8, fruit (× $^3/_4$); 9, 10, seed, two views (× $1^1/_2$), 8–10 from *Nel* 78. Drawn by Pat Halliday. Reproduced with permission from Flora of Southern Africa.

Dialium mossambicense Steyaert in Bull. Soc. Roy. Bot. Belg. **84**: 39 (1951). —Torre in Mendonça, Contrib. Conhec. Fl. Moçamb. **2**: 75 (1954). Type: Mozambique, Niassa Prov., Mecrusse Forest, *Gomes e Sousa* 2302 (BR holotype).

Dialium schlechteri sensu Steyaert in Bull. Soc. Roy. Bot. Belg. **84**: 37 (1951) in part for *Busse* 2587 ex Tanzania, non Harms.

Tree up to 7–20(25) m high; bark smooth, grey or grey-brown. Young branches ± glabrous. Leaves: petiole and rachis together 7–18 cm long, usually puberulous; leaflets (9)11–15, alternate or the lower ones ± opposite, (2)2.5–5.5(7) × (1)2–3(3.5) cm or the terminal one somewhat larger, the lower ones ± broadly ovate and rounded at the base, the upper ones elliptic and cuneate at the base, all weakly to strongly asymmetric at the base and with an obtusely acuminate apex, glabrous or rarely minutely puberulous beneath; petiolule 2.5–4 mm long, puberulous. Flowers up to 5 mm across, in panicles up to 30 × 15 cm; inflorescence axes appressed-pubescent; pedicels c. 1 mm. Buds and sepals up to 2.5 mm long, densely covered in fine, closely appressed, grey hairs outside; sepals white or cream inside (sometimes described as yellow in Tanzania). Petals 0 (rarely 1). Stamens 2, both on same side of flower; anthers 1.3–1.8 mm long; filaments 1.5–2.3 mm long. Ovary c. 1 mm long, brown tomentose; style c. 2 mm, brown hairy in lower part. Fruits 1.2–2 cm long, 1–1.5 cm diameter, spherical or usually slightly smoothly rounded at apex with style base not showing, elongate, densely dark brown-puberulous. Seeds 1 or 2, brown, shiny, 6–8 × 6–7 × 4 mm and lenticular when only one developing, 7–9 × 4–6 × 3 mm and elongate with a ± triangular cross-section when 2 per fruit.

Mozambique. N: Erati, near Namapa, fr. 17.viii.1948, *Barbosa* 1797 (BR, K, LISC, LMA); between Diaca and Mocimboa da Praia, c. 30 km from Mocímboa da Praia, fl. 25.iii.1961, *Gomes e Sousa* 4664 (K).

Conservation notes: Probably Lower Risk, Near Threatened.

Also in eastern Tanzania. In lowland forest and open woodland; up to 800 m.

As noted by Brenan in F.T.E.A., this species is closely allied to *D. bipindense* Harms from Gabon and Cameroon. Steyaert (1951) distinguished *D. mossamibicense* from *D. holtzii* mainly because it supposedly had a stipitate ovary, but I am unable to see any significant difference.

25. CHAMAECRISTA Moench

Chamaecrista Moench, Meth. Pl. Hort. Bot. Marburg.: 272 (1794). —Irwin & Barneby in Mem. New York Bot. Gard. **35**: 636–895 (1982).

Cassia subgen. *Absus* (DC.) Symon in Trans. Roy. Soc. S. Australia **90**: 77 (1966).

Annual or perennial herbs, sometimes woody at the base. Stems usually hairy with short curved (crisped) hairs and/or straight spreading hairs. Leaves paripinnate, usually with numerous leaflets; stipules narrowly triangular, appressed, persistent; petiole usually with a gland towards the top; rachis channelled above, the margins of the channel sometimes fused to form a ridge or crest; leaflets usually oblong, asymmetric. Inflorescences axillary or supra-axillary, few-flowered; pedicels slender, often elongating after flowering, bracteate near the apex. Calyx lobes subequal, narrowly ovate, acuminate. Petals yellow, delicate, obovate to oblong. Stamens 5–10, subequal, dehiscing by slits. Ovary oblong, hairy, with many ovules. Pods linear, elastically dehiscent. Seeds brown, shiny, rhomboid, flattened, usually with lines of small pits on the testa.

These plants were formerly placed as part of *Cassia*. Lock (Kew Bull. **43**: 333–342, 1988) followed Irwin & Barneby (Mem. New York Bot. Gard. **35**: 1–918, 1982) in recognising the segregate genera *Cassia* sensu stricto, *Senna*, and *Chamaecrista*. These can readily be separated using the characters in the key. It is worth mentioning that all species of *Chamaecrista* so far investigated appear to be nodulated while no species of *Senna* appears to be.

The species of *Chamaecrista* are often not easy to identify and two keys are given here. The first is a multi-access key that will allow quicker identification of the majority of specimens, although conclusions reached from it should always be checked by reference to the descriptions. The second is an orthodox dichotomous key that will probably allow 90% of specimens to be named with reasonable certainty.

Occasionally plants are seen with much-branched terminal inflorescences with many small flowers. These appear to be galled. *Chamaecrista absus* and *C. mimosoides* have both been seen in this state.

Chamaecrista rotundifolia (Pers.) Greene has been recorded from Zimbabwe by Drummond (Kirkia **8**: 219, 1972). Introduced from South America, it is widespread in coastal West Africa. It is a prostrate annual or short-lived perennial herb, slightly woody at the base. Each leaf is made up of only two leaflets.

Various characters are used in the naming of species of *Chamaecrista* and further details of some of these are given here.

1. *Habit.* This is usually clear from good specimens that include the plant base. If there is a single rooted base then the plant is probably annual (although occasionally perennials flower in their first year). Perennials often show the base of previous years' stems beside the current one, or a rhizome, or a substantial woody base. If in doubt, try both alternatives.

2. *The leaf rachis.* The leaf rachis in most species is furrowed between the leaflets. The lateral ridges fringing the central furrow usually have ciliate margins and these cilia may interlock like the teeth on the leaves of a Venus Fly-trap (*Dionaea*), partly obscuring the furrow. The ridges may be spreading, exposing the furrow, or they may be upright, leaving only a narrow slit between them. In some species the two ridges may fuse, producing a solid ridge or crest. This is usually developed only between the leaflets so that in lateral view the crest resembles blunt saw teeth ('crenate-crested'). When checking this character it is useful to remove a leaf, take off some of the leaflets to expose the rachis, and examine under a good lens or binocular microscope. The view directly downwards on to the upper surface of the rachis is the most informative.

3. *Petiolar gland.* This arises on the upper side of the petiole, usually close to the insertion of the first leaflet pair. It is usually darker than the petiole. There is usually a single gland but in a few species there may be two. Glands are of three intergrading types: sessile, circular or elliptic and button-like (the last sometimes appearing sunk into the petiole); shortly stalked, circular and raised, often with an expanded head, so resembling a short-stalked mushroom; and long-stalked, resembling short pins. There are intermediates, but most fall fairly clearly into one of these categories. Glands should be examined with a lens or binocular microscope both from above and from the side. Many are shown in FIGURE 3.2.**29**.

4. *Leaflets: size and midrib position.* The length of leaflets should be measured at the middle of a mature leaf. In most species the midrib of the leaflets is strongly excentric and the leaflet base is very asymmetric, with one side rounded and the other cuneate. The number of lateral veins arising from the base of the leaf on each side is given in the descriptions. The prominence of the veins varies to some extent between species.

5. *Indumentum.* Most species have an indumentum made up of two kinds of hairs: short, curved and ± appressed, and long and straight. The distribution of these varies between parts; in *Chamaecrista stricta*, for instance, the stems bear only short curved hairs but the pedicels have both.

Quick synoptic key to species of Chamaecrista *in the Flora area.*

Score the character states as indicated and look for a match in the table beneath. Read the notes above for help in interpreting character states.

Characters:
Habit: A: plant annual; B. plant perennial.
Rachis: C: crenate-crested; D: channelled or smooth.
Petiolar gland: (if no petiolar gland, see *C. absus* and *C. capensis*). E: long-stalked, the stalk longer than the diameter of the gland; F: short-stalked, the stalk shorter than the diameter of the gland, with a constriction between base and head; G: sessile.
Midrib position: H: strongly asymmetrical, narrow side of lamina less than half the width of the wide side; J: weakly asymmetrical or symmetrical, narrow side more than half the width of the wide side.
Petals: K: longest petal more than 10 mm long; L: longest petal less than 10 mm long.
Leaflet length: M: most leaflets more than 10 mm long; N: most leaflets less than 10 mm long.
Stem indumentum: P: at least some long straight spreading hairs present, usually also with shorter crisped hairs; Q: hairs short and crisped only, or (occasionally) stems glabrous.

Countries: Bo = Botswana; Mw = Malawi; Mz = Mozambique; Zm = Zambia; Zw = Zimbabwe.

C. absus and *C. capensis* are not included. Asterisked pairs have identical codings.

ACEHLNQ	*gracilior*	Mw, Mz, Zm, Zw	Flowers very small (2.5–4 mm). Petals pale
(A/B)C(F/G) HLNQ	*mimosoides*	Bo, Mw, Mz, Zm, Zw	Variable; usually annual; rachis markedly crenate-crested
ACGH(K/L)NP	*polytricha*	Mw, Mz, Zm, Mw	Gland sessile, saddle-shaped; leaflets long-pilose
ADFHLNQ	*fenarolii*	Zm, Zw	Flowers 6.5–7 mm
ADGHL(M/N)P	*falcinella*	Zm	Stipules prominent, asymmetric, semi-auriculate at base
ADGHLNQ	*stricta*	Mw, Mz, Zm, Zw	9–10 stamens
ADGJKMP	*kirkii**	Mw, Zm, Zw	Gland sessile, elliptic, up to 1.5 mm long
ADGJKMP	*wittei**	Mw, Mz, Zw	Gland sessile, circular, 0.5–0.6 mm diam.
ADGJLMP	*nigricans*	Zm	Whole plant in one plane. Flowers small in axillary groups
ADGJLNQ	*dimidiata*	Zw	4–5 stamens only
BCGHKNP	*plumosa*	Mz	Tufted perennial. Long pedicels
BCGHKNQ	*paralias*	Mz	Nerves of leaflets virtually invisible. Leaflets tightly imbricate
BCGH(K/L)NP	*polytricha*	Mw, Mz, Zm, Zw	Gland sessile, saddle-shaped. Leaflets long-pilose
BDEH(K/L)MQ	*grantii*	Mw, Mz	Glands between each leaflet pair. Broad leaflets. Prostrate
BDEJL(M/N)P	*zambesica*	Mz, Zw	Gland long-stalked. Midrib somewhat excentric
BDFHL(M/N)Q	*biensis*	Bo, Mz, Zm, Zw	Midrib strongly excentric, no distal nerves. Leaflet mucro long
BDFHLNQ	*katangensis*	Mw, Zm	Stems crisped-pubescent. Suffruticose habit. Gland like stalkless mushroom
BDGHKMP	*comosa*	Mw, Mz	Gland large, saddle-shaped, c. 2.5 × 1–1.2 mm. Flowers tend to be in axils of reduced leaves
BD(F/G)JK (M/N)Q	*robynsiana*	Zm	Vegetative parts virtually glabrous. Suffrutex of *Brachystegia* woodland
BDGJL(M/N)Q	*parva*	Mw, Zm, Zw	Gland large and flat. Pubescence short and crisped

Key to Chamaecrista *in the Flora Zambesiaca area.*

1. Leaflets in two pairs (if only one pair see *C. rotundifolia* above and aberrants of *C. absus*); whole plant viscid-glandular .... **1.** *absus*
– Leaflets in more than two pairs; plant not viscid-glandular 2
2. Petiole lacking a gland below the first pair of leaflets .... **7.** *capensis*
– Gland present on the petiole, below the first pair of leaflets 3
3. Glands present between most leaflet pairs, as well as on the petiole . . **2.** *grantii*
– Gland on the petiole only 4
4. Stamens 4–5 .... **3.** *dimidiata*
– Stamens 7–10 5
5. Leaflets less than 1 mm wide 6
– Leaflets more than 1 mm wide 10
6. Petiolar gland long-stalked, so that the gland resembles a short pin . . **11.** *gracilior*
– Petiolar gland sessile or short-stalked, not resembling a pin 7
7. Rachis crenate-crested 8
– Rachis furrowed or smooth 9
8. Perennial; leaflets thick-textured (nervation obscure); coastal Mozambique . . . . **12.** *paralias*
– Usually annual; leaflet nervation clearly visible; widespread ..... **13.** *mimosoides*
9. Petals c. 7 mm long; annual or short-lived perennial .... **10.** *fenarolii*
– Petals 12–14 mm long; perennial at low altitudes* .... **14.** *plumosa*
10. Whole plant flattened in one plane; flowers small, on short pedicels . . **5.** *nigricans*
– Plant not flattened in one plane; flowers larger, on distinct pedicels 11
11. Stems prostrate or decumbent; flowers small (petals 8 mm long or less) 12
– Stems erect; flowers large or small 13
12. Leaflets 8–14 pairs, 9–14 × 2.8–3.5 mm; petiolar gland 1.2–1.5 mm long, pin-headed .... **4.** *zambesica*
– Leaflets usually 17–23 pairs, 7–10.5 × 1.2–3 mm .... **6.** *biensis*
13. Flowers smaller (petals not more than 10 mm long) 14
– Flowers large (petals at least 10 mm long, usually more than 15 mm) 16
14. Petiolar gland shortly stalked; leaflets usually falcately curved, strongly asymmetric at base often with a distinct angular basal lobe .... **17.** *katangensis*
– Petiolar gland sessile; leaflets usually parallel-sided, weakly asymmetric at base without a distinct basal lobe 15
15. Leaflets up to 20 pairs; stems with both long straight hairs and short curved hairs; flowering pedicels at least 10 mm long .... **8.** *falcinella*
– Leaflets at least 25 pairs; stems with short curved hairs only; flowering pedicels up to 8 mm long .... **15.** *stricta*
16. Suffruticose perennial up to 30 cm tall, almost glabrous .... **16.** *robynsiana*
– Taller annuals or perennials; stems usually densely pubescent 17
17. Perennial with densely pubescent stems; leaflets 4–6 mm wide; petiolar gland sessile, more than 2 mm long, appearing sunk into the petiole; inflorescences tending to be concentrated towards stem apices .... **20.** *comosa*
– Annual or perennial; stems pubescent; leaflets usually narrower; petiolar gland smaller, often stalked; inflorescences axillary, scattered on the stems 18
18. Short-lived perennial or annual; flowers large (petals c. 15 mm long) . . **18.** *wittei*
– Annual or perennial; flowers smaller (petals up to 12 mm long, usually less) . . 19
19. Leaves with fewer than 20 pairs of leaflets; stipules falcately curved, prominently veined, overlapping and spreading at shoot apex .... **8.** *falcinella*

---

* If montane with intermediate-sized flowers see also 9, *C. polytricha*.

– Leaves with more than 20 pairs of leaflets; stipules straight, overlapping but not spreading at shoot apex 20
20. Leaves with 20–24 pairs of leaflets; leaflets 1.5–2.5 mm wide; petiolar gland sessile, flat . . . . **19.** *parva*
– Leaves with more than 24 pairs of leaflets; leaflets narrower or wider than above; gland sessile, flat or saddle-shaped 21
21. Annual or perennial; leaflets in 32–60 pairs, 0.8–1.4 mm wide; petiolar gland often saddle-shaped . . . . **9.** *polytricha*
– Annual; leaflets in 24–38 pairs, 1.8–4.5 mm wide; petiolar gland sessile, flat . . . . **21.** *kirkii*

1. **Chamaecrista absus** (L.) H.S. Irwin & Barneby in Mem. New York Bot. Gard. **35**: 664 (1982). —Lock, Leg. Afr. Check-list: 30 (1989). Type: Sri Lanka, Herb. Linnaeus 528.4 (LINN syntype). FIGURE 3.2.**30**.

Cassia absus L., Sp. Pl.: 537 (1753). —Klotzsch in Peters, Naturw. Reise Mossamb. **6**(1): 15 (1861). —Oliver in F.T.A. **2**: 279 (1871). —Taubert in Engler, Pflanzenw. Ost-Afrikas **C**: 201 (1895). —Harms in Engler, Pflanzenw. Afrikas **3**(1): 499 (1915). —Baker, Legum. Trop. Afr.: 639 (1930). —Brenan, Check-list For. Trees Shrubs Tang. Terr.: 98 (1949). —Keay in F.W.T.A., ed.2, **1**: 453 (1958). —Steyaert in F.C.B. **3**: 507 (1952). —Mendonça & Torre in C.F.A. **2**: 179 (1956). —Brenan in F.T.E.A., Legum.-Caesalp.: 81, fig.15 (1967). —Schreiber in Merxmüller, Prod. Fl. SW Afrika, fam. 59: 11 (1967). —Drummond in Kirkia **8**: 212 (1972); in **10**: 243 (1977). —Gordon-Gray in F.S.A. **16**(2): 95 (1977). —Irwin & Barneby in Mem. New York Bot. Gard. **30**: 277–283 (1977).

Annual erect sticky herb, sometimes woody at the base. Stems terete, lightly ridged, hirsute with hispid and glandular hairs of various lengths. Leaves paripinnate; rachis 3–5.5 cm long; stipules 3–6 × 0.5–1 mm, narrowly triangular, hirsute with glandular hairs; petiolar glands absent; leaflets in 2 pairs, 1.7–4.7 × 1.2–3.2 cm, broadly elliptic to broadly obovate, cuneate on proximal side, rounded on distal, apex rounded to obtuse or broadly acute, mucronulate, base asymmetric, sparsely appressed-pubescent on both surfaces; midrib median, 0–1 lateral nerves from base on proximal side, 1–2 on distal side. Inflorescences terminal, racemose, becoming leaf-opposed through growth of an axillary bud; bracts c. 2.5 × 1.5 mm, ovate, base cordate, apex acuminate; pedicels 0.3–1 cm long, with 1–2 minute bracteoles. Sepals 4.1–6.3 × 1.7–3 mm, glandular-hairy outside. Petals yellow, buff or orange, usually with darker markings, 6.5–8.7 × 3–4 mm, including the 1.5 mm long claw, broadly obovate. Stamens 5–6, subequal. Ovary 4 mm long, densely appressed-hairy; style c. 2 mm long, glabrous; stigma subterminal, oblique. Pods 4–5 × 0.6–0.8 cm, oblong, sparsely glandular-hairy. Seeds black, shiny with scattered surface pitting and brown dots, 5–6.5 × 3.5–5 mm, rhomboid to elliptic in outline, flattened.

Botswana. N: Ngamiland, Xamatshaa Is., fl. & fr. 18.ii.1975, *P.A. Smith* 1263 (K, SRGH). SW: Ghanzi, Eaton's Farm, fr. 2.v.1969, *Brown* s.n. (K). SE: Mahalapye village, fl. & fr. 16.xi.1963, *Yalala* 395 (K, SRGH). **Zambia.** B: Senanga Dist., Lilongo Forest Res., fl. & imm.fr., 6.ii.1952, *White* 2014 (BR, K). N: Chinsali Dist., 60 km SW of Isoka, fl. & fr. 15.iii.1975, *Hooper & Townsend* 770 (K). W: Ndola, fl. & fr. 22.iii.1954, *Fanshawe* 1010 (BR, K). C: Mt Makulu Research Station, 18 km (12 miles) S of Lusaka, fl. & fr. 29.iii.1957, *Angus* 1531 (K). E: Chipata Dist., Luangwa Valley, Mkhania, fl. & fr. 25.ii.1969, *Astle* 5517 (K). S: Mapanza Mission, fl. & fr. 22.iii.1953, *E.A. Robinson* 143 (K). **Zimbabwe.** N: Mt Darwin Dist., Chiswiti Reserve, fl. 18.i.1960, *Phipps* 2332 (BR, K, SRGH). W: Hwange (Wankie), fl. & fr. ii.1955, *Levy* 1175 (K, PRE). C: Makoni Dist., Chiduku, fl. & fr. iv.1955, *Davies* 1168 (K, SRGH). E: Mutare (Umtali) Dist., The Grove, Mt Sheni, fl. & fr. 5.iii.1950, *Chase* 1988 (K, SRGH). S: Masvingo Dist., Makoholi Experimental Farm, fl. 23.iii.1949, *D.A. Robinson* 303 (K, LISC, SRGH). **Malawi.** N: Mzimba Dist., Katete Mission, fl. 19.iv.1974, *Pawek* 8393 (K,

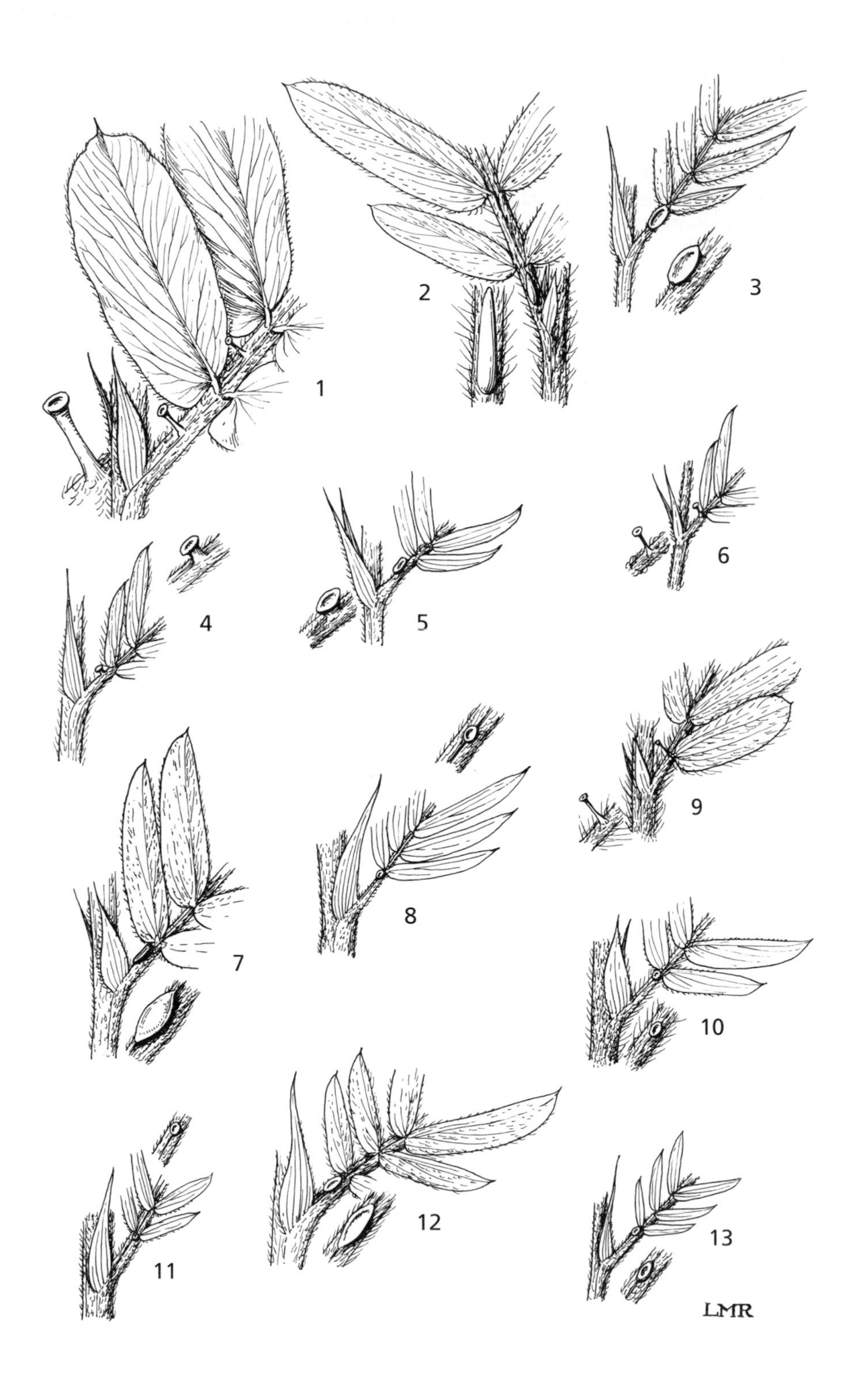
1
2
3
4
5
6
7
8
9
10
11
12
13
LMR

MAL, MO, SRGH). C: near Chitala, fl. 12.ii.1959, *Robson* 1546 (K, LISC). S: Mangochi Dist., near Nkopola, 19 km NW of Mangochi (Fort Johnston), fl. & fr. 2.iii.1970, *Brummitt* 8848 (K). **Mozambique.** Z: Lugela, Namagoa, fl. 19.iii.1949, *Faulkner* Kew series 402 (BR, K). MS: Chemba, crossroads to Tambara, fl. & fr. 4.iv.1962, *Balsinhas & Macuácua* 549 (K, LISC). GI: between Nhachengo and Vilankulo (Vilanculos), 48 km from Nhachengo, fl. & fr. 22.iii.1952, *Barbosa & Balsinhas* 4980 (K). M: Maputo (Lourenço Marques), fl. & fr. 15.ii.1920, *Borle* 552 (K).

Widespread in tropical Africa; also in S Asia and Australia. Woodland, open and wooded grassland, generally in disturbed places along roads and in cultivated ground; near sea-level to c. 1500 m.

According to Irwin & Barneby (1977), the Old World plants belong to var. *absus*, and normally have 5–7 stamens, while the New World plants, with 3–4 stamens, belong to var. *meonandra* (H.S. Irwin & Barneby) H.S. Irwin & Barneby. The Old World variety also occurs in the New World as an introduction.

Fanshawe 8411, from Kitwe in W Zambia, is much-branched, some of the leaves are bi- or trifoliolate, and many of the leaflets are narrower than usual. No other similar material has been seen and the plant is probably galled, as suggested by the collector. *Peter* 45368 from Victoria Falls has long many-flowered inflorescences and is also apparently an abberant individual.

2. **Chamaecrista grantii** (Oliv.) Standl. in Smithsonian Misc. Coll. **68**(5): 5 (1917). — Lock, Leg. Afr. Check-list: 31 (1989). Syntypes: Mozambique, 'Maravi Country' (W of Lake Malawi), *Kirk* (K syntype), and Tanzania, Morogoro Dist., Mbuiga, *Grant* s.n. (K syntype). FIGURE 3.2.**29**/1.

Cassia grantii Oliv. in F.T.A. **2**: 279 (1871); in Trans. Linn. Soc., London **29**: 64, fig.35 (1872). —Taubert in Engler, Pflanzenw. Ost-Afrikas **C**: 201 (1895). —Harms in Engler, Pflanzenw. Afrikas **3**(1): 500 (1915). —Baker, Legum. Trop. Afr.: 639 (1930). —Steyaert in F.C.B. **3**: 528 (1952). —Mendonça & Torre in C.F.A. **2**: 186 (1956). —Brenan in F.T.E.A., Legum.-Caesalp.: 83 (1967).

Perennial decumbent or prostrate suffruticose herb from a slender woody rootstock. Stems apparently annual, terete, pubescent with short curved hairs. Leaves paripinnate, 4–6 × 1.8–2 cm; stipules c. 5 × 2 mm, narrowly ovate, apex acuminate, base rounded, symmetric, surfaces glabrous, margins ciliate; petiolar gland single, stalked, pinhead-like, with further similar glands between all or most of the leaflet pairs; rachis channelled, not crested; leaflets in (6)8–13 pairs, 10–20 × 5–8 mm, oblong, almost symmetrical, proximal side of base rounded, distal cuneate, apex mucronate, surfaces glabrous, margins ciliate, margins inrolled over base of apical mucro; midrib 2–3 mm from distal margin, secondary nerves prominent beneath, indistinct above, 2–3 lateral nerves from base on proximal side of midrib, 4–5 on distal side. Inflorescences axillary, usually 2–4-flowered; pedicels 3.3–4.5 cm long in flower, apparently not elongating in fruit. Flowers large; sepals acuminate, appressed-pubescent outside; petals yellow, up to 10 × 6 mm. Stamens 10. Ovary densely appressed-white-pubescent; stigma terminal, fringed with hairs. Pods to 5 × 0.5 cm, sparsely pubescent. Seeds c. 3.5 × 2 mm, rhomboid, flattened.

Fig. 3.2.**29**. CHAMAECRISTA. Lower part of leaf (× 2) & petiolar gland (× 4) of various species. 1, C. GRANTII, from *Mgaza* 345; 2, C. NIGRICANS, from *Tanner* 1383; 3, C. PARVA, from *Lynes* I.h.107; 4, C. FENAROLII, from *Burtt* 2015; 5, C. KATANGENSIS, from *Bullock* 2017; 6, C. GRACILIOR, from *Milne-Redhead & Taylor* 9285; 7, C. COMOSA, from *Milne-Redhead & Taylor* 8257A; 8, C. STRICTA, from *Milne-Redhead & Taylor* 9666; 9, C. ZAMBESICA, from *Tanner* 2882; 10, C. WITTEI, from *Richards* 8814; 11, C. DIMIDIATA, from *Milne-Redhead & Taylor* 10049; 12, C. KIRKII var. KIRKII, from *Milne-Redhead & Taylor* 9073; 13, C. MIMOSOIDES, from *Milne-Redhead & Taylor* 9667. Drawn by Lura Ripley. Adapted from F.T.E.A.

Malawi. N. Namitawa (Namwitawa) Hill, lower slopes, fl. ix.1903, *McClounie* 94 (K). C: Dowa, 3 km (2 miles) W of Lake Nyasa Hotel, fl. 4.viii.1950, *Chase* 3856 (K, SRGH). **Mozambique.** N: Região de Ribáuè, fl. 16.x.1948, *Andrada* 1405 (K, LISC). Z: between Gilé and Alto Ligonha, 65.1 km from Gilé, fl. 11.x.1949, *Barbosa & Carvalho* 4379 (K, LISC). T: Angónia Dist., near Ulongue, fl. & fr. 21.xi.1980, *Macuácua* 1283 (K, LISC, LMA).

Also in Angola, Tanzania, Kenya and Uganda. Open *Brachystegia* and *Acacia* woodlands; grasslands and cultivated land; 400–1000 m.

3. **Chamaecrista dimidiata** (Roxb.) Lock in Kew Bull. **43**: 336 (1988); Leg. Afr. Check-list: 30 (1989). Type from India. FIGURE 3.2.**29**/11.

Senna dimidiata Roxb., Fl. Ind., ed. 2, **2**: 352 (1832).

Cassia hochstetteri Ghesq. in Bull. Jard. Bot. État **9**: 155 (1932). —Brenan in F.T.E.A., Legum.-Caesalp.: 96 (1967). Type: Ethiopia, Tigray, Mt. Scholoda, 1837, *Schimper* I: 66 (BR lectotype, K).

Annual erect (or prostrate) herb. Stems terete with slight pale ridges decurrent from the stipule bases, pubescent with appressed curved hairs. Leaves paripinnate, 4.4–4.9 × 0.8–1.2 cm; stipules 5.2–6 × 1.2–1.6 mm, narrowly triangular, glabrous, apex long-acuminate, base rounded to subcordate, asymmetric; petiolar gland dark, single, 0.75–0.9 × 0.5–0.6 mm, elliptic, sessile, shallowly cup-shaped; rachis channelled, not crested, margins of channels pilose; leaflets in (9)24–2(31) pairs, 4.1–5.1 × 0.8–1.1 mm, narrowly ovate-oblong, asymmetrical, base asymmetrically truncate, apex acuminate-mucronate, surfaces glabrous to sparsely appressed-pubescent, margins sparsely ciliate, cilia ascending; midrib 0.2–0.3 mm from distal margin, 0–1 lateral nerves from base on proximal side of midrib, 3–4 on distal side. Inflorescences supra-axillary, 1–2-flowered; pedicels 0.4–0.7 cm long in flower. Flowers small; sepals narrowly ovate, acute, sparsely appressed-pubescent outside; petals yellow, up to 6 × 4 mm. Stamens 4–5; if 5 then one smaller than the rest. Ovary densely appressed-pubescent; stigma terminal, fringed with hairs. Pods [from elsewhere] (15)22–43 × 3.5–6 mm. Seeds brown.

Zimbabwe. E: Nyanga (Inyanga), Juliasdale, fl. 21.ii.1972, *Corby* 2213 (K, LISC, SRGH).

A single record from the Flora area; 'roadside'. Elsewhere in Africa it occurs in grasslands and old cultivation, usually in montane areas. Also in Ethiopia and Tanzania, and in South and SE Asia. Recorded from Madagascar by Brenan (1967) but these specimens are attributed to *C. stricta* by Du Puy & Rabevohitra (Legumin. Madagascar: 102, 2002).

It is tempting to regard this and other specimens of this plant as abnormal forms of another taxon, but the small flowers on short pedicels and the crisped stem pubescence are fairly distinctive.

4. **Chamaecrista zambesica** (Oliv.) Lock in Kew Bull. **43**: 338 (1988); Leg. Afr. Check-list: 33 (1989). Type: Mozambique, Shamwara, *Kirk* s.n. (K holotype). FIGURE 3.2.**29**/9.

Cassia zambesica Oliv. in F.T.A. **2**: 280 (1871). —Harms in Engler, Pflanzenw. Afrikas **3**(1): 500 (1915). —Baker, Legum. Trop. Afr.: 640 (1930) as *zambesiaca*. —Brenan in F.T.E.A., Legum.-Caesalp.: 85 (1967). —Drummond in Kirkia **8**: 213 (1972).

Usually perennial prostrate herb from a somewhat woody base. Stems with slight pale ridges decurrent from the stipule bases, pubescent with long and short straight hairs. Leaves paripinnate, 5 × 2.3 cm; stipules narrowly ovate-triangular; petiolar gland single, dark, 1.2–1.5 mm long, pinhead-shaped; rachis channelled, not crested, occasionally with one or more glands similar to or smaller than the one on the petiole; leaflets in 8–14 pairs, 9.1–14 × 2.8–3.5 mm, oblong, asymmetrical, proximal side of base rounded, distal cuneate, apex mucronate, margins inrolled at apex, glabrous to sparsely pubescent above, more densely so beneath; midrib

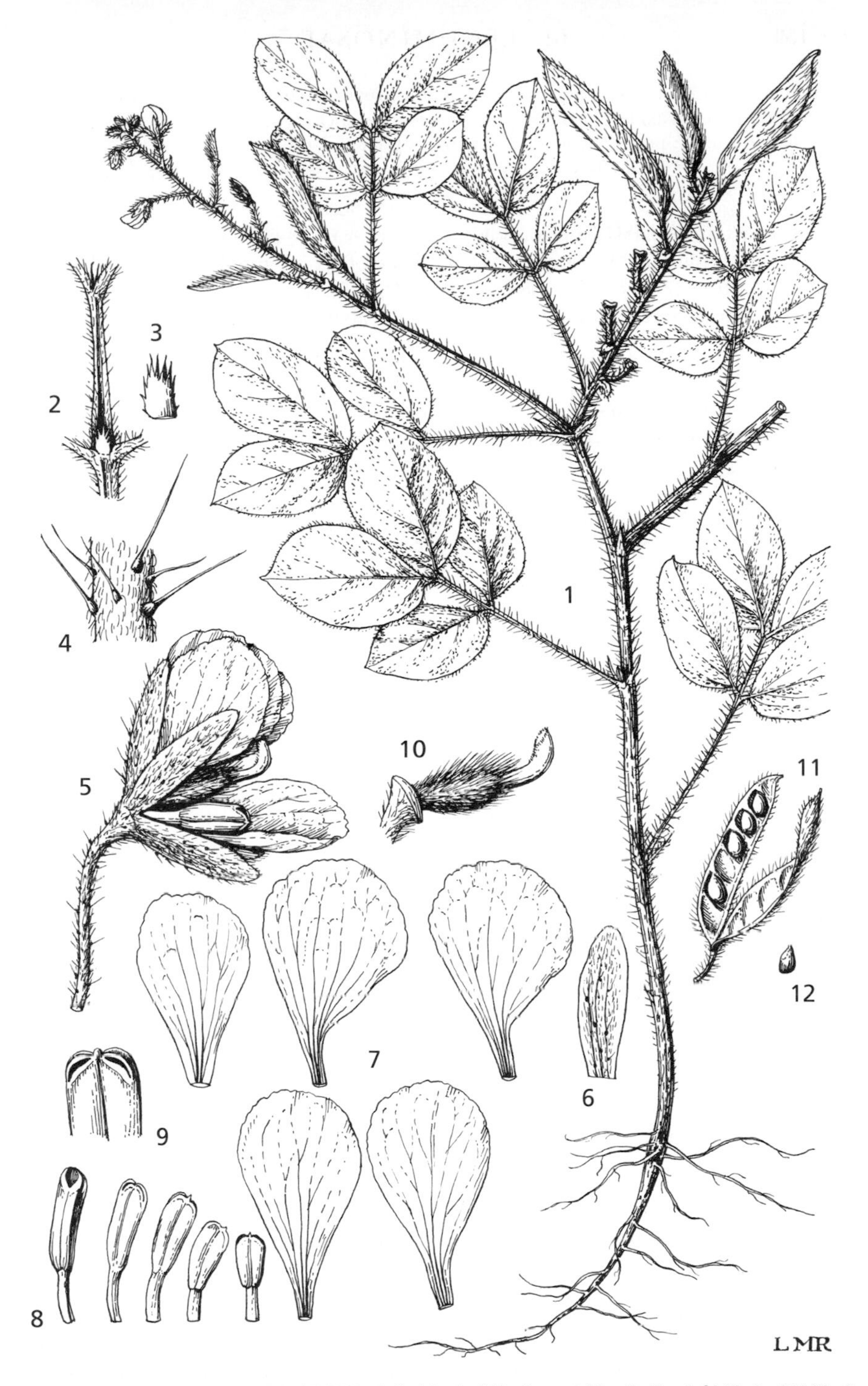

Fig. 3.2.**30**. CHAMAECRISTA ABSUS. 1, habit (× ²⁄₃), from *Milne-Redhead & Taylor* 10102; 2, gland on leaf rachis (× 4); 3, gland (× 8); 4, hairs on young stem (× 8); 5, flower (× 4); 6, sepal (× 4); 7, petals (× 4); 8, stamens (× 4); 9, tip of anther (× 8); 10, ovary (× 4), 2–10 from *Polhill & Paulo* 1223; 11, pod (× ²⁄₃); 12, seed (× ²⁄₃), 11 & 12 from *Milne-Redhead & Taylor* 10102. Drawn by Lura Ripley. From F.T.E.A.

strongly excentric; 4–6 lateral nerves from base on proximal side of midrib, 1–2 on distal side. Inflorescences supra-axillary, 2–4-flowered; pedicels 1–1.5 cm long in flower, not elongating in fruit. Flowers small; petals yellow, c. 7 × 4 mm. Stamens 10. Ovary densely appressed-white-pubescent. Pods 3.8 × 0.4 cm, pubescent with long hairs. Seeds dark brown, c. 2.5 × 2.5 mm, rhombic, flattened.

Zimbabwe. E: Mutare Dist., Gimbokki Farm, fl. & fr. 10.iii.1957, *Chase* 6360 (EA, K, SRGH). **Mozambique.** MS: Manica, Mavita, edge of Munhinga R. by road to Chimoio (Vila Pery), fl. & fr. 25.iv.1948, *Barbosa* 1559 (K, LISC).

Also in Kenya and Tanzania. Grasslands and abandoned cultivation; sea level to 950 m.

5. **Chamaecrista nigricans** (Vahl) Greene in Pittonia **4**: 30 (1899). —Lock, Leg. Afr. Check-list: 32 (1989). Type: Yemen, Wadi Surdûd, *Forsskål* (C holotype). FIGURE 3.2.**29**/2.

Cassia nigricans Vahl, Symb. Bot. **1**: 30 (1790). —Oliver in F.T.A. **2**: 280 (1871). —Harms in Engler, Pflanzenw. Afrikas **3**(1): 500 (1915). —Baker, Legum. Trop. Afr.: 641 (1930). —Steyaert in F.C.B. **3**: 518 (1952). —Mendonça & Torre in C.F.A. **2**: 181 (1956). —Keay in F.W.T.A., ed. 2, **1**: 452 (1958). —Brenan in F.T.E.A., Legum.-Caesalp.: 81 (1967).

Upright herb, sometimes woody towards the base, probably usually annual, 30–100 cm tall. Stem smooth, pubescent with short curled and long straight yellowish hairs. Leaves paripinnate, tending to be held in a single plane, 5–8.5 × 3.5–4.2 cm; stipules 5.5–6.5 × 1.5 mm, narrowly triangular, rounded or truncate at the base, pubescent with long and short hairs; petiolar gland oval, cushion-shaped, depressed; rachis channelled; leaflets in (7)11–14 pairs, 15–24 × 3.5–5 mm, oblong-elliptic, slightly asymmetric, proximal side of base rounded to cuneate, distal cuneate, apex mucronate, mucro 0.8–1.2 mm long, margins inrolled above into mucro, yellowish pubescent above and beneath; midrib slightly excentric to almost central; secondary nerves moderately prominent beneath, indistinct above, 2–3 lateral nerves from base on proximal side of midrib, 1–2 on distal side. Inflorescences supra-axillary, an additional axillary one sometimes arising later, 4–7-flowered; pedicels 1.5–3 mm long, apparently not elongating in fruit. Flowers small; petals 3.5–4.5 × 1.5–3.5 mm. Stamens 10. Ovary densely appressed-white-pubescent. Pod blackish, 2–2.5 × 0.4–0.5 cm, pubescent. Seeds dark brown, c. 2.5 × 1.8 mm, rhomboid, flattened.

Zambia. N: Mbala Dist., Mpulungu, fl. & imm.fr. 4.v.1952, *Richards* 1604 (K).

Drier regions of tropical Africa from Senegal to Angola and Tanzania, also in Yemen, Saudi Arabia and peninsular India. Usually in tall or medium grasslands, often in abandoned cultivation; 860 m.

This is the only record from the Flora area and East African material has been used to fill out the description. The small flowers on very short pedicels, and the almost symmetrical leaflets are characteristic. The plant is distinctive in life because the leaves all tend to be held in a single plane.

6. **Chamaecrista biensis** (Steyaert) Lock in Kew Bull. **43**: 335 (1988); Leg. Afr. Check-list: 30 (1989). Type: Angola, Bié Prov., *Gossweiler* 9 (BR holotype).

Cassia katangensis var. *biensis* Steyaert in Bull. Jard. Bot. État **20**: 260 (1950).

Cassia biensis (Steyaert) Mendonça & Torre in Bol. Soc. Brot., Sér. 2, **29**: 33, fig.1C (1955); in C.F.A. **2**: 184, fig.38C (1956). —Schreiber in Merxmüller, Prod. Fl. SW Afrika, fam. 59: 11 (1967). —Drummond in Kirkia **8**: 212 (1972). —Ross, Fl. Natal: 195 (1973). —Gordon-Gray in F.S.A. **16**(2): 100 (1977).

Perennial prostrate to decumbent suffruticose herb from a woody rootstock. Stems generally annual, terete with slight ridges decurrent from the stipule bases, glabrous. Leaves paripinnate, 4–5.2 × 1.5–2.2 cm; stipules 6–9 × 1.2–1.5 mm, narrowly triangular, spreading pubescent, apex long-acuminate, base rounded to subcordate, asymmetric; petiolar gland single, 0.4–0.5 × 0.4–0.5 mm, subcircular, shortly stalked; rachis channelled, not crested; channel margins ciliate; leaflets in (9)17–23(25) pairs, 7–10.5 × 1.2–3 mm, narrowly oblong, asymmetrical, base asymmetrically truncate, apex abruptly acuminate-mucronate, margins ciliate, often densely so, cilia ascending, surfaces glabrous to sparsely appressed-pubescent; midrib very close to distal margin, 0–1 lateral nerves from base on proximal side of midrib, c. 4 on distal side. Inflorescences axillary, (1)2-flowered; pedicels 1.7–4 cm long in flower, longer in fruit. Flowers small; sepals narrowly ovate, acute, mucronate, appressed-pubescent outside; petals yellow, up to 10 × 7 mm. Stamens 10. Ovary densely appressed-white-pubescent; style glabrous; stigma terminal, densely fringed with hairs. Pods 33–48 × 4–5 mm, sparsely appressed-pubescent, particularly near the margins. Seeds dull ochre, c. 2 × 1.2 mm, rhomboid in outline, flattened, sparsely pitted.

Botswana. N: Odiakwe, fl. & fr. 9.iii.1965, *Wild & Drummond* 6825 (K, SRGH). SW: 70 km (45 miles) N of Kang, fl. & fr. 18.ii.1960, *Wild* 5048 (K, SRGH). SE: Kgatleng Dist., Masama Ranch, fl. 10.xi.1978, *Hansen* 3545 (C, GAB, K, PRE, SRGH, UPS). **Zambia.** B: Sesheke Dist., fl. iv.1911?, *Gairdner* 480 (K). **Zimbabwe.** W: Nyamandhlovu, fl. & fr. 18.x.1953, *Plowes* 1642 (K, SRGH). C: Gweru (Gwelo), fl. 4.ix.1966, *Biegel* 1491 (K, SRGH). S: Mwenezi (Nuanetsi), between Tswiza and Nyala sidings, fl. 27.iv.1962, *Drummond* 7765 (K, SRGH). **Mozambique.** M: Maputo (Lourenço Marques), between Moamba and Pessene, fl. & fr. 19.ii.1948, *Torre* 7363 (LISC).

Also in Angola, Namibia and South Africa. In grasslands and open woodlands, almost always on Kalahari sand or other sandy soils; up to 1400 m.

Generally distinct in its prostrate habit, short leaves, small number of leaflets each with the midrib very close to the distal margin, relatively long pedicels and small flowers. The Mozambique plants are rather more densely pubescent than others.

7. **Chamaecrista capensis** (Thunb.) E. Mey., Comment. Pl. Afr. Austr. **1**: 158 (1836). Syntypes: South Africa, Western Cape Prov., between Loerie (?Luri) and Sundays (Sontags) R., *Thunberg* s.n. (UPS holotype).

Cassia capensis Thunb., Prodr. **1**: 79 (1794). —Harms in Engler, Pflanzenw. Afrikas **3**(1): 500 (1915). —Ross, Fl. Natal: 195 (1973). —Gordon-Gray in F.S.A. **16**(2): 98 (1977).

Prostrate herb, probably perennial, drying blackish. Stems terete, pubescent with short and long curved hairs. Leaves paripinnate, 3–4.5 × 1.5–2 cm. Stipules narrowly ovate-triangular, up to 5 mm long. Petiolar gland absent. Rachis channelled, the margins of the channel ciliate. Leaflets in 14–19 pairs, 8–10 × 1.5–1.9 mm, narrowly oblong, asymmetrical, ciliate on the margins, otherwise glabrous; proximal side of base rounded, distal cuneate; apex mucronate; midrib strongly excentric; 4–5 nerved from base on proximal side of midrib, 1–2 on dorsal side; leaflet venation almost parallel. Inflorescence slightly supra-axillary, 2–3-flowered. Pedicels 8–10 mm long, elongating slightly in fruit. Flowers medium to large; petals 7–8 × 4–5 mm, probably yellow. Stamens 10. Ovary densely appressed white-pubescent. Pods up to 6 × 0.5 cm, very sparsely appressed-pubescent. Seeds not seen.

Mozambique. M: Inhaca Is., Delagoa Bay, fl. & fr. 31.viii.1959, *Watmough* 314 (K, SRGH).

Also in South Africa. Abandoned cultivation; near sea level.

Conservation notes: Data Deficient. No judgement can be made on the basis of the single specimen seen from the Flora area.

This is the only specimen seen from the Flora area. It would appear to belong to var. *flavescens* (E.Mey.) Vogel, if Gordon-Gray's account in F.S.A. is followed. Its dark colour on drying, and the complete absence of a petiolar gland, makes it a distinctive plant. The taxonomy of the group (or species) needs further attention.

8. **Chamaecrista falcinella** (Oliv.) Lock in Kew Bull. **43**: 336 (1988); Leg. Afr. Check-list: 30 (1989).

Cassia falcinella Oliv. in F.T.A. **2**: 281 (1871). —Taubert in Engler, Pflanzenw. Ost-Afrikas **C**: 201 (1895). —Baker, Legum. Trop. Afr.: 641 (1930). —Ghesquière in Bull. Jard. Bot. État **9**: 162 (1932) excl. var. *longifolia*. —Steyaert in F.C.B. **3**: 520 (1952). —Brenan in Kew Bull. **14**: 178 (1960); in F.T.E.A., Legum.-Caesalp.: 90 (1967). —Schreiber in Merxmüller, Prod. Fl. SW Afrika, fam. 59: 11 (1967). —Drummond in Kirkia **8**: 212 (1972). —Gordon-Gray in F.S.A. **16**(2): 101 (1977). Type: Tanzania, Bukoba Dist., Karagwe, *Grant* 445 (K holotype).

Upright or ascending annual herb, sometimes slightly woody at the base. Stems terete, pubescent with short crisped and long straight spreading hairs. Leaves paripinnate, 4–5.5 × 1–1.5 cm; stipules 9–14.5 × 2.5–4.2 mm, narrowly falcate-triangular, apex acuminate, base strongly asymmetric, the proximal side narrowly cuneate, the distal side rounded to cordate and decurrent, outer surface hairy or not, margins ciliate, venation very prominent; petiolar gland single, elliptic to circular, 0.7–1.1 mm in diameter, sessile; rachis channelled above, ciliate on the channel margins and pilose beneath; leaflets in 9–19 pairs, 4–14 × 2.5–3.7 mm, narrowly oblong-ovate, almost straight, somewhat asymmetric, base asymmetric, proximal side rounded, distal side narrowly cuneate, apex mucronate, sparsely hairy above, more densely so beneath; margins ciliate; venation very prominent beneath; midrib excentric; 4–5 lateral nerves from base on proximal side, 0–1 on distal side. Inflorescences supra-axillary, 2–3-flowered; pedicels 1–2 cm long in flower and fruit, spreading pilose. Sepals 5–9 × 1.5–2 mm, narrowly ovate, acuminate, pilose outside. Petals yellow, 4–10 × 3–6 mm, broadly obovate to elliptic. Stamens 10. Ovary c. 5 mm long, densely appressed-pubescent; style c. 1.2 mm long, hooked, proximally sparsely appressed-pubescent, distally glabrous; stigma terminal, ciliate-margined. Pods 3.5–4 × 0.4 cm, sparsely pilose. Seeds dark brown, 2.5–3 × 1.5–1.8 mm, rhomboid, with lines of minute pits on the testa.

There are three varieties, of which two have been collected in the Flora area. The nominate variety occurs in East Africa.

Var. **intermedia** (Brenan) Lock in Kew Bull. **43**: 336 (1988). Type: Tanzania, Ufipa Dist., L. Kwela, 19 km from Mpui, *McCallum Webster* T1 (K holotype).

Cassia falcinella var. *intermedia* Brenan in Kew Bull. **14**: 179 (1960); in F.T.E.A., Legum.-Caesalp.: 91 (1967).

Petals c. 7–10 mm long. Stipules pubescent on the outer surface.

Zambia. N: Mbala Dist., Mbala (Abercorn) sandpits, fl. 13.ii.1964, *Richards* 19010 (K).

Also in Congo, Uganda, Kenya and Tanzania. Miombo woodland and associated secondary bushland and grassland, often in sandy and sometimes damp places; 1500–1750 m.

Var. **parviflora** (Steyaert) Lock in Kew Bull. **43**: 336 (1988). Type: Rwanda, Gabiro, *Becquet* 613 (BR holotype).

Cassia falcinella var. *parviflora* Steyaert in Bull. Jard. Bot. État **20**: 251 (1950); in F.C.B. **3**: 521 (1952). —Brenan in Kew Bull. **14**: 179 (1960); in F.T.E.A., Legum.-Caesalp.: 91 (1967). —Drummond in Kirkia **8**: 212 (1972). —Gordon-Gray in F.S.A. **16**(2): 101 (1977).

Petals 4–6 mm long.

Botswana. N: Maun, Thamalakane R., fl. & fr. 23.ii.1967, *Lambrecht* 11 (K, LISC, SRGH). **Zambia.** B: Masese, fl. & fr. 15.v.1962, *Fanshawe* 6822 (K, NDO). C: Luangwa Valley, Mfuwe, near Mwamba R., fl. & fr. 6.iii.1967, *Astle* 5055 (K). S: Livingstone Dist., near Maramba R., fl. & fr. 11.iii.1952, *White* 2235 (FHO, K). **Zimbabwe.** N: Gokwe Dist., Sengwa Research Station, fl. & fr. 8.iii.1976, *Guy* 2394 (K, SRGH). W: Hwange Nat. Park, near Shapi Camp, fl. & fr. 22.ii.1967, *Rushworth* 86 (K, SRGH). C: Charter

Estate, fl. & fr. 19.v.1966, *Corby* 1615 (K, SRGH). S: Masvingo Dist., Makoholi Research Station, fr. 15.iii.1978, *Senderayi* 245 (K, SRGH).

Also in Congo, Rwanda, East Africa, Angola, Namibia and South Africa. Grassy places in mopane and mixed woodland, flood plains and dambos; 600–1200 m.

The prominently veined and somewhat falcate stipules overlap at the shoot tips to give a distinctive appearance to the species. Material from East Africa has been used to fill out the description.

9. **Chamaecrista polytricha** (Brenan) Lock in Kew Bull. **43**: 337 (1988); Leg. Afr. Check-list: 32 (1989). Type: Tanzania, Songea Dist., 29 km E of Songea, *Milne-Redhead & Taylor* 9402 (K holotype).

Cassia polytricha Brenan in Kew Bull. **14**: 185 (1960); in F.T.E.A., Legum.-Caesalp.: 99 (1967). —Drummond in Kirkia **8**: 212 (1972).

Annual or perennial herb. Stems terete, hairy; hairs mostly straight and spreading. Leaves paripinnate, 3.8–6 × 0.8–1.5 cm; stipules 12–16 × 1.4–1.8 mm, very narrowly triangular, apex long-acuminate, base asymmetrically rounded to subcordate, margins and outer surface hairy; petiolar gland single, 0.4–0.6 mm in diameter, circular, sessile, sometime with the margins turned down so that the gland appears saddle-shaped; rachis weakly crenate, usually ciliate on the ridge margins, hairy beneath; leaflets in 32–60 pairs, 4.5–7.5 × 0.8–1.4 mm, narrowly oblong, weakly asymmetric, base asymmetric, proximal side rounded, distal side rounded-cuneate, apex acute, scattered-hairy mainly on the margins; midrib excentric; 4–5 lateral nerves from base on proximal side, 0–1 on distal side. Inflorescences supra-axillary, 1–3-flowered; pedicels 1.2–2.4 cm long in flower, elongating slightly in fruit, spreading hairy. Sepals 6.2–10.5 × 1.4–3.8 mm, narrowly ovate, acuminate, spreading hairy outside. Petals yellow, 6.3–12.2 × 4–8 mm, broadly obovate to elliptic. Stamens 10, four longer than the rest. Ovary 4–6 mm long, densely appressed-pubescent; style 4–6 mm long, almost glabrous; stigma terminal. Pods 4.5–5 × 0.4–0.5 cm, sparsely hairy. Seeds mid-brown, shiny, 2.5–3 × 1.3–1.5 mm, oblong-rhomboid.

There are three varieties of which two have been recorded from the Flora area. The third, var. *puchella* (Brenan) Lock, occurs in S Tanzania only, although Brenan (F.T.E.A) states 'doubtfully in Rhodesia'. I have not traced the specimen on which this statement is based. Var. *pulchella* is annual, with 2–3 flowers in each inflorescence.

Var. **polytricha**

Perennial. Petals 8–13 mm long.

Zimbabwe. E: Chimanimani Dist., Chimanimani Mts, Dragon's Tooth, fl. 28.viii.1966, *Bisset* CH55 (K, SRGH). **Malawi.** C: Nkhota-kota, fl. 2.v.1963, *Verboom* 890 (K). S: Zomba Dist., lower slopes of Zomba Mt. near Mlunguzi stream, fl. & fr. 6.iii.1977, *Brummitt & Seyani* 14803 (K, MAL). **Mozambique.** N: S face of Serra de Ribáuè, fl. 19.xii.1962, *Leach & Schelpe* 11409 (K, SRGH). Z: Guruè, 26 km from Mutuali to Lioma, fl. 10.ii.1964, *Torre & Paiva* 10501 (LISC). MS: Dombe Dist., E bank of Makuripini R., 5 km above confluence with Haroni R., fl. & fr. 6.i.1969, *Bisset* 11 (K, SRGH).

Also in Tanzania. Grasslands, often montane; 400–1700 m.

Material from the Chimanimani Mts. is more obviously subshrubby, often with bare stems and rather smaller leaves. These are probably montane ecotypes.

Var. **pauciflora** (Brenan) Lock in Kew Bull. **43**: 338 (1988). Type: Tanzania, Songea Dist., Kimarampaka stream, *Milne-Redhead & Taylor* 8488 (K holotype).

Cassia polytricha var. *pauciflora* Brenan in Kew Bull. **14**: 185 (1960); in F.T.E.A., Legum.-Caesalp.: 99 (1967). —Drummond in Kirkia **8**: 212 (1972).

Annual. Inflorescences 1(2)-flowered.

Zambia. E: Fort Young, fl. & fr. ix.1896, *Nicholson* s.n. (K). **Zimbabwe.** E: Nyanga Dist., M'kango R., Hondi Valley, fl. & fr. 7.iii.1966, *Corby* 1557 (K, SRGH). **Malawi.** S: Kasupe Dist., Chikala Hills, W side, fl. 17.ii.1975, *Brummitt, Banda, Seyani & Patel* 14355 (K, MAL). **Mozambique.** N: 25 km on Marrupa to Lichinga road, fl. 8.viii.1981, *Jansen, de Koning & de Wilde* 143 (K, WAG). MS: Chimoio, Amatongas, fl. 14.ii.1948, *Andrada* 1035 (LISC).

Also in Tanzania. Grasslands, often in disturbed places or on shallow soils over rock, 600–1100 m.

Specimens of this taxon often have a distinctive appearance because the leaflets do not seem to fold up as readily as those of other taxa so that the leaves are often open on the sheet, and the upper leaflets are much shorter than the lower.

10. **Chamaecrista fenarolii** (Mendonça & Torre) Lock in Kew Bull. **43**: 337 (1988); Leg. Afr. Check-list: 31 (1989). Type: Angola, Bié, near Chinguar, *Fenaroli* 1198a (Herb. Fenaroli holotype). FIGURE 3.2.**29**/4.

Cassia fenarolii Mendonça & Torre in Bol. Soc. Brot. Sér.2, **29**: 34, fig.1b (1955); in C.F.A. **2**: 186, fig.38b (1956). —Brenan in F.T.E.A., Legum.-Caesalp.: 86 (1967). —Drummond in Kirkia **8**: 212 (1972).

Annual or short-lived perennial usually prostrate woody-based herb. Stems terete, pubescent with crisped hairs. Leaves paripinnate, 2.2–5.5 × 0.4–0.9 cm; stipules 3.2–5.5 × 0.5–0.8 mm, narrowly triangular, apex long-acuminate, base cuneate to subcordate, surfaces and margins glabrous; petiolar gland single, dark, 0.3–0.5 mm in diameter, circular, shortly (0.3 mm) stalked, shallowly cup-shaped; rachis channelled, but the ciliate rims of the channel can be very closely appressed so that it appears ridged, but not crenate-crested; leaflets in 24–40 pairs, 2.9–6 × 0.5–0.9 mm, narrowly triangular-oblong to falcate, asymmetric, base asymmetric, proximal side rounded, distal side narrowly cuneate, apex mucronate, margins ciliate, cilia ascending, surfaces glabrous to sparsely appressed-pubescent; midrib c. 0.6 mm from proximal margin, 0.2 mm from distal; c. 3 somewhat obscure lateral nerves from base on proximal side, 0–1 on distal side. Inflorescences axillary to somewhat supra-axillary, 1(2)-flowered; pedicels 1.2–1.7 cm long in flower, similar in fruit, pubescent with crisped and long straight hairs. Sepals ovate-elliptic, acute, appressed-pubescent outside. Petals yellow, up to 7 × 3 mm. Stamens 10. Ovary densely appressed-pubescent. Pods to 3 × 0.4 cm, sparsely appressed-pubescent. Seeds uniformly brown, c. 2 ×1.5 mm, rectangular-rhomboid, scattered-punctate.

Zambia. W: Solwezi Dist., fl. & fr. 15.v.1969, *Mutimushi* 3311 (K). **Zimbabwe.** C: Shurugwi (Selukwe), fl. xi.1899, *Cecil* 120 (K). S: Mberengwa Dist., Mt. Buhwa, fl. & fr. 3.v.1973, *Biegel, Pope & Simon* 4274 (K, SRGH).

Also in Angola and Tanzania. Miombo woodland, and on roadsides; 900–1500 m.

Closest to *C. gracilior*, from which it differs in the larger flowers and the very shortly stalked glands. Differs from prostrate forms of *C. mimosoides* in the smooth or weakly ridged (not crenate-crested) rachis. The base is generally somewhat woody, but none of the few specimens that I have seen shows the bases of old shoots from previous years, so that it is probably at best a short-lived perennial.

11. **Chamaecrista gracilior** (Ghesq.) Lock in Kew Bull. **43**: 336 (1988); Leg. Afr. Check-list: 30 (1989). Lectotype selected by Steyaert (1950): Congo, Katanga, Kiambi, *de Witte* 263 (BR lectotype). FIGURE 3.2.**29**/6.

Cassia gracilior (Ghesq.) Steyaert in Bull. Jard. Bot. État **20**: 248 (1950); in F.C.B. **3**: 517 (1952). —Mendonça & Torre in C.F.A. **2**: 186 (1956). —Brenan in F.T.E.A., Legum.-Caesalp.: 87 (1967). —Drummond in Kirkia **8**: 212 (1972).

Annual, occasionally short-lived perennial, usually erect herb. Stems terete, pubescent with crisped hairs. Leaves paripinnate, 2.4–6.5 × 0.4–0.5 cm; stipules 5–6 × 1 mm, very narrowly triangular, apex long-acuminate, base cuneate to subcordate, surfaces and margins glabrous; petiolar gland single, dark, 0.1–0.3 mm in diameter, circular, stalked, the stalk 0.3–0.6 mm long, the whole trumpet-shaped; rachis weakly crenate-crested, not winged, hairy only at the insertions of the leaflets above, sparsely crisped-pubescent beneath; leaflets in 30–40 pairs, 2–4 × 0.5–0.7 mm, narrowly triangular-oblong to falcate, asymmetric, base asymmetric, proximal side rounded, distal side narrowly cuneate, apex mucronate, margins sparsely ascending-ciliate, surfaces virtually glabrous; midrib 0.3–0.4 mm from proximal margin, 0.2 mm from distal; 2–3 lateral nerves from base on proximal side, 0–1 on distal side. Inflorescences axillary to distinctly supra-axillary, 1–2-flowered; pedicels 1–1.5 cm long in flower, a little longer in fruit, pubescent with crisped appressed hairs. Sepals 5–5.5 × 3 mm, narrowly ovate, acuminate, sparsely appressed-pubescent outside. Petals cream to yellow, drying pinkish, up to 4.5 × 3 mm. Stamens 10. Ovary c. 2.5 mm long, densely appressed-pubescent; style c. 1 mm long, glabrous; stigma oblique. Pods (1.5)2.5–3 × 0.3–0.35 cm, sparsely appressed-pubescent. Seeds (immature) pale brown, rectangular-elliptic, c. 2.5 × 1.5 mm.

Zambia. N: Mbala Dist., Mbala (Abercorn), near Kilema Farm, fr. 12.v.1962, *Richards* 16464 (K). W: Kasepa Dist., 11 km W of Chizera, fl. & fr. 23.iii.1961, *Drummond & Rutherford-Smith* 7201 (K, SRGH). C: Chakwenga headwaters, 100–129 km E of Lusaka, fl. & fr. 5.iii.1965, *Robinson* 6392, (K). E: 4 km (2.5 miles) W of Kachalola on Great North Road, fl. & fr. 17.iii.1959, *Robson* 1740 (BM, K, SRGH). S: Choma Dist., Siamambo, fl. & fr. 20.iii.1958, *Robinson* 2816 (K, SRGH). **Zimbabwe.** N: Lomagundi, Mpinga–Mtorashanga road, Lone Cow Estate, fl. 2.iii.1965, *Corby* 1245 (K, SRGH). C: Harare (Salisbury), fl. ii.1920, *Eyles* 2084 (K, PRE). **Malawi.** N: Mzimba Dist., 10 km (6 miles) N of Mzambazi, 16 km (11 miles) S of Mperembe, fl. 10.iii.1978, *Pawek* 13968 (K, MAL, MO, SRGH). C: Lilongwe Dist., Dzalanyama Forest Res., 4 km from Katete Gate, fl. & fr. 26.iii.1977, *Brummitt, Seyani & Patel* 14974 (K). S: Mangochi Dist., Monkey Bay, fl. 1.iii.1970, *Brummitt & Eccles* 8810 (K). **Mozambique.** N: Malema, Chefe Elema, base of Serra Mancuni, 20 km from Mutuáli, fl. & fr. 17.iii.1964, *Torre & Paiva* 11231 (LISC).

Also in Congo, Tanzania and Angola. Open and disturbed places in miombo woodland and around dambos, often on sandy soils; 500–1800 m.

Similar to *C. fenarolii*, from which it differs in the smaller flowers and in the markedly stalked petiolar gland. The stalk may be elongated so that the whole gland appears trumpet-shaped, or it may be shorter, but there is always a marked constriction below the apex of the gland.

12. **Chamaecrista paralias** (Brenan) Lock in Kew Bull. **43**: 337 (1988); Leg. Afr. Check-list: 32 (1989). Type: Mozambique, Inhambane, Pomene, *Gomes e Sousa* 1691 (K holotype, LISC).

Cassia paralias Brenan in Kew Bull. **14**: 187 (1960).

Perennial erect woody herb or shrublet from a woody rootstock. Stems perennial, terete with slight pale ridges decurrent from the stipule bases, appressed-pubescent. Leaves paripinnate, 2.5–5.5 × 0.6–0.8 cm; stipules c. 30 × 10 mm, narrowly ovate-triangular, apex acuminate, base rounded to subcordate, asymmetric, surfaces glabrous but appressed-pubescent towards the apex, almost black above the base; petiolar gland single, dark, 0.2–0.4 mm in diameter, sessile, circular, shallowly cup-shaped; rachis crenate-crested, sparsely appressed-pubescent; leaflets in (20)30–58 pairs, 3.3–3.8 × 0.7–0.9 mm, narrowly oblong, asymmetrical, proximal side of base rounded, distal narrowly cuneate, apex rounded, often mucronate, margins ascending-ciliate, surfaces sparsely appressed-pubescent; midrib c. 0.2 mm from distal margin, lateral nerves obscure. Inflorescences axillary, 1–2-flowered; pedicels 1–2 cm long in flower, longer in fruit. Flowers large; sepals acuminate, appressed-pubescent outside; petals yellow, 11–12 × 6.5–8 mm. Stamens 10; three larger than the rest. Ovary densely and shortly appressed-pubescent; style c. 3 mm long, with scattered appressed hairs; stigma terminal, fringed with hairs. Pods 3.5–5 × 0.5 cm, sparsely appressed-pubescent. Seeds dull brown with black dots, 3 × 1.8 mm, oblong.

Mozambique. GI: Inhambane, Massinga, Pomene, fl. & fr. 22.ix.1980, *De Koning & Jansen* 8453 (K, LMU).

Known only from a small area of coastal Mozambique. Sand dunes; 0–20 m.

Distinctive in its small, thick-textured tightly imbricate leaflets. Perhaps a very highly specialised derivative of *C. mimosoides*, which it resembles in its crenate-crested rachis.

Brenan (1960), mentions *Pedro & Pedrogão* 4644 (Niassa, Quinga, old dunes, 25.vii.1948 (EA)), as being very similar to this but with the stems glabrous when young. I have not seen this material.

13. **Chamaecrista mimosoides** (L.) Greene in Pittonia **4**: 27 (1899). —Lock, Leg. Afr. Check-list: 32 (1989). —van Wyk & Malan, Wild Fl. Highveld, ed.2: 122 (1997). Type: Sri Lanka, *Herb. Hermann* 2: 13, 78 (BM syntype). FIGURE 3.2.**29**/13.

Cassia mimosoides L., Sp. Pl.: 379 (1753). —Klotzsch in Peters, Naturw. Reise Mossambique **6**(1): 15 (1861). —Oliver in F.T.A. **2**: 280 (1871). —Taubert in Engler, Pflanzenw. Ost-Afrikas **C**: 201 (1895). —Harms in Engler, Pflanzenw. Afrikas **3**(1): 500 (1915). —Baker, Legum. Trop. Afr.: 642 (1930). —Brenan, Check-list For. Trees Shrubs Tang. Terr.: 96 (1949). —Steyaert in F.C.B. **3**: 514, fig.37 (1952). —Mendonça & Torre in C.F.A. **2**: 181 (1956). —Keay in F.W.T.A., ed.2, **1**: 453 (1958). —Brenan in F.T.E.A., Legum.-Caesalp.: 100 (1967). —Schreiber in Merxmüller, Prod. Fl. SW Afrika, fam. 59: 11 (1967). —Drummond in Kirkia **8**: 212 (1972). —Ross, Fl. Natal: 195 (1973). —Gordon-Gray in F.S.A. **16**(2): 104 (1977).

Annual erect or decumbent herb, sometimes woody at the base. Stems terete, sparsely to densely pubescent with curved appressed hairs. Leaves paripinnate, 7–11(13) × 0.8–1(1.4) cm; stipules 6–11 × 1–2 mm, very narrowly triangular, apex long-acuminate, base rounded, asymmetric, surfaces and margins glabrous; petiolar gland single, dark to pale, 0.3–0.6 × 0.3–0.9 mm, elliptic to circular, sessile to shortly stalked (to 0.3 mm), usually shallowly cup-shaped; rachis crenate-crested; leaflets in (31)40–59 pairs, 3.8–5(8) × 0.7–1.2(1.4) mm, narrowly oblong, asymmetric, base asymmetrically truncate, apex acute to obtuse, shortly mucronate; margins sparsely ascending-ciliate, surfaces glabrous or with at most a few scattered hairs; midrib 0.2 mm from distal margin; 0–2 lateral nerves from base on proximal side of midrib, 2–4 on distal side, often indistinct. Inflorescences supra-axillary, usually 2–3-flowered; pedicels 0.25–2 cm long in flower, 0.5–2.4 cm in fruit, appressed-pubescent. Sepals 5–7.7 × 2–2.8 mm, ovate to narrowly ovate, appressed (or rarely spreading) pubescent outside. Petals yellow, broadly obovate, 4.5–9.5 × 3–7 mm. Stamens 9–10, four larger than the rest. Ovary densely appressed-pubescent; stigma terminal, fringed with hairs. Pods 3.7–5.8 × 0.3–0.5 cm, linear, sparsely pubescent with short stiff appressed hairs or with longer more flexuous hairs. Seeds pale brown, sometimes with small darker spots, 2.2–3 × 1.3–1.8 mm, rounded-rhomboid.

Caprivi Strip. Okavango R. at Diyona Camp, 3 km (2 miles) E of Nyangana Mission Station, fl. & fr. 18.ii.1956, *De Winter & Marais* 4759 (K, PRE). **Botswana.** N: Mutsoi, NE of Nokaneng, fl. & fr. 23.iii.1967, *Lambrecht* 89A (K, SRGH). **Zambia.** B: Mongu, fl. & fr. 20.ii.1966, *Robinson* 6841 (K). N: Mbala Dist., Ndundu to Kawimbe road, near Ndundu, fl. 7.iv.1967, *Richards* 22178 (K). W: Mufulira, fl. 13.ii.1949, *Cruse* 486 (K, LISC). C: Mt Makulu Research Station, 3 km (2 miles) W of Chilanga, fl. & fr. 21.xii.1958, *Robson & Angus* 976 (K). E: Katete, St Francis's Hospital, fl. 28.ii.1957, *Wright* 165 (K). S: Mazabuka, Magoye Forest Res., fl. 16.i.1952, *White* 1931 (K, FHO). **Zimbabwe.** N: Hurungwe Dist., Msukwe R. bank, fl. 3.iii.1954, *Lovemore* 391 (K, SRGH). W: Matobo Dist., Farm Besna Kobila, fl. ii.1954, *Miller* 2157 (K, SRGH). C: Marondera Dist., Marondera (Marandellas), Delta Farm, fl. & fr. 18.ii.1967, *Corby* 1767 (K, SRGH). E: Nyanga Dist., 8 km (5 miles) N of Nyanga Village, fl. & fr. 28.iv.1967, *Rushworth* 899 (K, SRGH). S: Gwanda Dist., Tuli Expt. Station, fl. & fr. 18.i.1965, *Norris-Rogers* 605 (K, SRGH). **Malawi.** N: Nkhata Bay Dist., Old Bandawe, 53 km (33 miles) S of Nkhata Bay junction, fl. & fr. 25.vi.1977, *Pawek* 12811 (K, MAL,

MO). C: Nkhota Kota Dist., 8 km (5 miles) N of Nkhota Kota, fl. & fr. 17.vi.1970, *Brummitt* 11521 (K). S: Zomba Dist., Shire Valley 23 km W of Zomba, fl. & fr. 12.iii.1977, *Brummitt & Patel* 14835 (K, MAL). **Mozambique.** N: Namapa/Erati Dist., Namapa, fl. 27.iii.1961, *Balsinhas & Marrime* 317 (K, LISC). Z: between Namacurra and Macuze, 14.1 km from Namacurra, fl. & fr. 28.viii.1949, *Barbosa & Carvalho* 3852 (K). T: Mágoè, from Mágoè Velho towards Cachomba, 10 km, fl. 28.ii.1970, *Torre & Correia* 18141 (LISC, LMA, LMU). MS: between Gondola and Nhamatanda (Vila Machado), fl. & fr. 22.iii.1960, *Wild & Leach* 5218 (K, SRGH). GI: Inhambane–Velho (Estevam), 24 km E of Inhambane town, fl. & fr. n.d., *Gomes e Sousa* 2011 (K). M: Maputo, fl. & fr. 1.x.1930, *Gomes e Sousa* 364 (K).

Widespread in tropical Africa; also in South Asia, but assessment of the extent of its occurrence there must await critical studies of this and related Asian taxa.

Many references to this taxon pre-1965 will include several taxa and have mostly been excluded. No attempt has been made to check the numerous specimens that might clarify these references.

Chamaecrista mimosoides is a very variable taxon, although perhaps less so in the Flora Zambesica area than in the F.T.E.A. area where Brenan distinguished 7 forms, one of which has now been distinguished as a good species, *Chamaecrista telfairiana* (Hook. f.) Lock (see Polhill in Fl. Mascarenes, fam. 80: 21, 1990). To the south, in the F.S.A. area, Gordon-Gray distinguished 3 forms and most of the material from the F.Z. area can be placed in one or other of these. Her forms 2 and 3 are widespread, but her form 1 seems to be confined largely to the coasts and islands of Mozambique. It can be distinguished by the large gland that overlaps the sides of the petiole. These plants also tend to have short peduncles and to be hairier than most other material, often with the hairs more flexuous.

The species is almost always annual. A few plants probably persist from one wet season to the next but no new basal stem is produced, unlike in true perennials where new stems tend to be produced from the rootstock in each wet season.

Most plants are erect but some, mainly from roadsides and trampled areas, are more decumbent. Plants can persist when grazed or mown and may then have a low dense and very atypical growth form.

The indumentum is distinctive; the stems and most of the other parts bear short curved appressed hairs and lack long straight ones. The stipules are usually entirely glabrous. The ovary and pod usually bear a sparse stiff rather coarse appressed indumentum, but in some forms the hairs are longer and more flexuous. The leaf rachis is pubescent beneath but the leaflets themselves are virtually glabrous apart from marginal cilia.

The rachis is crenate-crested. In some similar species, particularly *C. fenarolii* and *C. stricta*, the two margins of the furrow on the upper surface of the rachis may be enlarged and may mimic a crest, but examination at × 20 will show that the two margins are not fused, so that there is a deep groove between them. In some forms of *C. mimosoides* the top of the crest may be flattened, sometimes with two rows of hairs, but the margins are firmly fused so that there is no groove between them.

Chamaecrista mimosoides is most easily confused with *C. fenarolii* and *C. stricta*. For distinctions see previous paragraph. *Chamaecrista stricta* generally has larger leaflets and shorter leaves, while *C. fenarolii* is smaller in all its parts, particularly the leaves and stipules. *Chamaecrista polytricha* is also often confused with *C. mimosoides* but the hairs on the stems are straight and spreading and the stipules are hairy. *Chamaecrista plumosa* differs in its perennial habit, usually longer pedicls, and in the ciliate margins of the stipules. *Chamaecrista gracilior* has smaller flowers and the petiolar gland is long-stalked so that the gland resembles a pin.

14. **Chamaecrista plumosa** E. Mey., Comment. Pl. Afr. Austr.: 159 (1836). —Lock, Leg. Afr. Check-list: 32 (1989). Type: South Africa, Eastern Cape Province, between Bashe R. and Umtata, 1500 ft., *Drège* s.n. (K).

Cassia plumosa (E. Mey.) Vogel, Syn. Gen. Cass.: 64 (1837). —Ross, Fl. Natal: 195 (1973). —Gordon-Gray in F.S.A. **16**(2): 103 (1977).

Perennial erect to decumbent herb from a woody rootstock. Stems generally annual except for the extreme bases, terete with slight pale ridges decurrent from the stipule bases, sparsely appressed-puberulous. Leaves paripinnate, 4–7 × 0.7–0.9 cm; stipules 3.5–6 × 0.7–0.9 mm, narrowly triangular, apex long-acuminate, base truncate to subcordate, surfaces glabrous, margins ascending-ciliate or not; petiolar gland single or occasionally 2 superposed, dark, 0.4–0.5 × 0.3–0.4 mm, elliptic, very shortly stalked or at least constricted below, shallowly cup-shaped; rachis ridged to crenate-crested, sparsely appressed-pubescent; leaflets in 21–43 pairs, 4.2–6.1 × 0.7–1.1 mm, narrowly triangular oblong, asymmetrical, base asymmetrically truncate, apex acuminate-mucronate, margins sparsely ciliate, cilia ascending, surfaces glabrous to very sparsely appressed-pubescent; midrib 0.2–0.3 mm from distal margin, 0–1 lateral nerves from base on proximal side of midrib, 2–4 on distal side. Inflorescences supra-axillary, 1–3-flowered; pedicels 2–3 cm long in flower, longer in fruit. Flowers large; mature buds 7.5–9.5 mm long; sepals ovate, acuminate, very sparsely appressed-pubescent outside; petals yellow, 12–14 × 6–8 mm. Stamens 10. Ovary densely appressed-pubescent; style glabrous; stigma terminal, fringed with hairs. Pods 3.6–4.4 × 0.3–0.4 cm, very sparsely appressed-pubescent. Seeds yellow-brown with darker spots, 3–3.5 × 1.5 mm.

Two varieties are recognised of which only var. *plumosa* occurs in the Flora area. Var. *erecta* (Schorn & Gordon-Gray) Lock occurs in KwaZulu-Natal and Mpumalanga. It is a taller plant, not normally prostrate, and also differs in its ecology, being less confined to coastal sands.

Var. **plumosa**

Stems prostrate, forming a mat, or semierect and definitely branched to form a compact rounded bush up to c. 40 cm high.

Mozambique. MS: Dombe, E bank Makurupini, 5 km above Haroni confluence, fl. 14.i.1969, *Bisset* 68 (SRGH). GI: Inhambane, região de Chidinguel, Zavala, fl. & fr. 11.xii.1944, *Mendonça* 3379 (K, LISC). M: Maputo, 33 km N of Maputo (Lourenço Marques) near Marracuene, fl. 25.i.1946, *Gomes e Sousa* 3361 (K).

Mozambique to eastern Cape Province of South Africa. Generally in grasslands on sandy soil near the sea.

Phipps 248 from Zimbabwe (E), Chimanimani, Upper Haroni, fl. & fr. 31.i.1957, resembles this but the rachis is grooved rather than crenate-crested.

Chamaecrista plumosa can be confused with *C. mimosoides* and *C. paralias*, all of which share the crenate-crested rachis. The crests can sometimes be low and inconspicuous in *C. plumosa*, although often more prominent towards the distal end of the leaf. *Chamaecrista plumosa* is always perennial, and its flowers are larger than those of most forms of *C. mimosoides* and appear long-pedicelled. The leaflets of *C. paralias* are rounded at the tips and thick-textured so that it is difficult to see the nerves, which are always clearly visible in the acute leaflets of *C. plumosa*.

15. **Chamaecrista stricta** E. Mey., Comment. Pl. Afr. Austr.: 159 (1836). —Lock, Leg. Afr. Check-list: 33 (1989). Type: South Africa, Eastern Cape Province, Bashe R., *Drège* s.n. (K isotype). FIGURE 3.2.**29**/8.

Cassia kirkii Oliv. var. *quarrei* Ghesq. in Bull Jard. Bot. État **9**: 153 (1932). Type: Congo, Katanga, Étoile, *Quarré* 380 (BR lectotype, chosen by Steyaert 1950).

Cassia capensis var. *humifusa* Ghesq. in Bull Jard. Bot. État **9**: 164 (1932), for lesser part to include *Robyns* 2448 (BR syntype).

Cassia quarrei (Ghesq.) Steyaert in Bull Jard. Bot. État **20**: 264, fig.26 (1950); in F.C.B. **3**: 521 (1952). —Brenan in F.T.E.A., Legum.-Caesalp.: 95 (1967). —Drummond in Kirkia **8**: 213 (1972). —Ross, Fl. Natal: 195 (1973). —Gordon-Gray in F.S.A. **16**(2): 102 (1977).

Cassia sparsa Steyaert in Bull Jard. Bot. État **21**: 359, fig.100 (1951) in part including holotype only; Steyaert in F.C.B. **3**: 523, fig.45 (1952). Type: Congo, Katanga, Keyberg, 8 km SW of Lubumbashi, *Schmitz* 3485 (BR holotype).

Annual or perhaps short-lived perennial herb or softly woody shrublet. Stems terete, pubescent with crisped hairs. Leaves paripinnate, 3.5–10 × 1–2 cm; stipules 5.5–8.5 × 1.2–2 mm, narrowly triangular, apex long-acuminate, base asymmetrically auriculate, glabrous; petiolar gland single, 0.7–1.1 mm in diameter, circular, sessile, just below the first leaflet pair; rachis weakly channelled above, sparsely ciliate on the channel margins and crisped pubescent beneath; leaflets in 25–36 pairs, 7.3–10.2 × 1.4–2.3 mm, narrowly oblong, almost straight, asymmetric, base asymmetric, proximal side rounded, distal side rounded-cuneate, apex mucronate, margins shortly ciliate or glabrous, usually glabrous on both surfaces; midrib excentric; 4–5 lateral nerves from base on proximal side, 1–2 on distal side. Inflorescences markedly supra-axillary, 2–3-flowered; pedicels 6–7 mm long in flower, 1.2–1.3 cm long in fruit, pubescent with both crisped and spreading hairs. Sepals 6.5–7.2 × 1.2 mm, narrowly ovate, acuminate, pubescent with crisped hairs outside. Petals yellow, 6–8 × 6–8 mm, broadly obovate to suborbicular. Stamens (7)8–9, with 3–4 longer than the rest; a small staminode sometimes present. Ovary 3.2–4 mm long, densely appressed-pubescent, distally glabrous; stigma terminal, ciliate-margined. Pods 5–6.3 × 0.3–0.5 cm, sparsely appressed-pubescent. Seeds pale yellow-brown with faint darker spots, 2.5–3.2 × 1.6–2 mm, rhomboid, flattened.

Zambia. N. Mbala Dist., Mbala (Abercorn) to Kawimbi Road, near Mbala, fl. 14.iii.1952, *Richards* 993 (K). W: Kitwe, fl. & fr. 29.iii.1969, *Mutimushi* 3016 (K, NDO). C: Kabwe (Broken Hill), Mpima Forest Station, fl. & fr. 8.iv.1962, *Morze* 350 (K). E: Petauke Dist., Nyimba to Luembe road, fl. 11.xii.1958, *Robson* 900 (BM, K). S: Mapanza Mission, fl. & fr. 22.iii.1953, *Robinson* 137 (K). **Zimbabwe.** W: Bulawayo, fl. ii.1939, *Hopkins* 7780 (SRGH). C: Harare Dist., near Gwebi R., fl. 22.ii.1961, *Rutherford-Smith* 556 (K, SRGH). E: near Chirinda, fl. 14.iii.1950, *Hack* 72/50 (SRGH). S. Masvingo Dist., Zimbabwe Nat. Park, fr. 29.iii.1973, *Chiparawasha* 651 (SRGH). **Malawi.** C: Lilongwe Dist., Agricultural Research Station, fl. & fr. 11.iii.1966, *Salubeni* 426 (K). S: Blantyre, near Ndirande Forest Plantation, fl. 12.iii.1969, *Msinkhu* 7 (K, SRGH). **Mozambique.** N: Malema, Mutuáli, base of Mt. Cucuteia, fl. & fr. 16.iii.1964, *Torre & Paiva* 11196 (LISC). MS: Báruè, Serra de Choa, 9 km from Catandica (Vila Gouveia) to Serra, fl. 24.v.1971, *Torre & Correia* 18599 (LISC, LMA, LMU).

Also in Congo, Kenya, Tanzania, Swaziland and South Africa. Open places in woodland, river banks, disturbed places and a weed of cultivation, often forming extensive local populations; 750–1500 m.

16. **Chamaecrista robynsiana** (Ghesq.) Lock in Kew. Bull. **43**: 338 (1988); Leg. Afr. Check-list: 33 (1989). Type from Congo.

Cassia robynsiana Ghesq. in Bull. Jard. Bot. État **9**: 149, fig.1/29–31 (1932). —Steyaert in F.C.B. **3**: 517 (1952).

Perennial erect suffruticose herb from a slender woody rootstock. Stems generally annual, terete with slight pale ridges decurrent from the stipule bases, glabrous. Leaves paripinnate, 6–9 × 1.3–1.8 cm; stipules c. 60 × 22 mm, narrowly ovate-triangular, apex acuminate, base rounded to subcordate, asymmetric, surfaces glabrous, margins ciliate with hairs bent near the base at right angles towards the stipule apex; petiolar gland single or occasionally 2 superposed, dark, 1 × 0.4 mm, elliptic, sessile or very shortly stalked, shallowly cup-shaped; rachis channelled, not crested; leaflets in (12)18–24 pairs, 8–14 × 2–4.5 mm, narrowly triangular-oblong, asymmetrical, base asymmetrically truncate, apex acuminate-mucronate, margins ciliate, cilia ascending, surfaces glabrous to sparsely appressed-pubescent; midrib 0.5–0.7 mm from distal margin, 1–2 lateral nerves from base on proximal side of midrib, 3–6 on distal side. Inflorescences axillary, usually 1–2-flowered but in robust plants each pedicel may bear two flowers so that the inflorescence becomes 4-flowered; pedicels 2.2–6 cm long in flower, longer in fruit. Flowers large; sepals long-acuminate, appressed-pubescent outside; petals yellow, up to 18 × 10 mm. Stamens 10. Ovary densely appressed-pubescent; stigma terminal, fringed with hairs. Pods sparsely appressed-pubescent; mature pods and seeds not seen.

Zambia. W: Solwezi Dist., Chisera stream on Solwezi–Mwinilunga road, fl. 19.ix.1952, *Angus* 467 (BR, FHO, K). C: Mumbwa, 1911–12, *Macaulay* 97 (K).

Also in Congo (Katanga). In *Brachystegia* (chipya) woodland; sometimes in seasonally flooded valleys (dambos) or copper-induced clearings; 1200–1400 m.

17. **Chamaecrista katangensis** (Ghesq.) Lock in Kew Bull. **43**: 337 (1988). Type: Congo, Katanga, Welgelegen, *Corbisier* 577 (BR syntype). FIGURE 3.2.**29**/5.

Cassia fallacina var. *katangensis* Ghesq. in Bull. Jard. Bot. État **9**: 145 (1932).

Cassia katangensis (Ghesq.) Steyaert in Bull. Jard. Bot. État **20**: 258 (1950); in F.C.B. **3**: 529 (1952). —Brenan in F.T.E.A., Legum.-Caesalp.: 88 (1967).

Perennial herb producing annual upright shoots from a thin woody rootstock. Stems terete, thinly to densely pubescent with crisped hairs and often also longer straight hairs. Leaves paripinnate, 5–12 × 0.5–2.8 cm; stipules 6–12 × 0.7–2.3 mm, narrowly triangular, apex acuminate, base rounded to subcordate, margins and outer surface hairy; petiolar gland single, 0.4–0.7 mm in diameter, circular, stalk 0.3–0.5 mm long; rachis channelled above, ciliate on the channel margins, glabrous to hairy beneath; leaflets in 26–42 pairs, 15.5–24 × 4–6 mm, narrowly oblong, falcate, asymmetric, base strongly asymmetric, proximal side rounded but at least some leaflets usually with a distinct angular basal lobe, distal side narrowly cuneate, apex acute; margins ciliate, glabrous to hairy on both surfaces; midrib very excentric; 3–5 lateral nerves from base on proximal side, 0–1 on distal side. Inflorescences axillary to supra-axillary, 2–3-flowered; pedicels 1.5–2 cm long in flower, pubescent with curved appressed and/or spreading erect hairs. Sepals 5.5–6 × 1.5–3 mm, narrowly ovate, acuminate, pubescent with spreading hairs outside. Petals yellow. c. 8 × 6 mm, broadly obovate. Stamens 10, four longer than the others. Ovary c. 3 mm long, densely appressed-pubescent; style c. 2 mm long, curved, almost glabrous; stigma terminal, fringed with cilia. Pods 3.5–5.5 × 0.4–0.5 cm, sparsely appressed-pubescent. Seeds brown, 2.2–2.3 × 1.7–1.8 mm, rhombic-quadrate in outline.

Var. **katangensis**

Leaves relatively small, 5–7 cm long; leaflets usually less than 10 mm long; indumentum of curved appressed hairs, sometimes with a few longer erect hairs.

Zambia. N: Mbala Dist., Kambole, path to Katenga Falls, fl. & fr. 21.ii.1957, *Richards* 8296 (K). W: Mufulira, fl. 17.i.1948, *Cruse* 155 (K). **Malawi**. N: Rumphi Dist., Chisenga, fl. & fr. 17.ii.1961, *Richards* 14377 (K).

Cruse 155 was cited by Steyaert as a type of *Cassia* (*Chamaecrista*) *meelii*, but as Brenan has pointed out on a note on the sheet, it does not agree with the protologue and should be excluded as a type. The only specimen cited by both Ghesquière and Steyaert is *Corbisier* 577.

Var. **nuda** (Steyaert) Lock in Kew Bull. **43**: 337 (1988). Type from Congo.

Cassia katangensis var. *nuda* Steyaert in Bull. Jard. Bot. État. **20**: 258 (1950).

Leaves usually 8–12 cm long; leaflets 15–19 × 2–4 mm, usually almost glabrous.

Zambia. W: Misaka Forest Res., fr. 7.v.1969, *Mutimushi* 3054 (K, NDO). **Zimbabwe.** W: Hwange (Wankie), Gwai-Lutope Junction, fl. 26.ii.1968, *Wild* 6006 (SRGH). **Malawi.** N: Mzimba Dist., Champira, Katete Mission, fl. & fr. 24.iv.1974, *Pawek* 8501A (K, MAL, MO, SRGH).

Var. **A**

Indumentum of all parts rather dense, mainly of long erect hairs.

Zambia. C: 100–129 km E of Lusaka, Chakwenga headwaters, fl. & fr. 5.iii.1955, *E.A.Robinson* 6418 (K). N: top of Chikwalala Ridge, W aspect, fl. & fr. v.1955, *Lawton* 186 (K).

These are very hairy plants, arising from a short upright rootstock. Plants similar to these from Katanga (e.g. *Schmitz* 3435 (BR)) were named as *C. katangensis* by Steyaert so they fell within his concept of the species.

18. **Chamaecrista wittei** (Ghesq.) Lock in Kew Bull. **43**: 338 (1988); Leg. Afr. Check-list: 33 (1989). Type: Congo, Katanga, Kasiki, *de Witte* 443 (BR lectotype, BM). FIGURE 3.2.**29**/10.

Cassia wittei Ghesq. in Bull. Jard. Bot. État **9**: 154, fig.2/11–15 (1932). —Brenan, Check-list For. Trees Shrubs Tang. Terr.: 96 (1949); in F.T.E.A., Legum.-Caesalp.: 94 (1967). —Steyaert in F.C.B. **3**: 521 (1952). —Drummond in Kirkia **8**: 213 (1972).

Annual or short-lived perennial erect herb, often woody at the base, to 1.6 m tall. Stems terete, pubescent with crisped hairs and at least some longer spreading hairs. Leaves paripinnate, 6–9 × 1.5–2 cm; stipules 9–15 × 2–2.5 mm, narrowly triangular, apex acuminate, base rounded to subcordate, surfaces and margins glabrous; petiolar gland single, 0.5–0.6 mm in diameter, circular, sessile; rachis channelled above, crisped-pubescent on the channel margins and with scattered longer hairs beneath; leaflets in 20–43 pairs, 10.5–13 × 1.8–2.7 mm, narrowly oblong, weakly falcate, asymmetric, base asymmetric, proximal side rounded, distal side rounded to cuneate, apex mucronate, margins shortly ascending-ciliate, surfaces virtually glabrous; midrib slightly excentric; 3–4 lateral nerves from base on proximal side, 2–3 on distal side. Inflorescences supra-axillary, 3–4-flowered; pedicels 1.2–1.5 cm long in flower, 1.5–2 cm long in fruit, pubescent with crisped appressed and also with at least some longer spreading hairs. Sepals 10–12 × 4–6 mm, ovate or elliptic, acuminate, pubescent with spreading hairs outside. Petals yellow, c. 15 × 10–12 mm, broadly obovate to elliptic. Stamens 9, four longer than the others; a small staminode sometimes present. Ovary c. 8 mm long, densely appressed-pubescent; style c. 4 mm long, proximally appressed-pubescent, distally glabrous; stigma oblique. Pods 5–7 × 0.5–0.7 cm, sparsely spreading-pubescent. Seeds pale brown, 3.5–3.7 ×1.8–2.1 mm, oblong-elliptic, indistinctly pitted.

Zimbabwe. E: Nyanga (Inyanga), Pungwe Rest Huts, fl. 25.i.1951, *Chase* 3666 (K, SRGH). **Malawi.** N: Mzimba Dist., South Viphya Plateau, 11 km S of Chikangawa on link road to Luwawa, fl. & fr. 8.v.1970, *Brummitt* 10453 (K). C: Lilongwe Dist., Dzalanyama Forest Res., a few km S of ranch house, fl. 26.iii.1977, *Brummitt, Seyani & Patel* 14924 (K). S: Zomba Dist., Zomba Plateau below road to summit opposite Malosa saddle, fr. 2.viii.1970, *Brummitt & Banda* 12390 (K). **Mozambique.** MS: Border Farm (Quinta da Fronteira), fl. & imm.fr. 19.iii.1961, *Chase* 7444 (K, SRGH).

Also in Cameroon, Congo, Ethiopia, Uganda and Tanzania. Upland grassland; 1000–2000 m.

A distinctive large-flowered species. The indumentum is variable. Short crisped hairs are always present, and there are usually at least some longer spreading hairs. In some specimens these are numerous giving the young parts an almost furry appearance. They are often yellowish.

19. **Chamaecrista parva** (Steyaert) Lock in Kew Bull. **43**: 337 (1988); Leg. Afr. Check-list: 32 (1989). Type: Congo, Upemba Nat. Park, Lusinga, *de Witte* 2476 (BR holotype). FIGURE 3.2.**29**/3.

Cassia parva Steyaert in Bull. Jard. Bot. État **20**: 266 (1950); in F.C.B. **3**: 524 (1952). —Brenan in F.T.E.A., Legum.-Caesalp.: 88 (1967). —Drummond in Kirkia **8**: 212 (1972).

Perennial erect or ascending suffruticose herb from a woody rootstock. Stems generally annual, terete, pubescent with appressed curved hairs, with or without longer spreading hairs. Leaves paripinnate, 5–8.6 × 0.9–1.3 cm; stipules 6.5–11.5 × 1.5–2.2 mm, narrowly triangular, apex acuminate, base rounded, asymmetric, surfaces glabrous, margins ciliate; petiolar gland single, towards the apex, 0.9–2 × 0.4–1 mm, elliptic, saucer-shaped; occasionally a second, smaller gland nearer the petiole base; rachis channelled, the channel margins ciliate; leaflets in (16)20–24(31) pairs, 5.5–11.5 × 1.5–2.5 mm, narrowly oblong, curved, asymmetric, base asymmetric, proximal side rounded-auriculate, distal margin narrowly cuneate, apex acuminate-mucronate, margins ciliate, cilia ascending, surfaces glabrous to sparsely pubescent; midrib 0.5–0.8 mm from distal margin; 1–2 nerves from base on proximal side of midrib, 3–5 on distal side. Inflorescences axillary to supra-axillary, usually 2–3-flowered; pedicels 0.8–2.5 cm long in flower, similar in fruit, spreading-hairy. Sepals 7.5–9 × 3.2–4.8 mm, elliptic, acuminate, spreading-hairy outside. Petals yellow, 10–13 × 6.5–9 mm, broadly obovate. Stamens 10, 4 longer than the rest. Ovary appressed white-hairy; stigma terminal, fringed with hairs. Pods 4–5.7 × 0.4–0.5 cm, sparsely hairy; mature pods and seeds not seen.

Zambia. N: Mbala Dist., hill above Ndundu, fl. 11.iii.1962, *Richards* 16243 (K). W: Mwinilunga Dist., Kalene Hill, fl. 15.xii.1963, *Robinson* 6066 (K). E: Nyika, fl. 30.xii.1962, *Fanshawe* 7343 (K, NDO). **Zimbabwe.** N: Makonde Dist., Mtorashanga Impshi Mine road, fl. 18.ii.1969, *Corby* 2079 (K). C: Marondera Dist., fl. i.1931, *Rattray* 231 (K). E: Nyanga Dist., 4 km (2.5 miles) along Bonda road from Sanyatwe P.O., fl. 31.xii.1964, *Corby* 1206 (K). **Malawi.** N: Rumphi Dist., Nyika Plateau, Chelinda Camp, fl. 3.ii.1978, *Pawek* 13730 (K, MAL, MO). C: Dedza Dist., foot of Chongoni Mt., fl. 17.i.1959, *Robson & Jackson* 1236 (K). S: Zomba Mt., fl. & fr. 25.i.1959, *Robson* 1312 (K).

Also in Kenya, Tanzania and Congo. Upland grassland and wooded grasslands; 1450–2400 m.

Chamaecrista parva and *C. katangensis* form a species pair. Both are generally small suffructices from a slender woody rootstock. The main distinction lies in the petiolar gland — large, sessile and saucer-shaped in *C. parva* and shortly stalked in *C. katangensis*. In both species there are tall vigorous forms (distinguished as var. *nuda* in *C. katangensis*). It is tempting to consider them as forms of a single taxon but for the time being I prefer to keep them separate. Brenan (in F.T.E.A) remarks on the similarity between the petiolar glands of *C. parva* and *C. comosa* but the latter (as I understand it in the Flora area) is generally a very much more vigorous plant, always densely hairy, and tending to have the flowers concentrated towards the apices of the stems.

20. **Chamaecrista comosa** E. Mey., Comment. Pl. Afr. Austr.:160 (1836). —Lock, Leg. Afr. Check-list: 30 (1989). —van Wyk & Malan, Wild Fl. Highveld, ed.2: 122 (1997). Type from South Africa, Eastern Cape Province between Umzimvubu and Umsikaba rivers, *Drège* s.n. (?B† holotype). FIGURE 3.2.**29**/7.

Cassia comosa (E. Mey.) Vogel, Syn. Gen. Cass.: 65 (1837). —Brenan in F.T.E.A., Legum.-Caesalp.: 89 (1967). —Ross, Fl. Natal: 195 (1973). —Gordon-Gray in F.S.A. **16**(2): 97 (1977).

Cassia comosa var. *lanata* Steyaert in Bull. Jard. Bot. État **20**: 252 (1950); in F.C.B. **3**: 525 (1952). Type from South Africa.

Perennial herb with usually annual upright shoots from a woody vertical rootstock. Stems terete, densely pubescent with spreading hairs. Leaves paripinnate, 9.5–14 × 2–2.5 cm; stipules 8–13 × 2–4 mm, narrowly triangular, apex acuminate, base rounded to subcordate, margins and outer surface hairy; petiolar gland single or occasionally paired, 2.3–2.5 × 1–1.2 mm, elliptic, sessile; rachis channelled above, ciliate on the channel margins and hairy beneath; leaflets in 23–25 pairs, 15.5–24 × 4–6 mm, narrowly oblong, weakly falcate, asymmetric, base asymmetric, proximal side rounded, distal side rounded-cuneate; apex mucronate, margins hairy, scattered-hairy beneath; midrib excentric; 4–5 lateral nerves from base on proximal side, 2–3 on distal side. Inflorescences supra-axillary, 4–6-flowered, tending to be aggregated towards the shoot apices; pedicels 1.2–1.5 cm long in flower, 1.5–2 cm long in fruit, pubescent with spreading hairs. Sepals 10–12 × 4–6 mm, ovate, acuminate, pubescent with spreading hairs outside. Petals yellow, c. 13 × 6–9 mm, broadly obovate to elliptic. Stamens 9, four longer than the others; a small staminode sometimes present. Ovary c. 9 mm long, densely appressed-pubescent; style c. 6 mm long, proximally appressed-pubescent, distally glabrous; stigma oblique. Pods (immature) 6–6.5 × 0.5 cm, spreading pubescent. Mature seeds not seen.

Malawi. C: Lilongwe Dist., Dzalanyama Forest Res., a few km S of ranch house, fl. & imm.fr. 26.iii.1977, *Brummitt, Seyani & Patel* 14923 (K). S: Blantyre township area, fl. 3.iii.1948, *Faulkner* Kew series 214 (K). **Mozambique.** Z: flanks of Milange Mts, fl. 24.ii.1943, *Torre* 4827 (K, LISC).

Also in South Africa (KwaZulu-Natal). The typical variety occurs in South Africa, and the var. *capricornia* Steyaert in Congo, Tanzania and South Africa. Upland grasslands and open woodlands; up to 1600 m.

Differs from *C. wittei* in the perennial habit, the larger leaves and leaflets, the spreading (not crisped) indumentum and in the hairy margins to the stipules.

21. **Chamaecrista kirkii** (Oliv.) Standl. in Smithsonian Misc. Coll. 68(5): 5 (1917). — Lock, Leg. Afr. Check-list: 31 (1989). Type: Malawi, Manganja Hills, *Kirk* s.n. (K holotype).

Annual erect herb, sometimes becoming woody at the base. Stems terete with slight pale ridges decurrent from the stipule bases, with both short curved appressed and longer spreading hairs. Leaves paripinnate, 8.5–12 × 2–2 cm; stipules 8.5–12 × 1.2–2.1 mm, narrowly triangular, apex long-acuminate, base truncate, asymmetric, surfaces sparsely hairy, margins ciliate; petiolar gland single, dark, 0.9–1.5 × 0.5–0.9 mm, elliptic, sessile, saucer-shaped; rachis channelled, pilose; leaflets in 24–38 pairs, 8–19 × 1.8–4.5 mm, narrowly ovate-oblong, asymmetrical, base asymmetrically truncate, apex acuminate-mucronate, margins ciliate, cilia ascending, surfaces glabrous to pubescent; midrib 0.8 mm from distal margin, 1–2 lateral nerves from base on proximal side of midrib, 4–5 on distal side. Inflorescences supra-axillary, usually 4-flowered; peduncle up to 1 cm long, bracteate; pedicels 1.5–2.5 cm long in flower, slightly longer in fruit. Flowers large; sepals narrowly ovate, spreading-pubescent outside; petals yellow, 10–14 × 6–11 mm. Stamens 10. Ovary densely appressed-pilose; style sparsely pilose; stigma terminal, fringed with hairs. Pods 8–9 × 0.5–0.6 cm, sparsely appressed-pubescent. Seeds yellow-brown, c. 3 × 2 mm, rhombic-quadrate in outline, flattened, lightly pitted.

Three varieties have been distinguished on the basis of stem pubescence. The typical variety is the commonest in the Flora area. The specimen here called var. *guineensis* has a few long spreading hairs among the short ones on the stem and is somewhat intermediate. A third variety, var. *glabra* Steyaert, occurs in Uganda and eastern Congo.

Chamaecrista kirkii is most likely to be confused with *C. comosa*, but the latter is perennial, the inflorescences tend to be concentrated towards the end of the stems, in the axils of reduced leaves, and the petiolar gland is larger (2.3–2.5 mm long).

Var. **kirkii**. FIGURES 3.2.**29**/12 & 3.2.**31**.

Cassia kirkii Oliv. in F.T.A. **2**: 281 (1871). —Taubert in Engler, Pflanzenw. Ost-Afrikas **C**: 201 (1895). —Harms in Engler, Pflanzenw. Afrikas **3**(1): 500 (1915). —Baker, Legum. Trop. Afr.: 642 (1930). —Brenan, Check-list For. Trees Shrubs Tang. Terr.: 96 (1949). —

Steyaert in F.C.B. **3**: 525, fig.38 (1952). —Mendonça & Torre in C.F.A. **2**: 183 (1956). —Keay in F.W.T.A., ed. 2, **1**: 452 (1958). —Brenan in F.T.E.A., Legum.-Caesalp.: 92, fig.16 (1967). —Drummond in Kirkia **8**: 212 (1972).

Stems with spreading hairs and usually also densely appressed-pubescent; leaflets pubescent or not.

Zambia. N: Mbala Dist., M'bulu R., fl. 2.ii.1962, *Richards* 15983 (K). **Zimbabwe.** E: Nyanga (Inyanga), near Nyamingura R., fl. & fr. 24.iv.1958, *Phipps* 1237B (K, SRGH). **Malawi.** N: Nkhata Bay Dist., 8 km (5 miles) E of Mzuzu, fl. & fr. 7.viii.1973, *Pawek* 7311 (K). S: Manganja Hills, *Kirk* s.n. (K).

Widespread in tropical Africa from Nigeria through East Africa south to Zimbabwe. Usually in wet places in woodland and grassland, often by rivers or pools. The most vigorous plants come from the mud of pool and river margins.

Var. **guineensis** (Steyaert) Lock in Kew Bull. **43**: 337 (1988). Type: Congo, Yahila, *Louis* 11215 (BR holotype; K).

Cassia kirkii var. *guineensis* Steyaert in Bull. Jard. Bot. État. **20**: 256, fig.9 (1950); in F.C.B. **3**: 525, fig.38 (1952). —Brenan in F.T.E.A., Legum.-Caesalp.: 92 (1967).

Stems pubescent with short curved hairs; leaflets glabrous to sparsely pubescent.

Malawi. S: Blantyre Dist., 2 km N of Limbe, Upper Hynde Dam, fl. & fr. 14.ii.1970, *Brummitt* 8552 (K).

Tropical Africa from Sierra Leone and Cameroon to East Africa, south to Malawi. Habitat as the typical variety.

26. SENNA Mill.

Senna Mill., Gard. Dict., abr. ed. 4, **3** (1754). —Irwin & Barneby in Polhill & Raven, Adv. Legume Syst. **1**: 97–106 (1981); in Mem. New York Bot. Gard. **35**: 1–918 (1982).

Trees, shrubs or woody herbs, unarmed (in the Flora area). Leaves paripinnate; leaflets opposite or subopposite; stipules various but often persistent and prominent. Inflorescence racemose or paniculate; bracts often prominent but usually falling at or before anthesis; bracteoles absent. Calyx lobes 5, equal or subequal. Petals 5, yellow. Stamens 10, all with almost straight filaments which are not more than twice the length of their anther; anthers basifixed, dehiscing by terminal pores only. Ovary stipitate, curved; ovules numerous; stigma terminal or subterminal. Pods cylindrical, flattened or winged, often transversely septate inside, tardily dehiscent (not elastically) or indehiscent. Seeds numerous, areolate.

A tropical genus with about 250 species in the tropics of the Old and New World, most numerous and diverse in the latter, mainly in tropical seasonal vegetation but also in forest. A number of species are widely cultivated for their attractive flowers. Until the work of Irwin & Barneby (1981) *Senna* was placed, with *Chamaecrista* and *Cassia* sensu stricto, in *Cassia* sensu lato. None of the species in the genus is known to form root nodules (cf. *Chamaecrista* — see J. Sprent (2001), Nodulation in Legumes, RBG Kew).

The majority of the species recorded from the Flora area are either pantropical weeds or cultivated ornamentals. Only *Senna italica, S. petersiana, S. singueana* and perhaps *S. didymobotrya* are indigenous. All species of which material has been seen are included in the key but the decision as to which should receive a full account and which should be mentioned in passing is somewhat arbitrary. As a rule, species which appear to be either established in the wild or widely and prominently planted, have been given full accounts while others receive only a mention.

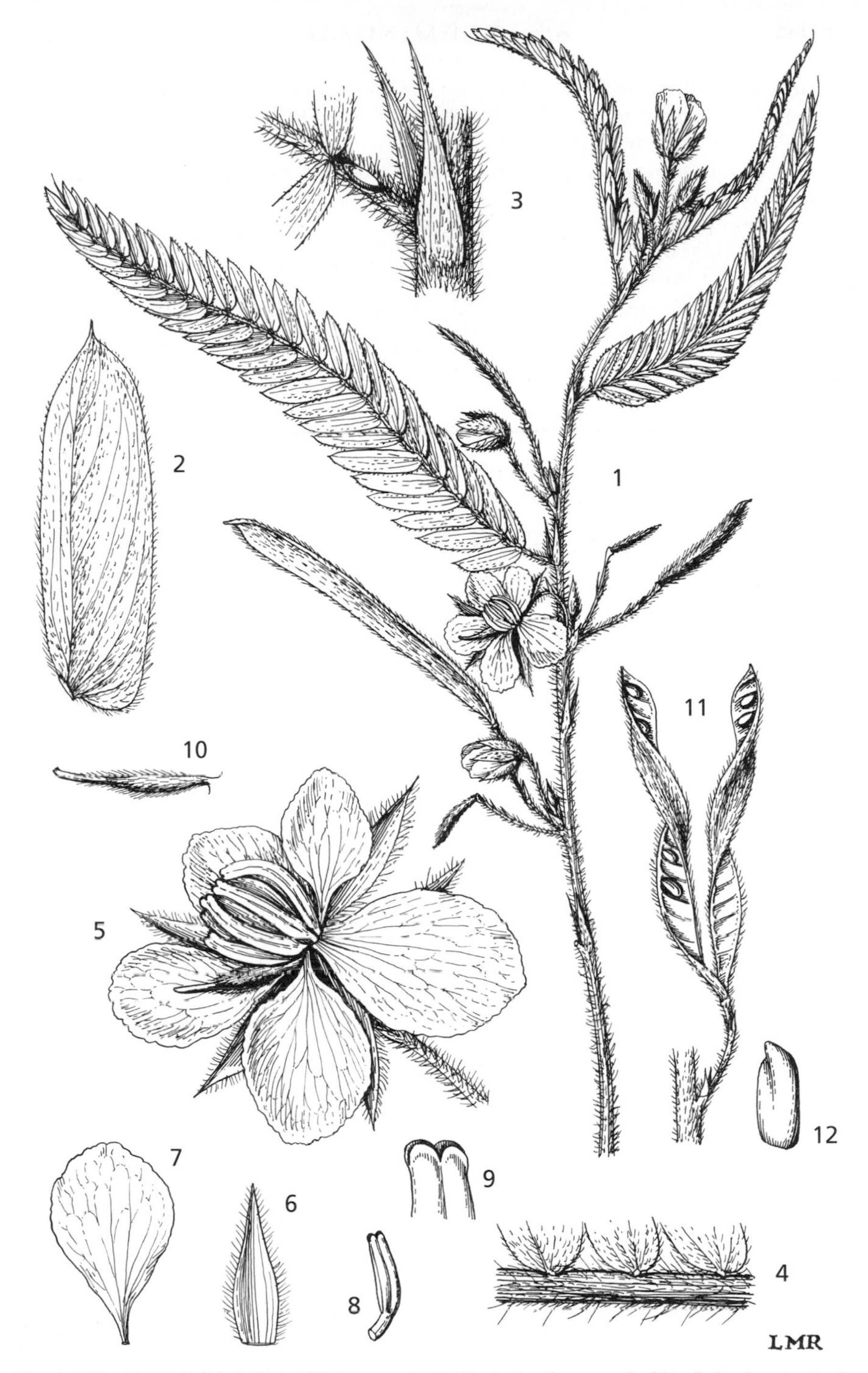

Fig. 3.2.**31**. CHAMAECRISTA KIRKII var. KIRKII. 1, fertile stem (× $^{2}/_{3}$); 2, leaflet (× 4); 3, petiolar gland (× 3); 4, part of leaf rachis, viewed from side (× 4), 1–4 from *Milne-Redhead & Taylor* 9637; 5, flower (× 2); 6, sepal (× 2); 7, petal (× 2); 8, stamen (× 2); 9, tip of anther (× 8); 10, gynoecium (× 2), 5–10 from *Rounce* 27; 11, pod (× $^{2}/_{3}$); 12, seed (× 4), 11 & 12 from *Drummond & Hemsley* 1492. Drawn by Lura Ripley. From F.T.E.A.

Senna alexandrina Mill. (*Cassia senna* L.; *Cassia angustifolia* Vahl; *Cassia acutifolia* Delile) is represented by a single collection from the Flora area. **Mozambique**. T: 'in the streets and rubbish heaps near Tette', fl. & fr. xi.1858, *Kirk* s.n. (K). The specimen belongs to var. *alexandrina*, which is the widely-used medicinal plant, and it seems certain from the habitat data that these plants arose from discarded seeds or pods. There is no evidence that the species has become permanently established. A note by Brenan at Kew mentions one other specimen that I have not seen — **Mozambique**. N: Niassa, Ilha do Ibo, 5.ix.1968, *Pedro & Pedrogão* 5044 (EA). *Senna alexandrina* differs from *S. italica* in the leaflets, which are narrowly elliptic or narrowly ovate with apiculate tips, and in the fruits, which are flattened, almost straight, larger than those of *S. italica* (c. 5 × 1.5 cm), and smooth.

Senna artemisioides (DC.) B.R. Randell (*Cassia artemisioides* DC.) is an Australian species which has been cultivated in the Flora area. **Zimbabwe**. C: Harare (Salisbury) Municipal Nursery, fl. 13.viii.1965, *Müller* 165 (K, SRGH). It is a low shrub with 4–5 pairs of linear leaflets. The whole plant is covered with appressed white hairs and there are prominent tufts of black hairs on the leaf rachis at the leaflet insertions. The flowers are yellow.

Senna multijuga (Rich.) H.S. Irwin & Barneby (*Cassia multijuga* Rich.), a native of South America, is cultivated in gardens in **Zimbabwe**, e.g. C: Harare (Salisbury), fl. 5.ii.1954, *Hughes* in SRGH 45375 (K, PRE, SRGH); E: Clogheen, Mutare (Umtali), fl. & fr. 26.ii.1958, *Pole-Evans* 5368 (K). It is a tree 10–15 m high with pinnate leaves with 12–16 pairs of oblong to obovate leaflets and large panicles of yellow flowers. The retuse (not acute) apices of the leaflets and the flattened (not terete) pods distinguish it from *Senna spectabilis*. All material seen appears to belong to the subsp. *multijuga*, the most widely cultivated of the subspecies.

Senna pendula (Willd.) H.S. Irwin & Barneby in Mem. New York. Bot. Gard. **35**: 378 (1982) (*Cassia pendula* Willd., *Cassia coluteoides* Collad.), a native of tropical South America, is recorded from the Flora area as a garden plant. It is a very variable taxon; Irwin & Barneby (1982) describe 19 varieties. It resembles *Senna bicapsularis* but has larger leaves with 3–5 pairs of leaflets, with glands between each pair of leaflets. It is also very similar to *S. surattensis* but that species has 10 fertile stamens. Specimens from Zimbabwe (e.g. *Wood* in SRGH 56270) have a densely woolly ovary and appear to belong to the var. *advena* (Vogel) H.S. Irwin & Barneby; *Biegel* 2479 (K, SRGH) from Mutare is similar and appears to have been collected outside a garden. *Balsinhas* 1846, from Maputo, Jardim Tunduru (Jardim Vasco da Gama), has an almost glabrous ovary and rather smaller leaflets. It is not clear to which of the varieties it should be attributed; indeed, it is doubtful whether cultivated plants far from their origin can necessarily be named to variety.

Senna sophera (L.) Roxb. (*Cassia sophera* L.) is very similar to *S. occidentalis*. It was described from Sri Lanka but is also widespread in the Neotropics, and its place of origin is uncertain although most likely to be in the Old World. No specimens have been seen from the Flora area but it may well occur. *S. sophera* has 4–10 pairs of leaflets (not 4–5 as in *S. occidentalis*), peduncles 0.8–2.5 cm (not 0.3–0.5 cm) long, and obtuse to subacute (not acute) bracts. It is usually weedy like *S. occidentalis* but has been tried as a cover crop in Tanzania.

Senna surattensis (Burm. f.) H.S. Irwin & Barneby (*Cassia surattensis* Burm. f.) has been recorded from South Africa as an escape from cultivation. One specimen from the Flora area might represent this taxon but is sterile. **Zimbabwe**. C: Marondera Dist., Marondera (Marandellas), near km peg 79.0 on Mutare (Umtali) Road, 7.xii.1977, *Biegel* 5564 (K, SRGH). *S. surattensis* is very similar to *S. bicapsularis* but differs in having 10, not 7, fertile stamens, and in its linear subpersistent stipules, pubescent petiolules, larger leaves with at least 4 pairs of leaflets, and in possessing glands between all leaflet pairs except the terminal one. Forms of *Senna pendula* are also similar but have only 7 fertile stamens and, in the most commonly cultivated form, a densely woolly ovary.

Key to Senna *in the Flora Zambesiaca area.*

All indigenous and several cultivated and/or naturalised species are included.

The key relies to a substantial extent on vegetative characters. The only *Senna* in the area that regularly flowers when leafless is *S. singueana*. However, beware of *Cassia abbreviata* and *Cassia angolensis*, both of which regularly flower when leafless.

1. Small cultivated shrub; leaflets linear, 1–2 mm wide, silvery-sericeous .... *artemisioides*
– Shrubs, trees or herbs; leaflets more than 5 mm wide, not silvery-sericeous .. 2
2. Shrubs or herbs; bracts of young inflorescence completely enclosing the young buds, broadly ovate, overlapping, giving the young inflorescence a cone-like appearance; stipules persistent at least for a few nodes below the shoot apex ... 3
– Trees, shrubs, or herbs; bracts of young inflorescence not as above; stipules, if persistent, either linear or cordate-auriculate 5
3. Herb; stems often prostrate and not woody except at the extreme base; pod flattened, curved, less than three times as long as broad .... **9.** *italica*
– Shrubs; stems erect, woody below; pods either 4-winged or flat, more than four times as long as broad 4
4. Leaflets oblong-elliptic, not more than 6.5 cm long and usually much less; flowers bright yellow with darker veins; pods not winged ...... **3.** *didymobotrya*
– Leaflets oblong, usually much more than 5 cm long; flowers deep yellow or orange; pods 4-winged .... **4.** *alata*
5. Leaflets acute to acuminate at the apex 6
– Leaflets obtuse to rounded or retuse at the apex (often mucronate) 13
6. Petiole with a cone-like or domed gland on the petiole near the base; rachis eglandular 7
– Petiole eglandular; rachis with glands between the junction of all or lower leaflets or eglandular 9
7. Stem and leaves long hairy (hairs visible to the naked eye) .... **6.** *hirsuta*
– Stem and leaves glabrous to inconspicuously puberulous 8
8. Leaflets 4–5 pairs; peduncles 0.3–0.5 cm long; bracts acute ...... **5.** *occidentalis*
– Leaflets 4–10 pairs; peduncles 0.8–2.5 cm long; bracts obtuse to subacute ..... *sophera*
9. Leaflets 6–12 pairs, pubescent at least beneath 10
– Leaflets 3–8 pairs, glabrous 11
10. Stipules linear, caducous; glands lacking; ovary glabrous; pods cylindrical ..... **7.** *spectabilis*
– Stipules prominent, persistent, cordate-auriculate to reniform; glands usually visible between each of proximal 1–4 pairs of leaflets though caducous; ovary hairy; pods laterally compressed .... **11.** *petersiana*
11. Leaf rachis eglandular; leaflets narrowly ovate to narrowly elliptic, at least four times as long as wide; rare weed .... *alexandrina*
– At least one gland on the rachis between the first pair of leaflets; leaflets relatively broader 12
12. Rachis with a gland between the proximal petiolule pair only; leaflets usually obtuse to rounded at the apex, only occasionally acute .... **8.** *bicapsularis*
– Rachis with a gland between each petiolule pair; leaflets acute to acuminate at the tip .... **10.** *septemtrionalis*

13. Tree; leaflets in 16–40 pairs; a botuliform gland present between each of at least the lowest two pairs of leaflets . . . . *multijuga*
– Herb, shrub or tree; leaflets in no more than 12 pairs 14
14. Trees; leaves almost always with more than 4 pairs of leaflets 15
– Herbs or shrubs; leaves with up to 4 pairs of leaflets 16*
15. Leaves 6–12-jugate; flowers dull yellow; pods flattened, irregularly twisted when ripe; fertile anthers 7, all of similar size but two with longer filaments than the rest; planted tree . . . . **1.** *siamea*
– Leaves 4–9-jugate; flowers bright golden yellow; pods weakly flattened, remaining straight when ripe; fertile anthers 7, three with anthers much larger than the rest; wild tree . . . . **12.** *singueana*
16. Inflorescences 1–3-flowered; pods linear, usually falcately curved, 0.5–0.7 cm wide . . . . **2.** *obtusifolia*
– Inflorescences more than 3-flowered; pods narrowly cylindrical, straight, 1–1.5 cm wide . . . . **8.** *bicapsularis*

1. **Senna siamea** (Lam.) H.S. Irwin & Barneby in Mem. New York. Bot. Gard. **35**: 98 (1982). —Lock, Leg. Afr. Check-list: 39 (1989). Type: cultivated plant in Réunion under the name of 'Siamois', *Commerson* s.n. (P-LA holotype).

Cassia siamea Lam., Encycl. Méth. Bot. **1**(2): 648 (1785). —Harms in Engler, Pflanzenw. Afrikas **3**(1): 497 (1915). —Baker, Legum. Trop. Afr.: 639 (1930). —Brenan, Check-list For. Trees Shrubs Tang. Terr.: 98 (1949). —Steyaert in F.C.B. **3**: 504, fig.36 (1952). —Mendonça & Torre in C.F.A. **2**: 178 (1956). —Keay in F.W.T.A., ed. 2, **1**: 453 (1958). —White, F.F.N.R.: 120 (1962). —Gordon-Gray in F.S.A. **16**(2): 77 (1977).

Tree to 20 m or more; bark smooth, grey. Branchlets dark brown, sparsely puberulous, somewhat ridged longitudinally. Leaves paripinnate; stipules not seen (minute, subulate according to Brenan), presumably early caducous; petiole 2–4.2 cm long, pubescent; rachis 7.5–23 cm long, pubescent; leaflets in 6–12 pairs, 2.4–6.8 × 1.1–2.9 cm, elliptic to oblong, base rounded to cuneate, apex obtuse to retuse, shortly mucronate, sparsely appressed-pubescent beneath, marginal nerve prominent; petiolules 2–3 mm long. Inflorescences terminal, paniculate; bracts 3–4 mm long, linear, pubescent, present at anthesis; pedicels 2.5–3.5 cm long at anthesis. Sepals 5, dimorphic, two c. 5 × 4 mm, three c. 10 × 8 mm, broadly ovate, glabrous to sparsely puberulous. Petals yellow, c. 12 × 10 mm, broadly obovate, claw 2 mm long. Stamens 10; seven fertile with anthers 4–5 mm long, two with filaments c. 10 mm long, five with filaments c. 5 mm long; three probably infertile with reduced anthers and short filaments. Ovary c. 5 mm long, puberulous; style 2–3 mm long, glabrous. Pod 15–23 × 0.9–1.3 cm, linear, glabrous, tardily dehiscent along both sutures. Seeds dark brown, c. 7 × 6 mm, discoid, flat.

Zambia. B: Mongu, Barotse National School, planted, fl. vii.1954, *Cooling* 84 (FHO). C: Luangwa Dist., Katondwe Mission, fl. 31.xii.1972, *Strid* 2741 (FHO). S: Livingstone Dist., Katambora, fl. & fr. 15.iii.1952, *White* 2272 (FHO, K). **Zimbabwe.** W: Hwange Dist., Victoria Falls, below Falls Hotel, fl. 7.ii.1977, *Moyo* 13 (K, SRGH). **Malawi.** N: Kisitu Bridge, Rumphi–Mzuzu road, fl. 12.ii.1976, *E. Phillips* 1173 (K, MO). S: Mangochi Dist., Monkey Bay, grounds of Marina Beach Hotel, fl. 2.iii.1970, *Brummitt* 8836 (K). **Mozambique.** N: Niassa, 37 km (23 miles) E of Cuamba (Nova Freixo), fl. & fr. 18.v.1961, *Leach & Rutherford-Smith* 10870 (K, LISC, SRGH). M: Magude, fl. 17.vii.1946, *Gerstner* 6647 (K, LISC, PRE).

Originally from SE Asia, now widely planted and sometimes naturalised.

Lamarck wrote that his plant was said by Commerson to be cultivated in Réunion

* If a cultivated shrub resembling *S. bicapsularis*, but with rather more leaflets, see also *S. pendula* and *S. surattensis* noted above.

"sous le nom 'Siamois'". The sheets in the Lamarck Herbarium have no locality or collector but are assumed to be the type material. Larsen & Larsen (Flore du Laos, Cambodge et du Viet-Nam) state that the type material was collected in India but do not give their reasons.

Very widely planted for poles and firewood; coppices well after cutting and regrows very rapidly. Certainly much more widespread than the available specimens would suggest. The inflorescences are rather diffuse and not particularly showy, and the old pods remain on the tree giving it a rather untidy appearance.

2. **Senna obtusifolia** (L.) H.S. Irwin & Barneby in Mem. New York Bot. Gard. **35**: 252 (1982). —Lock, Leg. Afr. Check-list: 38 (1989). Type a cultivated plant, grown from seed from Cuba, near Havana, *Herb. Dillenius* (OXF).

Cassia obtusifolia L., Sp. Pl.: 377 (1753). —Klotzsch in Peters, Naturw. Reise Mossambique **6**(1): 13 (1861). —Brenan in Kew Bull. **13**: 248–252 (1958); in F.T.E.A., Legum.-Caesalp.: 77 (1967). —Schreiber in Merxmüller, Prod. Fl. SW Afrika, fam. 59: 11 (1967). —Drummond in Kirkia **10**: 243 (1977).

Cassia tora sensu auct. mult. non L., e.g. Oliver in F.T.A. **2**: 275 (1871). —Harms in Engler, Pflanzenw. Afrikas **3**(1): 497 (1915). —Baker, Legum. Trop. Afr.: 636 (1930). —Steyaert in F.C.B. **3**: 512 (1952). —Mendonça & Torre in C.F.A. **2**: 180 (1956). —Keay in F.W.T.A., ed.2, **1**: 455 (1958). —White, F.F.N.R.: 120 (1962). —Gordon-Gray in F.S.A. **16**(2): 94 (1977).

Annual or short-lived perennial shrub or woody herb 0.5–2 m tall. Branchlets pale grey-brown, glabrous, longitudinally ridged when young, later terete. Leaves paripinnate; stipules 7–9 mm long, linear, with ciliate margins, caducous; petiole 2.5–3.9 cm long, sparsely pubescent; rachis 1.5–2.6 cm long, sparsely pubescent, glabrescent, with a terete reddish gland 1–1.5 mm long between the proximal leaflet pair and sometimes a similar but smaller gland between the middle leaflet pair; leaflets in (2)3 pairs, 2.9–5.7 × 1.8–2.9 cm (smaller in stunted or dry-season material), the distal largest, broadly elliptic to broadly obovate, base cuneate or rounded, slightly asymmetric, apex obtuse to rounded or retuse, often mucronate, glabrous except for the pubescent margin and scattered hairs on the midrib beneath; petiolules 1–2 mm long, pubescent. Inflorescences axillary, 1–2-flowered; bracts 3–4 mm long, linear, with ciliate margins, caducous; peduncle very short; pedicels 1.2–2 cm long at anthesis. Hypanthium with reddish glands. Sepals 5, two c. 8 × 5 mm, three 5–6 × 3–4 mm, elliptic to obovate, glabrous except for the ciliate margins which also bear scattered reddish glands. Petals orange to yellow, 14–16 × 7–9 mm, including a claw c. 2 mm long, obovate, slightly unequal. Stamens 10; three fertile with anthers 4–5 mm long and filaments 3–4 mm long, four fertile with anthers c. 2 mm long and filaments c. 2 mm long; two trumpet-shaped staminodes 2–3 mm long and one linear staminode c. 2 mm long. Ovary c. 15 mm long, terete, appressed-pubescent; style 2–2.5 mm long, glabrous; stigma terminal, oblique, trumpet-shaped. Pod 13–20 × 0.5–0.7 cm, linear, falcately curved or straight, elliptic in section, acuminate at apex, tardily dehiscent along both sutures. Seeds olive-brown, c. 5 × 2–3 mm, rhomboid, with a linear areole on each broad face.

Botswana. N: Linyanti R. floodplain at Hyaena Camp, fl. & imm.fr. 21.x.1978, *P.A. Smith* 2502 (K, SRGH). **Zambia.** B: Nangweshi, fl. & imm.fr. 19.vii.1952, *Codd* 7123 (K, PRE). N: Chiengi Dist., Luchinda R. Bridge, fr. 25.v.1961, *Astle* 676 (K). C: 38 km (24 miles) W of Lusaka on old Mumbwa Road, fl. & fr. 17.iv.1962, *Angus* 3128 (FHO, K). E: Chipata (Fort Jameson) Dist., Luangwa Game Reserve, Big Lagoon Camp, fl. & fr. 10.x.1960, *Richards* 13340 (K). S: Mapanza Mission, fl. & imm.fr. 22.iii.1953, *Robinson* 142 (K). **Zimbabwe.** N: Hurungwe Dist., Msuku R., fr. iv.1956, *Davies* 1896 (K, SRGH). W: Hwange Dist., Matetsi Safari Area, fl. 26.iv.1979, *Gonde* 240 (K, SRGH). C: Chegutu Dist., Poole, fl. 3.iv.1946, *Wild* 1007 (K, SRGH). E: Chimanimani Dist., Umvumvumvu River Gorge, fl. & fr. 21.iv.1963, *Chase* 7998 (K, LISC, SRGH). **Malawi.** N: Nkhata Bay Dist., Chinteche, lakeshore, fl. & imm.fr. 11.v.1970, *Brummitt* 10612

(K). C: Lilongwe Dist., 7 km NW of Malingunde, 15 km SW of Lilongwe, fl. & imm.fr. 25.iii.1977, *Brummitt, Seyani & Patel* 14921 (K). S: Chikwawa Dist., W. bank of Shire R. at Kasisi, 6 km N of Chikwawa, fl.& imm.fr. 21.iv.1970, *Brummitt* 10023 (K). **Mozambique.** N: Nampula, fl.& imm.fr. 3.iv.1937, *Torre* 1250 (COI, LISC). T: Estima acampamento do G.P.Z., fl. 27.iii.1972, *Macêdo* 5105 (K, LISC). MS: Buzi R., fr. iv.1962, *Goldsmith* 130/62 (COI, K, LISC, SRGH). M: Maputo (Lourenço Marques), Xinavane, near Palmeira, fl. 31.iii.1959, *Barbosa & Lemos* 8416 (COI, K, LISC).

A pantropical weed, probably of New World origin, widespread in tropical Africa. Usually in grasslands and bushlands, particularly in disturbed, trampled or overgrazed sites; up to 1200 m.

Conservation notes: A very widespread taxon; Lower Risk, Least Concern.

The distinction between this species and *Senna tora* (L.) Roxb. has been discussed at length by Brenan (Kew Bull. **13**: 248, 1958). *S. tora* has shorter pedicels (0.5–1 cm in flower and less than 1.5 cm in fruit), no abrupt constriction below the apex of the largest anthers, and seed areoles which are 1.5–2 mm wide, not linear. *S. tora* in this strict sense occurs mainly in SE Asia and I have not seen any material from the Flora area.

3. **Senna didymobotrya** (Fresen.) H.S. Irwin & Barneby in Mem. New York Bot. Gard. **35**: 467 (1982). —Lock, Leg. Afr. Check-list: 37 (1989). —Pooley, Trees Natal: 154 (1993). Type: Ethiopia, *Rueppell* s.n. (FR holotype). FIGURE 3.2.**32**.

Cassia didymobotrya Fresen. in Flora **22**: 53 (1839). —Oliver in F.T.A. **2**: 276 (1871). —Taubert in Engler, Pflanzenw. Ost-Afrikas **C**: 201 (1895). —Harms in Engler, Pflanzenw. Afrikas **3**(1): 498 (1915). —Baker in J. Linn. Soc., Bot. **40**: 62 (1911); Legum. Trop. Afr.: 638 (1930). —Brenan, Check-list For. Trees Shrubs Tang. Terr.: 97 (1949); in F.T.E.A., Legum.-Caesalp.: 66, fig.12 (1967). —Steyaert in F.C.B. **3**: 504, fig.36 (1952). —Mendonça & Torre in C.F.A. **2**: 177 (1956). —White, F.F.N.R.: 120 (1962). —Drummond in Kirkia **10**: 243 (1977). —Ross, Fl. Natal: 195 (1973). —Gordon-Gray in F.S.A. **16**(2): 79 (1977).

Cassia verdickii De Wild. in Ann. Mus. Congo Belge, Bot. **1**: 49, t.16 fig.6–11 (1902). —Harms in Engler, Pflanzenw. Afrikas **3**(1): 498 (1915). —Baker, Legum. Trop. Afr.: 638 (1930). Type from Congo.

Rather soft-wooded shrub, 2–4 m or more tall, unpleasant-smelling when fresh. Branchlets dark brown, sparsely pubescent. Leaves paripinnate; stipules 10–15 × 8–11 mm, broadly ovate, cordate at base, acuminate at apex, persistent; petioles 1.7–3.5(8.5) cm long, pubescent; rachis 10–25 cm long, pubescent; leaflets in 6–16 pairs, opposite, 21–55 × 6–20 mm, oblong, asymmetrically cuneate at base, acute and mucronate at apex, pubescent above and beneath. Inflorescence an axillary raceme, axis 22–34 cm long; bracts present in bud, falling before anthesis, 10–16 × 8–13 mm, broadly ovate, base cordate, apex acute, apiculate. Sepals 5, c. 13 × 5 mm, elliptic. Petals 5, yellow, with prominent darker veins particularly when dry, c. 20 × 10 mm, broadly elliptic to obovate. Stamens 10, seven of them probably fertile, two of these with filaments c. 2 mm long and anthers c. 10 mm long, five with filaments c. 2 mm long and anthers c. 5 mm long; three probably infertile with filaments c. 2 mm long and reduced anthers. Ovary c. 12 mm long, densely pubescent; stipe c. 5 mm long; style c. 12 mm long, glabrous. Pods blackish, 8–10.4 × 1.8–2.4 cm, oblong, corrugated on the flat surface, very tardily dehiscent along both sutures. Seeds dull brown, 6.5–7.5 × 3.5–5 mm, oblong, flattened, apiculate at the proximal end, with a narrowly obovate to oblong areole in the centre of each face.

Zambia. B: Barotse Plain, planted for ornament, fl. & fr. vi.1933, *Trapnell* 1297 (K). N: Mbala Dist., Vomo Gap, Fwambo side, fl. & fr. 27.ix.1960, *Richards* 13290 (K). W: Kitwe, fl. & fr. 5.vi.1967, *Fanshawe* 10097 (K). C: Lusaka, Leopard Hills Road, fl. & fr. 28.iv.1957, *Angus* 1562 (FHO, K). **Zimbabwe.** C: Chinamora Reserve, NNE of Harare (Salisbury), fl. 18.vi.1958, *Leach* 8245 (K, SRGH). E: St Trias Hill Mission, fl. ii.1917, *Mundy* 3161 (K). S: Bikita Dist., 73.5 km (46 miles) E of Masvingo (Fort Victoria), fl. 24.v.1959, *Leach* 9013 (K). **Malawi.** C: Dedza Dist., 5 km E of Dedza on road to Ncheu, fl. & fr. 1.iv.1970, *Brummitt* 9585 (K). S: Chiradzulu Dist., 3–5 km from Chiradzulu on

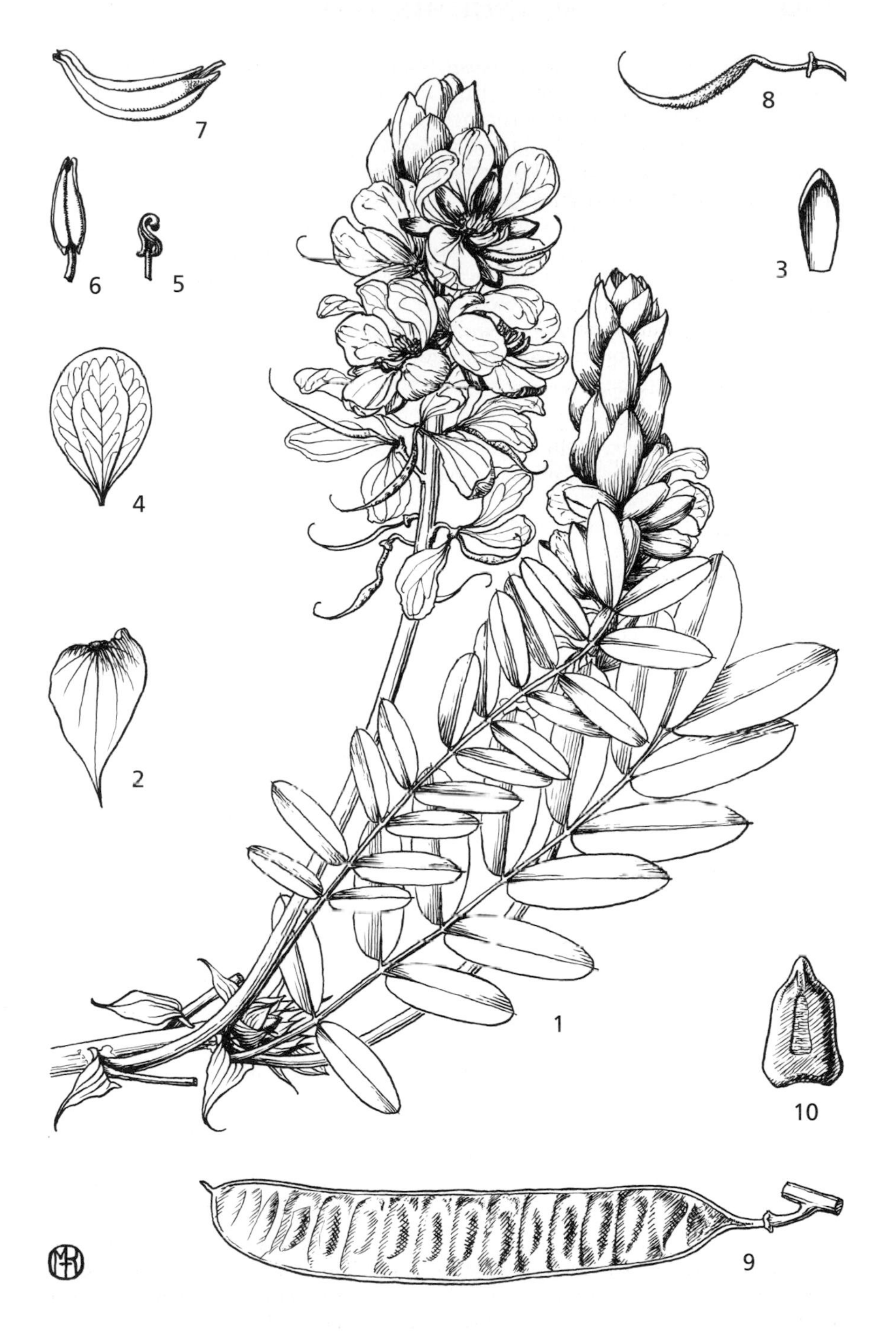

Fig. 3.2.**32**. SENNA DIDYMOBOTRYA. 1, flowering branchlet (× ⅔); 2, stipule (× 1); 3, sepal (1); 4, petal (× 1); 5–7, three different sorts of stamen (× 1); 8, gynoecium (× 1), 1–8 from *Milne-Redhead & Taylor* 10919; 9, pod (× ⅔); 10, seed (× 2), 9 & 10 from *T.H.E. Jackson* 304. Drawn by Olive Milne-Redhead. From F.T.E.A.

road to Limbe, fl. & fr. 11.iv.1970, *Brummitt* 9801 (K). **Mozambique.** N: Maniamba, at base of Jéci Mts, near Malulo, c. 60 km from Lichinga (Vila Cabral), fl. 3.iii.1964, *Torre & Paiva* 11013 (LISC). T: between Furancungo and Ulongue (Vila Coutinho), 85 km from Furancungo, fl. 15.vii.1949, *Barbosa & Carvalho* 3639 (LISC).

Widespread in tropical Africa. Probably indigenous in swamps and beside rivers in seasonal regions; also on roadsides and in waste places; widely planted as an ornamental and possibly because of its uses as medicine and as a fish poison (Steyaert 1952).

Conservation notes: A very widespread taxon; Lower Risk, Least Concern.

Irwin & Barneby (1982) doubted Steyaert's placement of *C. verdickii* in synonymy, pointing out that the illustration shows rather few (10) pairs of leaflets, and does not show the characteristic floral bracts. Some specimens from the Flora area have even fewer leaflets, and the bracts may be absent from older inflorescences. Steyaert's conclusion is followed here.

4. **Senna alata** (L.) Roxb., Fl. Ind., ed.2: 349 (1832). —Lock, Leg. Afr. Check-list: 36 (1989). Type a cultivated plant, probably originally from South America, *Herb. Clifford* (BM syntype).

Cassia alata L., Sp. Pl.: 378 (1753). —Oliver in F.T.A. **2**: 275 (1871). —Harms in Engler, Pflanzenw. Afrikas **3**(1): 497 (1915). —Baker, Legum. Trop. Afr.: 637 (1930). —Brenan, Check-list For. Trees Shrubs Tang. Terr.: 96 (1949). —Steyaert in F.C.B. **3**: 507 (1952). —Keay in F.W.T.A., ed.2, **1**: 452 (1958). —Brenan in F.T.E.A., Legum.-Caesalp.: 64 (1967).

Shrub to 3 m or more. Branchlets olive-brown, longitudinally ridged, usually minutely puberulous, glabrescent. Leaves paripinnate; stipules asymmetrically triangular, persistent; rachis generally 25–75 cm long; leaflets in (5)7–12 pairs, 6.5–8 × 3–3.5 cm (much larger in South American material), oblong, asymmetrically rounded at base, rounded and apiculate at apex, sparsely pubescent mainly on the veins beneath, glabrescent. Inflorescences racemose, axillary, to 25 cm long; bracts orange, c. 2.1 × 1.3 cm, broadly obovate, falling before anthesis but prominent earlier. Petals golden yellow with dark veins, particularly when dry, 13–15 × 10–12 mm, broadly elliptic. Stamens 10, two with anthers c. 10 mm long, five with anthers c. 4–5 mm long; three with anthers 1–2 mm long. Ovary c. 10–12 mm long, densely pubescent, longitudinally ridged; stipe 1–2 mm long; style 3–4 mm long, almost glabrous. Pod dark brown to black, shiny, 4-winged, 12–16 × 1.5–2.5 cm, very tardily dehiscent. Seeds brown, c. 7 × 4 × 2 mm, obovate, angular, ridged on each face, areolate on each of the narrow faces.

Malawi. S: Chowe Estate, c. 15 km E of Mangochi, fl. & fr. viii.1980, *Arnall* s.n. (K); Machinga Dist., Machinga East, Mwekuwa village, cultivated, fl. 30.iv.1982, *Patel* 880 (K, MAL).

Indigenous to South America; widely cultivated in the wetter parts of Africa for its showy flowers and medicinal properties ('Ringworm Plant'); sometimes becoming naturalised. Probably more widespread than the few collections suggest; recorded for Zimbabwe by Biegel but no material seen.

Easily recognised by the upright spikes of golden yellow flowers topped by the orange floral bracts, and the four-winged pods.

The seeds are flattened at right angles to those of *C. didymobotrya*, so that the areoles are on the narrow 'sides' of the seed, not on the broad face.

5. **Senna occidentalis** (L.) Link, Handb. **2**: 140 (1831). —Irwin & Barneby, Mem. New York Bot. Gard. **35**: 436 (1982). —Lock, Leg. Afr. Check-list: 38 (1989). Type a cultivated plant, *Herb. Clifford* (BM syntype). FIGURE 3.2.**33**.

Cassia occidentalis L., Sp. Pl.: 377 (1753). —Klotzsch in Peters, Naturw. Reise Mossambique **6**(1): 13 (1861). —Oliver in F.T.A. **2**: 274 (1871). —Taubert in Engler, Pflanzenw. Ost-Afrikas **C**: 201 (1895). —Harms in Engler, Pflanzenw. Afrikas **3**(1): 497 (1915). —Baker, Legum. Trop. Afr.: 635 (1930). —Brenan, Check-list For. Trees Shrubs

Tang. Terr.: 97 (1949); in F.T.E.A., Legum.-Caesalp.: 78, fig.14 (1967). —Steyaert in F.C.B. **3**: 513 (1952). —Mendonça & Torre in C.F.A. **2**: 181 (1956); —Keay in F.W.T.A., ed.2, **1**: 455 (1958). —White, F.F.N.R.: 119 (1962). —Schreiber in Merxmüller, Prod. Fl. SW Afrika, fam. 59: 11 (1967). —Drummond in Kirkia **10**: 243(1977). —Ross, Fl. Natal: 195 (1973). —Gordon-Gray in F.S.A. **16**(2): 85 (1977).

An erect herb, sometimes woody at the base, 0.5–1(2) m high. Stems ridged, subglabrous. Leaves paripinnate; petiole 2.8–6 cm long, with a single large dome- or cone-shaped gland close to the base; rachis 5–12.6 cm long, eglandular; leaflets in 4–5 pairs, 2.9–12 × 1.2–4.2 cm, the distal pair largest, ovate to ovate-elliptic, acute to acuminate, base asymmetric, upper side cuneate, lower side rounded, glabrous except for a few hairs and scattered small glands on the petiolules and margins. Inflorescences axillary racemes, very short, almost umbellate; peduncles mostly 3–5 mm long. Sepals 5, c. 7 × 3–5 mm, obovate, glabrous. Petals yellow, dark veined when dry, c. 8 mm in diameter with a claw 1–2 mm long, obovate to subcircular. Stamens 10; five probably fertile of which two with anthers 5–6 mm long and three with anthers c. 4 mm long, five apparently sterile with reduced flattened anthers. Ovary 8–9 mm long, appressed-pubescent; style 3–4 mm long, glabrous except for a few hairs near the tip. Pod 8–12 × 0.8–1 cm, straight or slightly curved, tardily dehiscent or indehiscent, many-seeded. Seeds buff, c. 4.5 mm in diameter, suborbicular, verruculose, with an elliptic areole on each face.

Caprivi Strip. Andara, banks and islands of Okavango R., fl. 8.iii.1958, *Merxmüller & Giess* 1960 (K, M). **Botswana.** N: Matlapaneng bridge, Thamalakane R., fl. 28.iii.1977, *P.A. Smith* 1959 (K). **Zambia.** B: Mongu Dist., Mongu Township, fl. & fr. 11.ii.1952, *White* 2052 (FHO, K). N: Kasama Dist., 13 km (8 miles) N of Kasama on Mbala (Abercorn) road, fl. & fr. 22.xii.1961, *Astle* 1144 (K). C: 40 km (25 miles) from Lusaka on old Mumbwa road, fl. & fr. 7.iv.1960, *Angus* 2202 (FHO, K). S: Kafue Flats, fl. & fr. v.1908, *Allen* 711 (K). **Zimbabwe.** N: Mt Darwin Dist., upper reaches of Nyatandi R., fl. & fr. 27.i.1960, *Phipps* 2427 (K, SRGH). W: Victoria Falls, S bank of Zambezi, fl. & fr. 9.ii.1912, *Rogers* 5729 (K). C: Harare (Salisbury), fl. & fr., n.d., *Agric. Dept.* 3120 (K). E: Chimanimani (Melsetter), Nyanyadzi, fl. 24.i.1964, *Corby* 1073 (K, SRGH). **Malawi.** N: Karonga Dist., 11 km N of Karonga, fl. 15.vii.1970, *Brummitt* 12137 (K). C: Lilongwe Dist., Lilongwe, new capital site, fl. & fr. 29.iii.1970, *Brummitt & Little* 9511 (K). S: Chikwawa Dist., by Ngabu Rest House, fl. & fr. 23.iv.1980, *Brummitt & Blackmore* 9798 (K). **Mozambique.** N: Cabo Delgado, Pemba (Porto Amélia), fl. & fr. 21.xi.1952, *Barbosa & Balsinhas* 5277 (K, LISC). Z: Mocuba area, fl. 1943, *Torre* s.n. (LISC). T: Cabora Bassa Dist., between Marueira and Songo, 2–3 km from Songo, fl. & fr. 4.iii.1972, *Macêdo* 4985 (LISC). MS: Gorongosa Nat. Park, SE Urema Plains, fl. & fr. iii.1972, *Tinley* 2453 (K, LISC, SRGH). GI: Gaza, near Chibuto, fr. 28.v.1957, *Montalvão* 102 (K). M: c. 5 km (3 miles) N of Maputo (Lourenço Marques), fl. & fr. 29.iii.1948, *Rodin* 4174 (K).

Pantropical; Irwin and Barneby very tentatively suggest an Old World origin but this is most uncertain. However, if introduced to Africa, it has been there a long time as it was collected by Welwitsch in Angola in 1853 and by Kirk in Mozambique in 1859, the latter noting: 'Always around villages'. Generally a species of disturbed ground or a weed of cultivation.

Conservation notes: A very widespread taxon; Lower Risk, Least Concern.

6. **Senna hirsuta** (L.) H.S. Irwin & Barneby in Phytologia **44**(7): 499 (1979); in Mem. New York Bot. Gard. **35**: 425 (1982). —Lock, Leg. Afr. Check-list: 37 (1989). Type a cultivated plant, *Herb. Clifford* (BM holotype).

Cassia hirsuta L., Sp. Pl.: 378 (1753). —Steyaert in F.C.B. **3**: 513 (1952). —Keay in F.W.T.A., ed. 2, **1**: 455 (1958). —Brenan in F.T.E.A., Legum.-Caesalp.: 80 (1967). —Drummond in Kirkia **8**: 212 (1972); in **10**: 243 (1977). —Ross, Fl. Natal: 195 (1973). —Gordon-Gray in F.S.A. **16**(2): 86 (1977).

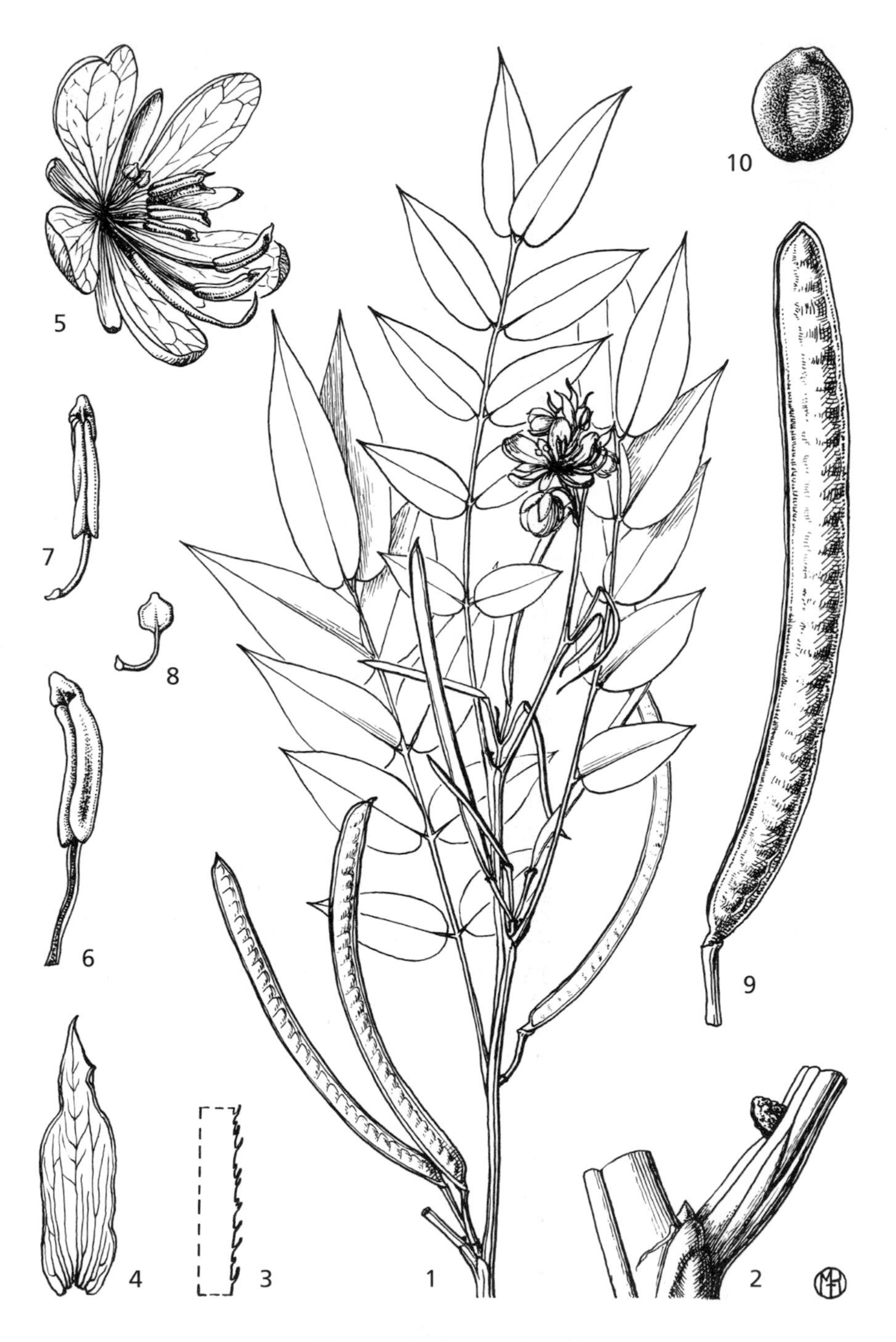

Fig. 3.2.**33**. SENNA OCCIDENTALIS. 1, fertile branch (× 1/2); 2, stem-leaf junction showing gland on petiole (× 4); 3, part of leaflet margin (× 4), 1–3 from *Fyffe* 131/24; 4, bract (× 2), from *Tweedie* 1033; 5, flower (× 2); 6, anticous stamen (× 4); 7, median stamen (× 4); 8, posticous stamen (× 4), 5–8 from *Fyffe* 131/24; 9, pod (× 1); 10, seed (× 4),9 & 10 from *Faulkner* 1648. Drawn by Olive Milne-Redhead. From F.T.E.A.

An erect herb, sometimes softly woody at the base, to 2 m high. Stems ridged, densely hairy. Leaves paripinnate; stipules 6–7 × 0.5 mm, linear, hairy, subpersistent; petiole 3–6.6 cm long, hairy, with a single large dome- or cone-shaped gland close to the base; rachis 6.9–12.2 cm long, hairy, eglandular; leaflets in 4–5 pairs, 2.8–10.6 × 1.4–3.5 cm, the distal pair usually largest, elliptic to ovate, acuminate, base symmetrical, cuneate, hairy on both surfaces. Sepals 5, dimorphic, the outer two c. 5 × 4 mm, broadly obovate, hairy, the inner three c. 10 × 4 mm, elliptic. Petals yellow, dark-veined when dry, 10–13 × 5–9 mm with a claw 1–2 mm long, obovate. Stamens 10; seven probably fertile of which two with anthers 6–7 mm long and three with anthers c. 4 mm long, three apparently sterile with reduced flattened anthers. Ovary 8–9 mm long, densely spreading-hairy; style 2–3 mm long, glabrous. Pod 6–13 × 0.5 cm, straight or slightly curved, hirsute, tardily dehiscent or indehiscent, many-seeded. Seeds grey-brown, 3–3.5 × 3–3.5 × 1 mm, suborbicular, with elliptic areoles c. 1 mm long on each edge.

Zimbabwe. E: Chipinga Dist., Zona Tea Estate, SE Chirinda Forest, fl. 15.v.1962, *Chase* 7717 (FHO, K, LISC, SRGH). **Malawi.** N: Rumphi Dist., Nchena Nchena, fl. 24.iv.1969, *Pawek* 2346 (K). S: Mulanje, Swazi [Estate], fl. 29.iii.1949 & fr. 18.vi.1949, *Faulkner* Kew series 395 (COI, K).

Weed of cultivation in the wetter parts of the Flora area, but apparently uncommon (or undercollected). Native of South America.

Irwin & Barneby (1982), distinguish 7 varieties. All specimens seen belong to var. *hirsuta*, which has hairs 0.6–2.8 mm long on both surfaces of the leaves and fairly straight pods 11–15 × 0.45–0.65 cm (see Irwin & Barneby, 1982: 428). This variety is a widespread weed in tropical Africa and the rest of the Old World.

7. **Senna spectabilis** (DC.) H.S. Irwin & Barneby in Mem. New York Bot. Gard. **35**: 600, figs.11 & 12 (1982). —Lock, Leg. Afr. Check-list: 40 (1989). Type: A cultivated plant, probably originally from Venezuela.

Cassia spectabilis DC., Cat. Pl. Hort. Bot. Monspel.: 90 (1813). —Baker, Legum. Trop. Afr.: 631 (1930) —Steyaert in F.C.B. **3**: 506 (1952). —Mendonça & Torre in C.F.A. **2**: 178 (1956). —White, F.F.N.R.: 120 (1962).

Tree to 10 m or more. Branchlets smooth, dark brown, puberulous to tomentellous, glabrescent. Leaves paripinnate; stipules 7–9 [4–10] mm long, linear, caducous; petiole 3.5–4 [1–3.5] cm long, pubescent, eglandular; rachis 13–28 [12–36] cm long, pubescent, eglandular; leaflets in 10–12 pairs, [(8)11–19(20)] 5.5–8 × 1.8–2.5 [3–9.5 × 1.1–2.7] cm, the lowest pairs somewhat smaller, ovate, acute, base symmetrical, cuneate, appressed-hairy beneath, sparsely appressed-puberulous above. Inflorescences large terminal panicles. Sepals 5, dimorphic, the outer two 5–6 × 4 mm, broadly ovate-elliptic, the inner three larger, 8–9 mm in diameter, suborbicular, almost glabrous. Petals dimorphic, golden-yellow, dark-veined when dry; four 17–19 × 14–16 mm, broadly obovate; one obovate, turned at right angles above the claw, larger than the rest. Stamens 10; seven fertile with anthers c. 5 mm long, and three apparently sterile with reduced flattened bilobed anthers. Ovary 12–15 mm long, glabrous; style glabrous. Pod black, 16–30 × 0.9–1.2 cm, terete or subquadrate, straight, glabrous, pendulous, very tardily dehiscent along one suture, many-seeded. Seeds brown, 5–7 × 3.5–5 mm, broadly obovate in outline, with elliptic areoles on each broad face. (Measurements and figures in square brackets are those for South America, from Irwin & Barneby 1982)

Zambia. C: Lusaka township, planted, fl. iii.1952, *White* 2145 (FHO). **Malawi.** N: Rumphi, fl. 29.iii.1976, *E.Phillips* 1543 (K, MO). S: Blantyre Dist., Lunzu Market, 15 km N of Blantyre, planted, fl. 21.iii.1970, *Brummitt* 9235 (K).

The specimens seen are all var. *spectabilis*. This, the most commonly planted variety, is indigenous to Central and South America, except for Brazil, where it is replaced by var. *excelsa* (Schrad.) H.S. Irwin & Barneby, which has a shrubbier habit, smaller leaves, and smaller leaflets which tend to be oblong-elliptic or elliptic rather than ovate.

A highly decorative and widely planted street and avenue tree in the wetter areas of tropical Africa. Like many cultivated species, probably very undercollected and much more common than the few records suggest. Readily distinguished by the almost cylindrical pendulous pods and the large upright panicles of golden-yellow flowers with asymmetrical corollas. The pods superficially resemble those of *Cassia fistula* and similar species, but are tardily dehiscent along one suture, and the seeds are areolate.

8. **Senna bicapsularis** (L.) Roxb., Fl. Ind., ed. 2, **2**: 342 (1832). —Lock, Leg. Afr. Check-list: 37 (1989). —Schmidt, Lötter & McCleland, Trees Shrubs Mpum. & Kruger Nat. Park: 192 (2004). Type: A cultivated plant, said to be from Sri Lanka but probably from the West Indies, *Herb. Linnaeus* 528.10 (LINN syntype).

Cassia bicapsularis L., Sp. Pl.: 376 (1753). —Baker, Legum. Trop. Afr.: 635 (1930). —Brenan, Check-list For. Trees Shrubs Tang. Terr.: 96 (1949). —Steyaert in F.C.B. **3**: 511 (1952). —Keay in F.W.T.A., ed. 2, **1**: 455 (1958). —White, F.F.N.R.: 120 (1962). —Brenan in F.T.E.A., Legum.-Caesalp.: 71 (1967). —Drummond in Kirkia **8**: 212 (1972); in Kirkia **10**: 243 (1977). —Ross, Fl. Natal: 195 (1973). —Gordon-Gray in F.S.A. **16**(2): 90 (1977).

Shrub 1–5 m tall. Branchlets olive, becoming grey, glabrous, longitudinally ridged. Leaves paripinnate; stipules c. 2 mm long, very narrowly triangular, early-caducous; petiole 0.6–2.3 cm long, glabrous; rachis 1–2.5 cm long, glabrous except for a patch of minute probably glandular hairs between each leaflet pair, with an ellipsoid gland c. 1 mm long between the proximal leaflet pair; leaflets in 3–4 pairs, 1.5–4 × 1.2–2 cm, the distal largest, broadly elliptic to broadly obovate or subcircular, base rounded to cuneate, slightly asymmetric, apex acute to rounded or retuse, often minutely mucronate, glabrous; petiolules c. 1 mm long. Inflorescences axillary, racemose; bracts 1–2 mm long, very narrowly triangular, early caducous; pedicels 4–5 mm long at anthesis. Sepals 5, elliptic, dimorphic, two c. 10 × 5 mm, three 5–7 × 3–4 mm, glabrous. Petals yellow, 12–14 × 6 mm, obovate. Stamens 9, trimorphic, two fertile with anthers c. 7 mm long and filaments c. 8 mm long, four fertile with anthers c. 5 mm long and filaments c. 1 mm long; three trumpet-shaped staminodes c. 4 mm long. Ovary c. 10 mm long, glabrous; style 4 mm long, glabrous. Pod 8–15 × 1–1.5 cm, oblong-linear, straight, cylindrical, rounded at apex, tardily dehiscent along both sutures, or indehiscent. Seeds olive-brown, 5–6 × 3.5–4.5 mm, broadly elliptic, without areoles.

Zambia. W: Ndola, police office, fl. & fr. 5.vii.1938, *Miller* A51/39 (FHO). C: Lusaka Forest Nursery, fl. 4.iii.1952, *White* 2183 (FHO). **Zimbabwe.** C: Gweru Dist., 6 km (4 miles) on Shurugwi (Selukwe) road, fl. 9.vi.1967, *Biegel* 2194 (K, LISC, SRGH). E: Chimanimani Dist., Mutambara Nature Res. by Umvumvumvu River, fl. & fr. 30.ix.1959, *Phipps* 2228 (K, SRGH). S: Mberengwa Dist., Buhwa Mt., near Buhwa School, fl. 2.v.1973, *Biegel, Pope & Simon.* 4268 (K, LISC, SRGH). **Malawi.** C: Dedza Dist., Chongoni Forest Res., fl. 12.vii.1968, *Salubeni* 1125 (K, LISC, SRGH). S: Chiradzulu Dist., 3–5 km from Chiradzulu on road to Limbe, fl. 11.iv.1970, *Brummitt* 9793 (K). **Mozambique.** M: region of Maputo, Salamanga, fl. 19.vii.1967, *Gomes e Sousa & Balsinhas* 4914 (COI, K).

Widely cultivated and readily naturalised. All four collections from Zambia at FHO are recorded as being cultivated, but elsewhere the species seems to be at least partially naturalised.

9. **Senna italica** Mill., Gard. Dict., ed. 8, no. 2 (1768). —Lock, Leg. Afr. Check-list: 37 (1989). —Pooley, Wild Fl. KwaZulu-Natal: 256 (1998). Type: Whereabouts unknown.

Cassia italica (Mill.) F.W. Andrews, Fl. Pl. Anglo-Egypt. Sudan **2**: 117 (1952). —Mendonça & Torre in C.F.A. **2**: 178 (1956). —Keay in F.W.T.A., ed. 2, **1**: 453 (1958). —Brenan in Kew Bull. **13**: 239 (1958); in F.T.E.A., Legum.-Caesalp.: 65 (1967). —Schreiber in Merxmüller, Prod. Fl. SW Afrika, fam. 59: 11 (1967). —Drummond in Kirkia **8**: 212 (1972). —Ross, Fl. Natal: 195 (1973). —Gordon-Gray in F.S.A. **16**(2): 81 (1977).

Prostrate or ascending shrublet or herb from a woody base, to 15–60 cm high. Branchlets pale grey-brown, pubescent, weakly longitudinally ridged. Leaves paripinnate; stipules 5–6 × 1 mm, narrowly triangular, pubescent, persistent; petiole 0.3–2 cm long, pubescent; rachis 5.5–8 cm long, pubescent, eglandular or with a very small reddish gland between the leaflet pairs; leaflets in (4)5–7(8) pairs, 1.6–3.6 × 0.6–2 cm, the distal largest, oblong to broadly obovate, base broadly cuneate or rounded, slightly asymmetric, apex obtuse to rounded or retuse, usually mucronate, glabrous to sparsely pubescent, the margin and midrib beneath more densely pubescent; petiolules up to 1 mm long, glabrous to pubescent. Inflorescences axillary many-flowered racemes, 7–26 cm long, including the peduncle; bracts c. 5 × 3 mm, broadly ovate, acuminate, pubescent, caducous; pedicels 2–3 mm long at anthesis. Sepals 5, elliptic to obovate, glabrous. Petals orange to yellow, obovate, slightly unequal. Stamens 10; two fertile with long arcuate anthers and very short filaments; five fertile with short straight anthers and straight filaments; three probably sterile with anthers curved into a semi-circle and straight filaments. Ovary flattened, glabrous to villous; style terete; stigma terminal. Pod weakly curved to almost straight, flattened, usually bearing distinct crests on the flat faces in the middle of each segment, mucronate (style-base), tardily dehiscent along both sutures. Seeds brown, 6–7 × 2–3 mm, obovate, flattened, with a narrowly elliptic areole c. 2 mm long on each broad face.

There are three virtually allopatric subspecies. Subsp. *italica* occurs in North Africa from the Cape Verde Islands to Egypt and Ethiopia, and in western Asia. It has inflorescences longer than the subtending leaf, a terete style which is not expanded at the apex, and large flowers with petals 9–20 mm long, large anthers 8–14 mm long, medium ones 4–6 mm long, and the smallest 1.5–2.5 mm long.

Subsp. **arachoides** (Burch.) Lock in Kew Bull. **43**: 339 (1988); in Leg. Afr. Check-list: 38 (1989). Type: South Africa, Northern Cape, Griqualand West, Asbestos Mtns., Kloof Village, *Burchell* 1680 (K holotype).

Cassia arachoides Burch., Trav. 1: 341 (1822). —Baker, Legum. Trop. Afr.: 636 (1930).

Cassia italica subsp. *arachoides* (Burch.) Brenan in Kew Bull. **13**: 242 (1958). —Schreiber in Merxmüller, Prod. Fl. SW Afrika, fam. 59: 11 (1967). —Gordon-Gray in F.S.A. **16**(2): 83 (1977); in J. S. Afr. Bot. **44**: 67–81 (1978).

Racemes 7–22 cm long, usually longer than the subtending leaf. Petals usually 10–12 mm long. Large anthers c. 8 mm long; medium 2–3 mm long; smallest c. 1 mm long. Style terete to the apex. Indumentum very variable from minutely puberulous to villous.

Botswana. SW: 64 km (40 miles) S of Tshane on road to Tshabong, fl. 11.i.1973, *Thompson* 1723 (K, PRE). SE: near Mahalapye Village, fl. 21.xii.1957, *De Beer* 534 (K, SRGH). **Zimbabwe.** W: Insiza Dist., Fort Rixon, fl.& imm.fr. 8.ii.1974, *Mavi* 1514 (K, SRGH). C: Gweru Dist., 42 km (26 miles) N of Gweru (Gwelo) on Lower Gwelo road at turn-off to 15 miles Hunter's Road, fl. 22.ii.1968, *Biegel* 2251 (K, SRGH). S: Masvingo Dist., road to Kyle Nat. Park from Masvingo (Fort Victoria), fl. 8.ii.1972, *Gibbs Russell* 1453 (K, SRGH). **Mozambique.** GI: Gaza, Massingir, fl. & fr. 22.vii.1982, *Matos* 5054 (LISC). M: between Maputo (L. Marques) and Namaacha, fl. & fr. 16.x.1940, *Torre* 1779 (K, LISC).

Also in Namibia, Swaziland and South Africa.

Conservation notes: A widespread taxon; Lower Risk, Least Concern.

A variable taxon; the variation has been discussed by Brenan (1958) and Gordon-Gray (1977, 1978). The most striking variant has villous-hairy ovaries and young pods. This is Brenan's 'Variant 1' and Gordon-Gray's 'Group 3' (1977) and 'Variant D' (1978). It has been collected mainly from northern parts of the former Transvaal, but three specimens from Botswana (SE) belong here: *E.J. Lugard* 229 (K), *Mitchison* A32 (K), and *Allen* 219 (K, PRE). The first two of these also have much larger flowers than is normal for the subspecies.

Subsp. **micrantha** (Brenan) Lock in Kew Bull. **43**: 339 (1988); Leg. Afr. Check-list: 38 (1989). Type: Kenya, Turkana Dist., *Padwa* 144 (K holotype).

Cassia italica subsp. *micrantha* Brenan in Kew Bull. **13**: 241 (1958); in F.T.E.A., Legum.-Caesalp.: 65 (1967). —Schreiber in Merxmüller, Prod. Fl. SW Afrika, fam. 59: 11 (1967). —Gordon-Gray in F.S.A. **16**(2): 83 (1977); in J. S. Afr.. Bot. **44**: 67–81 (1978).

Racemes 2–8 cm long, usually shorter than the subtending leaf. Petals 8.5–9 mm long. Anthers similar to or a little smaller than those of subsp. *arachoides*. Style ± trumpet-shaped at the apex.

Botswana. N: Kwebe Hills, fl. 1.iii.1898, *E.J. Lugard* 201 (K).

Senegal to Ethiopia and south to Tanzania; Angola and Namibia. Also in India.

Conservation notes: Only one record from the Flora area, but widespread at the global scale; Lower Risk, Least Concern, but possibly Endangered in the Flora area.

The remarkable difference in style form between this subspecies and the other two was pointed out by Gordon-Gray. It is surprising that there are no recent collections from Botswana; Gordon-Gray's map (1978: 73) shows that it is widespread in N Namibia although generally well west of the Botswana border.

10. **Senna septemtrionalis** (Viv.) H.S. Irwin & Barneby in Mem. New York Bot. Gard. **35**: 365 (1982). —Lock, Leg. Afr. Check-list: 39 (1989). Type a cultivated plant.

Cassia laevigata Willd., Enum. Pl. Hort. Berol.: 441 (1809). —Oliver in F.T.A. **2**: 275 (1871). —Harms in Engler, Pflanzenw. Afrikas **3**(1): 495 (1915). —Baker, Legum. Trop. Afr.: 634 (1930). —Steyaert in F.C.B. **3**: 511 (1952). —Keay in F.W.T.A., ed.2, **1**: 453 (1958). Type a cultivated plant in Berlin Bot. Garden.

Cassia floribunda sensu auct. mult. non Cav. —De Wit in Webbia **11**: 245 (1955). —Brenan in F.T.E.A., Legum.-Caesalp.: 70 (1967). —Drummond in Kirkia **8**: 212 (1972); in **10**: 243 (1977). —Ross, Fl. Natal: 195 (1973). —Gordon-Gray in F.S.A. **16**(2): 87 (1977).

Shrub or small tree to 5 m high. Branchlets pale olive-brown, glabrous, with slight longitudinal ridges. Leaves paripinnate; stipules 5–6 mm long, narrowly triangular, glabrous, subpersistent; petiole 3–7 cm long, glabrous; rachis 4.8–9.4 cm long, glabrous, with a conical reddish gland c. 1 mm long between each leaflet pair and terminating the rachis; leaflets in 3–4 pairs, 4.3–9.5 × 1.9–3.2 cm, the distal largest, ovate to elliptic, base cuneate, symmetrical, apex acuminate, occasionally acute, glabrous; petiolules up to 1 mm long, glabrous or with a few appressed hairs. Inflorescences simply racemose, axillary, 6–10-flowered, 3.5–9.5 cm long including the peduncle; bracts 5–8 × 1–2 mm, narrowly triangular, glabrous, caducous; pedicels 1.5–2.5 mm long at anthesis. Sepals 5, two larger c. 8 × 5 mm, three smaller 5–7 × 3–5 mm, elliptic to obovate, glabrous. Petals unequal, the largest 13–15 × 12–13 mm, broadly obovate, retuse at apex, the others broadly elliptic to obovate, acute, orange to yellow. Stamens 10, two fertile with arcuate anthers 7–8 mm long and filaments c. 10 mm long, five fertile with short straight anthers 2–4 mm long and straight filaments c. 1 mm long on 4 stamens and c. 4 mm long on the fifth, three probably sterile with anthers curved into a semicircle and straight filaments. Ovary c. 10 mm long, curved, terete, glabrous; style c. 4 mm long, terete; stigma terminal, recessed, surrounded by a fringe of hairs. Pod 8–9 cm long, c. 1 cm in diameter, straight, grooved, the sutures somewhat impressed, mucronate, glabrous, tardily dehiscent along both sutures. Seeds brown, in two rows, c. 5 × 3 mm, obovate, flattened, without areoles.

Zambia. N: Mbala Dist., Kambole, close to Mission House, fl. 30.i.1959, *Richards* 10790 (K). **Zimbabwe.** C: Harare (Salisbury), garden ornamental, fl. 29.xii.1975, *Biegel* 5193 (K, SRGH). E: Nyanga Dist., Gairezi R., fl. 15.iii.1964, *Corby* 1100 (K, SRGH). **Malawi.** N: Misuku Hills, Mughesse rain forest, fl. 29.xii.1970, *Pawek* 4234 (K, MAL). **Mozambique.** MS: Manica, Macequece, foot of Vumba Mts., fr. 6.iii.1948, *Barbosa* 1196 (LISC). M: near Cascata da Namaacha, fr. 22.xii.1944, *Torre* 6932 (BR, K, LISC. LMA).

Originally from Central America, but widely planted as an ornamental in the tropics. Most of the material from the Flora area is either cultivated or from close to gardens, but the species seems well-established in the Eastern Region of Zimbabwe, where it was first collected in 1906 (*Swynnerton* 199 (K)), and noted as being used medicinally by local people. Disturbed places; forest margins.

The notes on *Gilliland* Q1202 state that it is a tree to 30 ft. This is almost certainly a mistake.

There has been much confusion over the name of this taxon, but the account by Irwin & Barneby (see above) and the summary by Andrews & Knees (Kew Mag. **5**: 76–82, 1988) should clarify matters. The name *Cassia laevigata* was long used, but then De Wit incorrectly used the name *C. floribunda,* and this name came into general usage. Irwin & Barneby have now shown that the oldest name is *Cassia septemtrionalis,* and that the epithet *floribunda* should correctly be applied, as *Senna* × *floribunda* (Cav.) H.S. Irwin & Barneby, to the hybrid between *Senna septemtrionalis* and *S. multiglandulosa. Senna* × *floribunda* is quite widely cultivated and may occur in the Flora area; the leaflets are asymmetric at the base and the midrib, the lamina base and margin beneath and the petiole all bear weak subappressed hairs. These parts are glabrous in *S. septemtrionalis,* and the leaflet bases are symmetrical.

11. **Senna petersiana** (Bolle) Lock in Kew Bull. **43**: 340 (1988); Leg. Afr. Check-list: 39 (1989). —M. Coates Palgrave, Trees Sthn. Africa: 341 (2002). Type: "Mozambique, Querimba I. and Mozambique", *Peters* (B† holotype).

Cassia petersiana Bolle in Peters, Naturw. Reise Mossamb. **6**(1): 13 (1861). —Oliver in F.T.A. **2**: 272 (1871). —Taubert in Engl., Pflanzenw. Ost-Afrikas **C**: 200 (1895). —Baker in J. Linn. Soc. Bot. **40**: 61 (1911); Legum. Trop. Afr.: 633 (1930). —Harms in Engler, Pflanzenw. Afrikas **3**(1): 495, fig.264 (1915). —Brenan, Check-list For. Trees Shrubs Tang. Terr.: 97 (1949); in F.T.E.A., Legum.-Caesalp.: 72 (1967). —Steyaert in F.C.B. **3**: 508 (1952). —White, F.F.N.R.: 119 (1962). —Gomes e Sousa, Dendrol. Moçamb. Estudo Geral **1**: 260, fig.59 (1966). —Drummond in Kirkia **10**: 243 (1977). —Ross, Fl. Natal: 195 (1973). —Gordon-Gray in F.S.A. **16**(2): 92 (1977).

Cassia delagoensis Harvey in F.C. **2**: 272 (1862). Types: Mozambique, Delagoa Bay, *Forbes* s.n. (K syntype) and South Africa, Durban (Port Natal), *Hewittson* s.n. (K syntype).

Cassia petersiana var. *tomentosa* Baker f., Legum. Trop. Afr.: 634 (1930). Types: Zambia, Kabwe (Broken Hill), *Rogers* 8129 (K syntype) and Mumbwa, *Macaulay* 173 (K syntype).

Shrub or small tree 2–6 m high, sometimes described as scandent. Branchlets brown, puberulous to pubescent, longitudinally ridged when young. Leaves paripinnate; stipules 10–15 mm long, leafy, cordate-auriculate to reniform, often caudate-attenuate, pubescent, subpersistent; petiole 1.5–4.5 cm long, pubescent; rachis up to 20 cm long, appressed-pubescent, with scattered elongate reddish glands among the hairs, with a stalked botuliform often caducous reddish gland 1.5–2 mm long between each of the proximal 1–4 leaflet pairs; leaflets in 6–10 pairs, 4–7 × 1.5–3.5 cm, the middle ones usually the largest, ovate to ovate-elliptic, base cuneate to rounded or cordate, often slightly asymmetrical, apex acute to acuminate, mucronate, sparsely hairy above, more densely so beneath, with appressed hairs and reddish elongated glands; petiolules up to 1 mm long, appressed-pubescent. Inflorescences paniculate, terminal, made up of corymbose racemes; primary inflorescence branches (racemes) up to 9 cm long, subtended by reduced caducous leaves with few pinnae; stipules of these similar to those of foliage leaves, relatively persistent; proximal bracts of racemes very broadly ovate, acuminate, distal ones narrower, all usually densely pubescent, subpersistent; proximal pedicels up to 5 cm long at anthesis; distal ones shorter. Sepals 5, dimorphic, three c. 12 × 10 mm, two c. 7 × 5 mm, broadly ovate-elliptic, pubescent. Petals golden yellow, unequal, clawed, the largest up to 22 × 11 mm, oblong, somewhat asymmetrical, rounded at apex, the others broadly elliptic to ovate, rounded or retuse. Stamens 10, three large with arcuate fertile anthers 7–10 mm long and filaments 8–12 mm long, four fertile with slightly curved anthers 6–8 mm long and straight filaments c. 1 mm long, and three small, probably sterile, with reniform to subspherical anthers and straight filaments 2–4 mm long. Ovary 10–12 mm long, straight or curved, terete, densely appressed-pubescent with white hairs and reddish glands; style 4.5–5.5 mm long, terete, glabrous; stigma terminal, recessed, surrounded by a fringe of hairs. Pod black with brown margins, linear, 13–24 × 0.8–1.2 cm, straight or irregularly curved, chambered, glabrous or appressed-pubescent, indehiscent. Seeds brown, in a single row, 5–6 × 4 mm, obovate, flattened, with two narrowly elliptic areoles on the narrow faces.

Zambia. N: Mbesuma Ranch, fr. 20.vii.1961, *Astle* 814 (K). C: Lusaka Dist., 27 km (17 miles) N of Lusaka on Kabwe (Broken Hill) Road, fl. 15.iii.1961, *Drummond & Rutherford-Smith* 6910 (K, SRGH). S: Magoye, forest reserve, fl. 26.ii.1958, *Verboom* 318 (FHO, K). **Zimbabwe.** C: Chegutu (Hartley), fr. viii.1933, *Eyles* 7506 (K, SRGH). E: Chipinge Dist., Dotts Drift, c. 2.5 km (1.5 miles) E of Save R., fl. 18.iii.1960, *Johnstone* 115 (K, SRGH). S: Mwenezi Dist., Bubye R., near Bubye Ranch homestead, fl. 8.v.1958, *Drummond* 5699 (K, LISC, SRGH). **Malawi.** N: Mzimba Dist., near Emanyeleni, c. 19 km S of Rumphi on road to Ekwendeni, fl. 21.v.1970, *Brummitt* 10985 (K). C: Ntchisi Dist., Ntchisi Forest Res., fl. 19.vi.1970, *Brummitt* 11575 (K). S: Mangochi Dist., Kapinujoti, fl. 20.v.1954, *Jackson* 1328 (FHO, K). **Mozambique.** N: Marrupa, fl. 13.vi.1948, *Pedro & Pedrogão* 4301 (K). Z: Lugela Dist., Namagoa Plantation, fl. 30.v.1948 & fr. 20.xi.1948, *Faulkner* Kew series 275 (COI, K). T: between Fíngoè and Chicoa, 1.8 km from Fíngoè, fr. 29.vi.1949, *Barbosa & Carvalho* 3378 (K, LISC). MS: Mossurize (Algueirão), fl. 22.ii.1907, *Johnson* 135 (K). GI: Gaza, Xai-Xai (Vila de João Belo), Chipenhe, Régulo Chiconela, fl. 1.iv.1959, *Barbosa & Lemos* 8442 (COI, K, LISC). M: between Matutuíne (Bela Vista) and Catembe, 30.7 km from Bela Vista, fl. 17.ii.1952, *Barbosa & Balsinhas* 4737 (K).

Widespread in eastern Africa from Ethiopia, the Sudan and Uganda southwards to Mozambique and northern South Africa, and west to Cameroon and the Central African Republic. Also in Madagascar where probably introduced (see Du Puy, Legum. Madagascar: 91, 2002). Wetter bushland and thicket; forest margins; near sea-level up to 1650 m.

Conservation notes: A very widespread taxon; Lower Risk, Least Concern.

A very variable taxon, particularly in leaflet shape and indumentum. The leaflets vary from broadly ovate with a cordate base to narrowly elliptic with a cuneate base. In general, broader leaflets correlate with a denser indumentum (as in Baker's var. *tomentosa*) but this is not invariable and there are many intermediates. The density of the indumentum on the upper and lower surfaces of the leaflets is also not necessarily correlated; some plants have a sparsely appressed-puberulous lower surface and scattered hairs above; others have a shiny and virtually glabrous upper surface but a pubescent lower surface. Different forms may occur in the same area (e.g. Zimbabwe, Mutare: *Chase* 2287 with narrow very sparsely pubescent leaflets and caducous inflorescence bracts, and *Hopkins* in GHS 12470 with broader leaflets, denser pubescence and persistent inflorescence bracts), but there are also intermediates (e.g. *Eyles* 7076). It is just possible that the narrow-leaved plants (e.g. *Chase* 2287, *Wild* 3356, *Drummond* 5699 and *Loveridge* 85934), may be the result of hybridisation with another, perhaps introduced, species. See also hybrid with *S. singueana* on p. 163.

12. **Senna singueana** (Delile) Lock in Kew Bull. **43**: 340 (1988); Leg. Afr. Check-list: 39 (1989). —M. Coates Palgrave, Trees Sthn. Africa: 342 (2002). Type: Ethiopia, Singué (Jebel Singe), *Caillaud* s.n. (MPU holotype). FIGURE 3.2.**34**.

Cassia singueana Delile, Cent. Pl. Afr.: 28 (1826). —Brenan, Check-list For. Trees Shrubs Tang. Terr.: 98 (1949). —Keay in F.W.T.A., ed.2, **1**: 453 (1958). —Brenan in F.T.E.A., Legum.-Caesalp.: 73, fig.13 (1967). —Steyaert in F.C.B. **3**: 509 (1952). —Mendonça & Torre in C.F.A. **2**: 179 (1956). —O. Palgrave, Trees Central Africa: 97 (1957). —White,

F.F.N.R.: 120 (1962). —Schreiber in Merxmüller, Prodr. Fl. SW Afrika, fam. 59: 11 (1967). —Drummond in Kirkia **10**: 243 (1977).

Cassia goratensis Fresen. in Flora **22**: 53 (1839). —Klotzsch in Peters, Naturw. Reise Mossambique **6**(1): 13 (1861). —Oliver in F.T.A. **2**: 273 (1871). —Taubert in Engler, Pflanzenw. Ost-Afrikas **C**: 200 (1895). —Harms in Engler, Pflanzenw. Afrikas **3**(1): 495, fig.265 (1915). —Baker, Legum. Trop. Afr.: 634 (1930). Type: Ethiopia, *Rueppell* s.n. (FR holotype).

Cassia goratensis var. *flavescens* Baker f., Legum. Trop. Afr.: 634 (1930); —Brenan, Check-list For. Trees Shrubs Tang. Terr.: 98 (1949). Syntypes from Tanzania.

Cassia goratensis var. *glabra* Baker f., Legum. Trop. Afr.: 634 (1930). Type: no specimen cited.

Cassia tettensis Bolle in Peters, Naturw. Reise Mossamb. **6**(1): 14 (1861). —Oliver in F.T.A. **2**: 273 (1871). —Harms in Engler, Pflanzenw. Afrikas **3**(1): 495 (1915). —Baker, Legum. Trop. Afr.: 633 (1930). Type: Mozambique, "bei Sena und Tette" (?B† holotype).

Cassia singueana var. *glabra* (Baker f.) Brenan in Kew Bull. **4**: 77 (1949); Check-list For. Trees Shrubs Tang. Terr.: 99 (1949).

Cassia singueana Delile var. *flavescens* (Baker f.) Brenan in Kew Bull. **4**: 77 (1949).

Cassia sinqueana sensu auct. non Delile, orth. error —Gordon-Gray in F.S.A. **16**(2): 93 (1977) as *Cassia sinqueana.*

Usually a small tree 2–6 m high. Branchlets brown, lenticellate, glabrous to pubescent, longitudinally ridged when young, cracking and exfoliating when older. Leaves paripinnate; stipules 4–6 mm long, linear, caducous; petiole 2–4.5(6.5) cm long, pubescent; rachis (6)10–17.5(21) cm long, appressed-pubescent, with a stalked botuliform often caducous reddish gland 1.5–2 mm long between at least the proximal leaflet pairs; leaflets in 4–9 pairs, 3.2–4.8 × 1.2–2.5 cm, the lowest usually somewhat smaller than the rest, elliptic to oblong, base broadly cuneate to rounded, sometimes asymmetric, apex rounded to retuse (rarely acute), mucronate, glabrous to pubescent above, glabrous to more densely pubescent beneath, particularly on the midrib; petiolules up to 1 mm long, appressed-pubescent. Inflorescences racemose or paniculate, terminal and axillary, usually with at least some opposite branches; primary inflorescence branches (racemes) up to 5 cm long at first anthesis, later elongating; proximal bracts of racemes ovate, early caducous, with pairs of stalked glands persisting beside the scar; proximal pedicels up to 5 cm long at anthesis; distal ones shorter. Sepals 5, three larger c. 13 × 11 mm, three smaller c. 8 × 7 mm, broadly elliptic to subcircular, glabrous or nearly so. Petals golden yellow, unequal, clawed, the two largest up to 25 × 18 mm, oblong, one somewhat asymmetrical, rounded and undulate-dentate at apex, the others smaller, broadly elliptic to oblong, rounded at apex. Stamens 10, three large with arcuate fertile anthers 10–12 mm long and filaments 5–6 mm long, four fertile with almost straight anthers 6–8 mm long and straight filaments 1–2 mm long, and three small probably sterile with reniform to subsagittate anthers and straight filaments 2–4 mm long. Ovary 10–12 mm long, straight or curved, terete, glabrous to appressed-pubescent with white hairs; style 4–5 mm long, terete, glabrous; stigma terminal, recessed, surrounded by a fringe of hairs. Pod blackish, 10–25 × 0.8–1 cm, linear, straight or irregularly curved or twisted, chambered, glabrous, indehiscent. Seeds brown, in a single row, 5–6 mm in diameter, subcircular, flattened, with two elliptic 1–2 mm long pale areoles.

Zambia. B: 45 km (28 miles) S of Gonye Falls, fl. 18.vii.1952, *Codd* 7115 (K, PRE). N: Chambeshi R. terrace, fl. 25.vii.1961, *Astle* 823 (K, SRGH). W: Solwezi Dist., Mbulungu stream, W of R. Mutanda, fl. 9.vii.1930, *Milne-Redhead* 687 (K). C: Mt. Makulu Res. Station, 19 km (12 miles) S of Lusaka, fr. 1.xi.1956, *Angus* 1435 (FHO, K). E: Msoro, c. 80 km (50 miles) W of Chipata (Fort Jameson), fl.& imm.fr. 10.vi.1954, *Robinson* 853 (K). S: Mazabuka Dist., 37 km (23 miles) Magoye to Lusaka, fl. 14.iv.1952, *White* 2661 (FHO, K). **Zimbabwe.** N: Mutoko Dist., Mkota Reserve, fl. 18.iv.1951, *Lovemore* 22 in SRGH 32725 (K, SRGH). W: Binga Dist., Kavira Hot Springs, fl. vi.1958, *Davies* 2479 (K, SRGH). C: Harare (Salisbury), Shamva Road, fl. 22.v.1946, *Wild* 1097 (K, SRGH). E: Chimanimani Dist., Birirwiri R., fl. 23.vi.1968, *Chase* 8505 (K, LISC, SRGH). S: Chivi Dist., Messina to Masvingo (Fort Victoria) Road, near Lundi R., fr. 18.x.1930, *Fries, Norlindh & Weimarck* 2045 (K). **Malawi.** N: Karonga Dist., Chitimba to Chilumba, near Sangilo, fl. & fr. 16.ix.1975, *Pawek* 10105

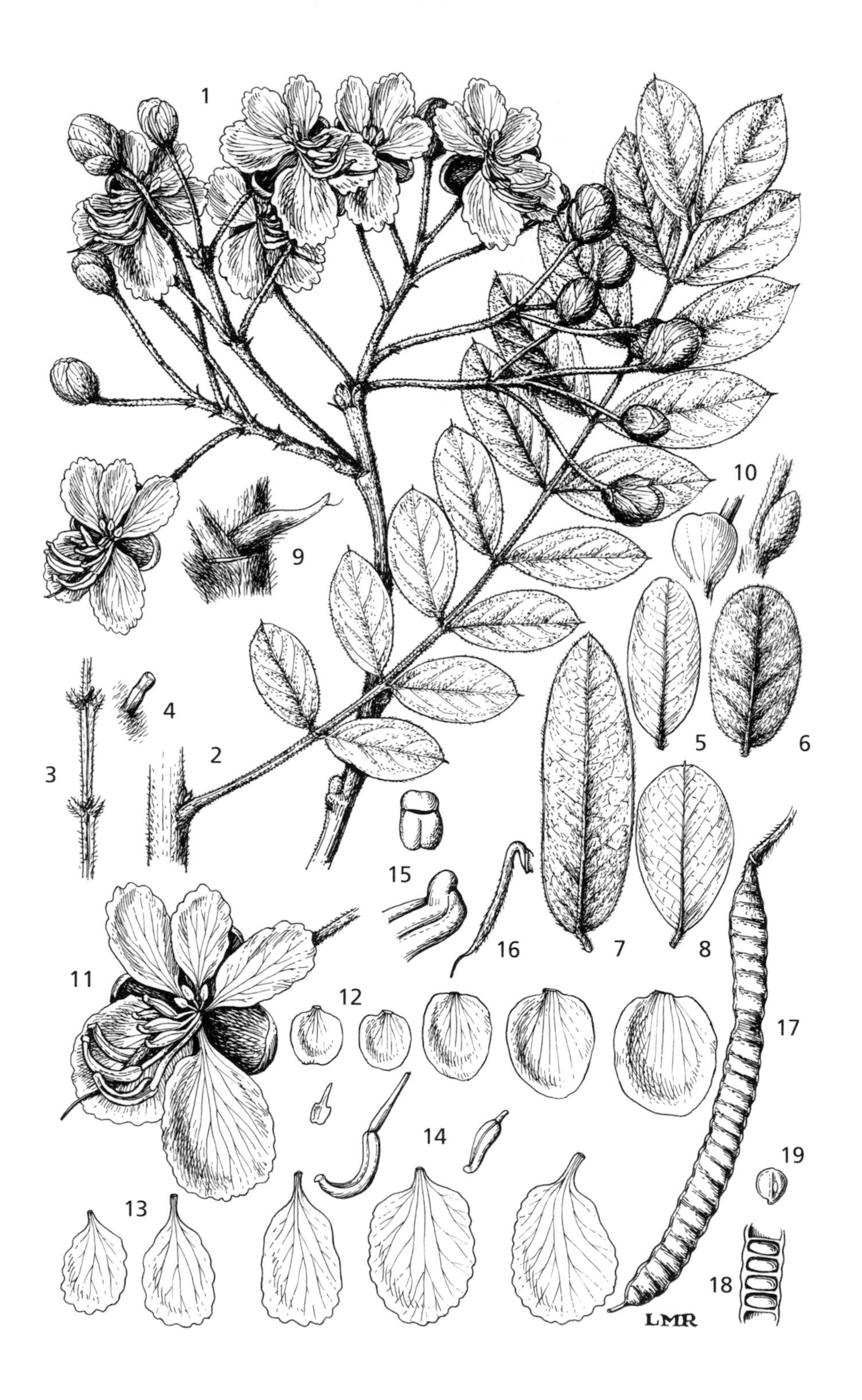

1
2
3
4
5
6
7
8
9
10
11
12
13
14
15
16
17
18
19
LMR

(K, MAL, MO, SRGH). C: Lilongwe, Agric. Res. Station, Mpatsocokolo to Chicewa, imm.fr. 26.ix.1951, *Jackson* 594 (K). S: Blantyre Dist., 10 km NW of Blantyre on road to Chileka, fl. 13.iv.1970, *Brummitt* 9821 (K). **Mozambique.** N: by Mutuali to Nyasa road, near Lurio R., fl. viii.1931, *Gomes e Sousa* 871 (COI, K). T: Moatize, Benga region, Zambezi valley, fl. 16.v.1948, *Mendonça* 4268 (BR, K, LISC).

Widespread in tropical Africa, but generally absent from the wetter, forested regions. Woodland, wooded grassland, bushland and thicket; 150–1750 m.

Conservation notes: A very widespread taxon; Lower Risk, Least Concern.

The leaflets are not particularly variable in size or shape, but range from completely glabrous to pubescent. Completely glabrous forms have been called var. *glabra*.

The label of *Peter* 47834 (K) from Harare (Salisbury), bears the name '*Cassia singueana* Delile var. *subgeabsens*'. This is presumably a corruption of '*subglabrescens*' but I have not been able to trace publication of this name or anything like it.

11 × 12. **Senna petersiana** × **Senna singueana**

This hybrid has been recorded from East Africa (see Brenan in F.T.E.A., Legum.-Caesalp.: 73 (1967)), and a specimen from Zambia (C: Mazabuka, fl. 27.ii.1963, *van Rensburg* 1500 (K)) has been so named by Brenan.

The leaflets are broader than is usual in *Senna petersiana*, and tend to be rounded to obtuse, not acute, at the apex. The stipules are narrower than in *S. petersiana* and tend to be persistent, not caducous as in *S. singueana*.

27. CASSIA L.

Cassia L., Sp. Pl.: 376 (1753); Gen. Pl., ed. 5: 178 (1754). —Irwin & Barneby in Polhill & Raven, Adv. Legume Syst. **1**: 97–106 (1981); in Mem. New York Bot. Gard. **35**: 1–918 (1982).

Trees or occasionally shrubs, unarmed. Leaves paripinnate; leaflets opposite or subopposite; stipules small, usually caducous. Inflorescence simply racemose, or paniculate; bracts and bracteoles small, caducous or subpersistent. Hypanthium small, obconical, infilled. Calyx lobes 5, equal, imbricate. Petals 5, large, yellow or white (pink in introduced species). Stamens 10, the three anticous ones long, sigmoidally curved at the base, often swollen near the middle; anthers dorsifixed, dehiscing by basal slits and often also by terminal pores. Ovary stipitate, curved; ovules numerous; stigma terminal or subterminal, often recessed forming a stigmatic chamber. Pods cylindrical, tardily dehiscent or indehiscent. Seeds numerous, without an areole, free in chambers within the fruit or surrounded by pulp.

A tropical genus with about 30 species in the tropics of the Old and New Worlds, mainly in tropical seasonal vegetation but also in forest. Several species have very attractive flowers and are widely cultivated as ornamentals. Until the work of Irwin & Barneby (1981, 1982), *Cassia* was generally regarded as including all those species now placed in *Senna* and *Chamaecrista*.

Fig. 3.2.**34**. SENNA SINGUEANA. 1, inflorescence (× ²/₃); 2, leaf (× ²/₃); 3, glands on leaf rachis (× 1); 4, detail of gland (× 6), 1–4 from *Milne-Redhead & Taylor* 11180; 5–8, undersurface of leaflets from 4 different specimens to show variation (× 2), from *Brasnett* 312, *Anderson* 928, *Tanner* 2906 & *Stolz* 2393; 9, gland on axis of inflorescence (× 8), from *Milne-Redhead & Taylor* 11180; 10, bracts, showing variation (× 2), from *Richards* 10138 & *Willan* 29; 11, flower (× 1); 12, sepals (× 1); 13, petals (× 1); 14, stamens, one of each size (× 1); 15, tip of anther viewed from above and from side (× 6); 16, gynoecium (× ²/₃), 11–16 from *Milne-Redhead & Taylor* 11180; 17, pod (× ²/₃), from *Bally* 7578; 18, part of pod cut longitudinally (× ²/₃); 19, seed (× ²/₃), 18 & 19 from *Lynes* I.g.22. Drawn by Lura Ripley. From F.T.E.A.

Cassia fistula L., Golden Shower, Indian Laburnum, is frequently planted. **Zimbabwe.** W: Hwange, cult. Victoria Falls Nat. Park, fl. 18.iii.1982, *Ncube* 85 (BM, SRGH). C: Harare, fl. 1.v.1956, *Grey* 13/56 in *SRGH* 68048 (K, SRGH). **Malawi.** S: Zomba Dist., Zomba Botanic Garden, fl. 1.xii.1977, *Brummitt & Seyani* 15240 (K). **Mozambique.** M: Magude, Delagoa Plantation, fl. 24.x.1944, *Mendonça* 3127 (BM, LISC). It differs from *C. afrofistula* in the inflorescence, which consists of groups of 1–3 pendulous unbranched racemes, in the leaflets, which are symmetrical at the base, have more lateral nerves, and are usually larger than those of *C. afrofistula*, in the glabrous (not pubescent) ovary, and in the pod, which has a thick rigid wall which cannot be cracked between finger and thumb.

Other species of *Cassia* sensu stricto are also planted in gardens and parks within the Flora area, but are not often collected; none appears to be naturalised. Several are listed by Biegel (Check List Ornam. Pl. Rhod. Parks & Gard., 1977). *Cassia fistula* (see above) is the only yellow-flowered species. Pink-flowered species include **Cassia agnes** (De Wit) Brenan from SE Asia. **Zimbabwe.** C: Harare, garden ornamental, fl. 15.xii.1975, *Biegel* 5181 (K, SRGH). E: Mutare (Umtali), fl. 23.xii.1956, *Chase* 5931 (K, SRGH). **Mozambique.** M: Maputo (Lourenço Marques), Jardim Tunduru [Vasco da Gama], fl. 29.iii.1971, *Balsinhas* 1814 (K). This has pink flowers and pods similar to those of *C. fistula.* **Cassia javanica** L., from SE Asia, resembles *C. agnes* but is pubescent throughout and has persistent auriculate stipules at least 1 cm long. **Zimbabwe.** C: Goromonzi Dist., cult. Ewanrigg Gardens, fl. 20.ii.1971, *Biegel & Müller* 3241 (K, SRGH). Biegel also lists *Cassia nodosa* Roxb. from SE Asia, which is similar to *C. javanica* but has narrow stipules to 5 mm long and usually acute leaflets, and *Cassia grandis* L.f., from northern South America, which also has pink flowers, but usually tinged orange, and a massive rough pod c. 5 cm in diameter with ridges along the sutures.

Note on the key: This includes indigenous species as well as the commonly planted yellow-flowered *C. fistula.* All indigenous species have yellow or white flowers. Introduced pink-flowered species are mentioned above.

1. Ovary glabrous or nearly so; fruits glabrous, rigid; seeds not enclosed in pulp . . 2
– Ovary pubescent; fruits glabrous or pubescent, rigid or not; seeds enclosed in pulp 3
2. Inflorescences erect or spreading, axis less than 20 cm long . . . . . . **1.** *angolensis*
– Inflorescences pendulous; axis usually 30 cm or more long; cultivated . . *fistula*
3. Petals white; inflorescences pendulous; filaments of three outer stamens abruptly swollen near the middle . . . . **2.** *burttii*
– Petals yellow; inflorescences erect or spreading; filaments of outer stamens swollen or not near the middle 4
4. Pedicels of lowest flowers longer than the simple inflorescence axis; bracts and bracteoles present and conspicuous at anthesis . . . . **3.** *abbreviata*
– Pedicels of lowest flowers much shorter than the usually branched inflorescence axis; bracts falling at or well before anthesis 5
5. Bracts persisting until anthesis . . . . **4.** *thyrsoidea*
– Bracts falling well before anthesis, while buds are young . . . . **5.** *afrofistula*

1. **Cassia angolensis** Hiern, Cat. Afr. Pl. Welw. **1**: 291 (1896). —Harms in Engler, Pflanzenw. Afrikas **3**(1): 493 (1915). —Baker, Legum. Trop. Afr.: 631 (1930). —Brenan, Check-list For. Trees Shrubs Tang. Terr.: 98 (1949); in F.T.E.A., Legum.-Caesalp.: 58 (1967). —Steyaert in F.C.B. **3**: 503 (1952). —Mendonça & Torre in C.F.A. **2**: 176 (1956). —Gordon-Gray in F.S.A. **16**(2): 107 (1977). —Lock, Leg.

Afr. Check-list: 27 (1989). —White, Dowsett-Lemaire & Chapman, For. Fl. Malawi: 306 (2001). Type: Angola, Cuanza Norte, Cazengo, Cacula, *Welwitsch* 1736 (LISU holotype, BM, K).

Tree to 20 m tall; bark smooth, grey. Branchlets very dark brown to almost black, longitudinally furrowed, pubescent to sparsely pubescent, glabrescent. Leaves paripinnate; stipules not seen, very early caducous; petiole 1.6–2 cm long, sparsely pubescent; rachis 9–14(18) cm long, sparsely pubescent; petiolules 1–3 mm long; leaflets opposite to subopposite, in 7–9 pairs, discolorous, 2.8–6.2 × 1.5–2.8 cm, ovate to ovate-elliptic, symmetrical, apex rounded or (usually) retuse, base rounded to broadly cuneate, glabrous; venation slightly prominent on both surfaces. Inflorescences axillary and terminal, racemose; axis puberulous; bracts and bracteoles not seen, presumably early caducous; pedicels (2.5)5–8 cm long, longitudinally ridged when dry, sparsely pubescent. Calyx lobes 5, 12–15 × 5–8 mm, ovate to elliptic, sparsely puberulous near the base outside, glabrous inside. Petals yellow, 28–30 × 16–27 mm, broadly elliptic. Stamens 10; 3 with filaments c. 40 mm long, glabrous, sigmoidally curved near the base, without a median swelling, anthers c. 5 mm long; 4 with straight filaments 10–15 mm long, anthers 5 mm long; 3 with filaments 12–15 mm long, curved towards the apex, anthers 3 mm long. Ovary multiovulate, c. 50 mm long and 2 mm in diameter, flattened-filiform, almost glabrous but very sparsely pubescent near the base; stigma terminal, recessed. Pod black, up to 70 cm long, 1.5–2 cm in diameter, terete, glabrous, many-seeded. Seeds buff, shiny, 8–9 × 5–6 × 4–5 mm, ellipsoidal, somewhat flattened, not embedded in pulp.

Zambia. S: Mazabuka Dist., W of Pemba, fl. 13.viii.1959, *West* 4027 (K, LISC, SRGH). **Malawi.** S: Mulanje Mt., forest beside Lukulezi R., fl. 7.x.1957, *Chapman* 453 (FHO, K). **Mozambique.** N: Cabo Delgado, região de Mueda, fl. 19.x.1942, *Mendonça* 954 (BM, K, LISC).

Also in Angola, Tanzania and Congo. Forest, dense thicket on Kalahari sands; 850–1500 m.

Conservation notes: A widespread taxon, but apparently always scarce or overlooked. Probably best treated as Data Deficient, but comes close to Vulnerable.

The seeds vary in colour within a single pod in *Bingham* 6552; some are buff, others, red-brown. *Semsei* 3188 (K) from Tanzania has subspherical seeds; possibly those seen from the Flora area and used for this account are immature.

The ovary is completely glabrous in Angolan material, and in the specimens from Zambia. Those from Malawi and Mozambique have sparsely pubescent ovaries, as do the East African specimens treated as this by Brenan in F.T.E.A. The material is all rather poor and further good collections and field observations are needed. The Malawi material all comes from submontane or montane forest; elsewhere the species grows in drier vegetation types such as thicket on Kalahari sands. It is possible that more than one taxon may be involved.

2. **Cassia burttii** Baker f. in J. Bot. **73**: 80 (1935). —Brenan, Check-list For. Trees Shrubs Tang. Terr.: 96 (1949); in F.T.E.A., Legum.-Caesalp.: 61 (1967). —Lock, Leg. Afr. Check-list: 28 (1989). Type: Tanzania, Morogoro Dist., Wami road, *B.D. Burtt* 5032 (BM holotype, EA, FHO, K).

Shrub or small tree, 1–6(8) m tall. Branchlets dark red-brown, longitudinally ridged, sparsely puberulous, glabrescent. Leaves paripinnate; stipules not seen, probably very early caducous; petiole 6–9 cm long, sparsely appressed-puberulous; rachis 13–25 cm long, sparsely appressed-puberulous; petiolules 3–5 mm long, sparsely puberulous; leaflets opposite, in 6–7 pairs, 4.8–9 × 3.2–5.3 cm, broadly ovate to elliptic, apex rounded or retuse, base rounded to broadly cuneate, sometimes slightly asymmetric, sparsely appressed-pubescent beneath, glabrescent; venation slightly prominent particularly beneath. Inflorescences axillary, racemose, 18–50 cm long, simple or branched, pendulous; axis sparsely appressed-pubescent; bracts and bracteoles not seen, early caducous; pedicels 24–27 mm long at anthesis, sparsely appressed-pubescent. Calyx lobes 5, c. 8 × 4 mm, elliptic to ovate, minutely puberulous inside and outside. Petals 5, white, 15–16 × 8–10 mm, elliptic to obovate. Stamens 10; filaments glabrous, three of them c. 17 mm long, sigmoidally curved near the base and abruptly swollen near the middle with anthers 3 mm long, the rest 5–7 mm long, straight, anthers 1–2 mm long. Ovary multiovulate, 15–17 mm long and 1–2 mm in diameter, filiform, densely sericeous-pubescent; stipe short; stigma terminal, recessed. Pod blackish, 40–60 × 1–1.5 cm, transversely septate, terete, glabrous. Seed shiny dark brown, c. 10 × 8 × 2–3 mm, broadly obovate-elliptic, flattened.

Mozambique. N: Cabo Delgado, between L. Nangade and Mocímboa [Moçimbua] do Rovuma, fl. 20.xii.1906, *Stocks* 100 (K).

Also in Tanzania. Habitat unrecorded, probably deciduous woodland; c. 200 m.

Conservation notes: Only known from the Flora area from a single very old collection. Not infrequent in Tanzania. Critically Endangered B1 within the Flora area; probably Lower Risk, Least Concern globally.

This remains the only collection of this distinctive and (when in flower) very conspicuous species from within the Flora area. The description has been prepared partly from Tanzanian material. The pendulous racemes and white petals are distinctive, although the yellow stamens have led some collectors to describe the flowers as pale yellow.

3. **Cassia abbreviata** Oliv. in F.T.A. 2: 271 (1871). —Taubert in Engler, Pflanzenw. Ost-Afrikas **C**: 200 (1895). —Harms in Engler, Pflanzenw. Afrikas **3**(1): 495 (1915). —Baker in J. Linn. Soc., Bot. **40**: 62 (1911); Legum. Trop. Afr.: 632 (1930). —Brenan, Check-list For. Trees Shrubs Tang. Terr.: 98 (1949); in Kew Bull. **13**: 231 (1958). —Steyaert in F.C.B. **3**: 502 (1952). —O. Coates Palgrave, Trees Central Africa: 93 (1957). —Gomes e Sousa, Dendrol. Moçamb. **1**: 70 (1958); Dendrol. Moçamb. Estudo Geral **1**: 261, fig.60 (1966). —White, F.F.N.R.: 120 (1962). —Schreiber in Merxmüller, Prod. Fl. SW Afrika, fam. 59: 10 (1967). —Brenan in F.T.E.A., Legum.-Caesalp.: 59 (1967). —Drummond in Kirkia **10**: 243 (1977). —Gordon-Gray in F.S.A. **16**(2): 73, figs.18.1, 20.1 (1977). —M. Coates Palgrave, Trees Sthn. Africa: 339 (2002). Types: Mozambique, near Lupata, *Kirk* s.n., near Tete, *Kirk* s.n.; Malawi, Manganja Hills, *Meller* s.n., and Lake Malawi (Nyasa), Cape Maclear, *Kirk* s.n. (K syntypes).

Tree 5–15 m tall. Branchlets brown, longitudinally ridged, pubescent to glabrous. Leaves paripinnate; stipules c. 3 mm long, narrowly triangular, early caducous; petiole 3–10 cm long, glabrous to pubescent; rachis (7.5)11–23 cm long, glabrous to pubescent; petiolules 3–5 mm long; leaflets opposite to subopposite, in 8–10 pairs, 1.7–8.1 × 1.1–4.4 cm (the lowest usually smaller and proportionately broader), ovate to elliptic, apex acute to rounded or retuse, base cuneate to rounded, glabrous or sparsely pubescent above, pubescent to almost glabrous beneath; venation slightly prominent on both surfaces. Inflorescences terminal, racemose; axis sparsely to densely pubescent; bracts 6–9 × 2–3 mm, ovate, acuminate, pubescent; bracteoles 3–5 × 2 mm, acuminate, pubescent; bracts and bracteoles present at anthesis. Calyx lobes 5, 11–13 × 4–6 mm, ovate-elliptic, sparsely to densely pubescent outside and inside. Petals obovate. Stamens 10; anthers and filaments glabrous; 3 stamens with filaments 35–50 mm long, sigmoidally curved in the basal third, swollen above the middle, anthers 5–6 mm long; 4 with straight filaments 7–10 mm long, anthers 5 mm long; 2 with filaments 9–10 mm long, swollen at the base, recurved above, anthers c. 2 mm long; one with straight filament 7–10 mm long, anther 2 mm long. Ovary 35–45 × 1–1.5 mm including stipe c. 10 mm long, linear, densely pubescent, multiovulate; stigmatic chamber subterminal, fringed with stiff hairs. Pod dark brown to black, 20–100 cm long, 1.5–2 cm in diameter, cylindrical, velvety-pubescent to glabrous, many-seeded. Seed dark brown, 12 × 10 × 3 mm, broadly obovate, flattened, enclosed in pulp.

Subsp. **abbreviata**; Brenan in Kew Bull. **13**: 231 (1958).

Leaves pubescent beneath with erect or crisped hairs, particularly near the midrib; inflorescence axis densely pubescent with erect hairs.

Zambia. N: Mbala Dist., valley below Kambole Escarpment, fl. 13.ix.1960, *Richards* 13245 (K). W: 9.5 km (6 miles) S of Mufulira, fl. 2.x.1948, *Cruse* 395 (K). C: Lusaka Dist., between Luangwa Bridge and Rufunsa R., fl. 6.ix.1947, *Brenan & Greenway* in *Brenan* 7822 (FHO, K). **Zimbabwe.** N: near Umvukwe Mts., 8 km N of Banket, fr. 23.iv.1948, *Rodin* 4401 (K). W: Bulawayo, fl. 1.xi.1921, *Borle* 352 (K). C: Marondera Dist., Marondera (Marandellas), Igava Road, fl. 30.ix.1966, *Corby* 1643 (K, LISC, SRGH). E: Mutare Dist., Mutare (Umtali) Commonage, fl. 26.ix.1948, *Chase* 1603 (BM, K, LISC, SRGH). **Malawi.** N: Rumphi Dist., Mzimba–Rumphi road, 32 km from Rumphi, fl. 14.x.1960, *Richards* 13344 (K). C: Dowa Dist., Chitala to Domera Bay, fl. 29.x.1941, *Greenway* 6380 (K). S: Chikwawa Dist., Lengwe Game Res., fr. 7.iii.1970, *Brummitt* 8947 (K). **Mozambique.** N: Niassa Prov., between Chai and Diaca, fl. 25.x.1960, *Gomes e Sousa* 4584 (COI, K). T: between Chicoa and Fingoé, 14.8 km from Chicoa, fr. 26.vi.1949, *Barbosa & Carvalho* 3287 (K, LISC). MS: Chemba, Chiou, Estação Experimental de CICA, fr. 21.iv.1960, *Lemos & Macuácua* 131 (K, LISC).

Also in Tanzania and Congo. Woodland and wooded grassland; 750–1400 m.

Conservation notes: A widespread taxon; Lower Risk, Least Concern.

Subsp. **beareana** (Holmes) Brenan in Kew Bull. **13**: 232 (1958); in F.T.E.A., Legum.-Caesalp.: 60 (1967). —Schreiber in Merxmüller, Prodr. Fl. SW Afrika, fam. 59: 10 (1967). —Drummond in Kirkia **10**: 243 (1977). —Gordon-Gray in F.S.A. **16**(2): 73 (1977). —Lock, Leg. Afr. Check-list: 27 (1989). —Van Wyk, Trees Kruger Nat. Park, ed.3: 88 (1994). Type from East Africa, uncertain locality, *O'Sullivan Beare* (London Pharmaceutical Soc. holotype, K fragm.).

Cassia beareana Holmes in Pharm. Journ. **68** (Ser. 4, 14): 42 (1902). —Harms in Engler, Pflanzenw. Afrikas **3**(1): 495 (1915). —Baker, Legum. Trop. Afr.: 631 (1930). —Brenan, Check-list For. Trees Shrubs Tang. Terr.: 98 (1949).

Cassia droogmansiana De Wild. in Ann. Mus. Congo Belge, Bot. sér. 4, **1**: 47, fig.17 (1902). —Harms in Engler, Pflanzenw. Afrikas **3**(1). 495 (1915). Type from Congo.

Cassia granitica Baker f. in J. Bot. **43**: 45 (1905). Type: Zimbabwe, near Bulawayo, *Eyles* 1080 (BM holotype).

Cassia bequaertii De Wild. in Repert. Spec. Nov. Regni Veg. **11**: 513 (1913). Type from Congo.

Cassia abbreviata var. *granitica* (Baker f.) Baker f., Legum. Trop. Afr.: 632 (1930). —Miller, Check-list For. Trees Shrubs Bech. Prot.: 24 (1948).

Cassia abbreviata var. *glabrifructifera* Steyaert in Bull. Jard. Bot. État. **21**: 357 (1951); in F.C.B. **3**: 502 (1952). Type: Congo, Kiniama, *A. Schmitz* 2834 (BR holotype).

Leaves pubescent beneath with usually sparse small appressed hairs, sometimes almost glabrous; inflorescence axis more shortly and sparsely pubescent than in the typical subspecies.

Botswana. N: Chobe Dist., Kasane, fl. 2.viii.1950, *Robertson & Elffers* 90 (K, PRE). SE: Tuli Block, 16 km (10 miles) E of Zanzibar, 160 km (100 miles) E of Palapye, fr. 15.xii.1976, *Mott* 1046 (K, SRGH). **Zambia.** N: Mansa Dist., L. Bangweulu, near Samfya Mission, fl. 20.viii.1952, *White* 3092 (BM, K, FHO). C: S Mumbwa Dist., 120 km (75 miles) W of Lusaka, N edge of Kafue Flats, 8 km (5 miles) ESE of Shimonjela's village, fl. 15.viii.1949, *Hoyle* 1134 (FHO). S: Mazabuka Dist., between Mazabuka and Sachenga, fl. & fr. 8.x.1930, *Milne-Redhead* 1229 (K). **Zimbabwe.** N: Hurungwe Dist., near Ruwe R. on Kariba Road, fl. 28.viii.1959, *Goodier* 601 (K, LISC, SRGH). W: Nyamandhlovu Pasture Res. Station, fl. 7.x.1953, *Plowes* 1641 (K, SRGH).

C: Kadoma (Gatooma), fr iv., *Eyles* 7050 (K, SRGH). E: 32 km (25 miles) N of Chipinge, fl. 7.x.1961, *Methuen* 282 (K). S: Mberengwa Dist., Sihanda R., st. 12.iv.1951, *Harvie* 10/51 in SRGH 33238 (K, LISC, SRGH). **Mozambique.** N: Rovuma R., 32 km (20 miles) from coast, imm.fr. 15.iii.1861, *Meller* s.n. (K). Z: Maganja da Costa, between R. Raraga and crossing to Régulo Muzo, 12.7 km from river, fl. 28.ix.1949, *Barbosa & Carvalho* 4246 (K, LISC). T: Cabora Bassa, c. 1 km from dam, N bank of R. Zambezi, fr. 23.iv.1972, *Pereira & Correia* 2240 (BM, LISC). MS: Madanda Forest, fl. ix–x.1911, *Dawe* 454 (K). GI: between Saúte and Funhalouro, fr. 19.v.1941, *Torre* 2693 (K, LISC). M: Magude near Mahele, fr. 29.xi.1944, *Mendonça* 3143 (BM, K, LISC).

Also in Congo, Kenya, Tanzania, South Africa and Namibia. Woodland and wooded grassland; probably usually in drier habitats than subsp. *abbreviata*; near sea level to 1400 m.

Conservation notes: A widespread taxon; Lower Risk, Least Concern.

Subsp. *kassneri* (Baker f.) Brenan (*Cassia kassneri* Baker f.) occurs in Kenya and N Tanzania.

4. **Cassia thyrsoidea** Brenan in Kew Bull. **13**: 234 (1958); in F.T.E.A., Legum.-Caesalp.: 62 (1967). —Lock, Leg. Afr. Check-list: 29 (1989). —White, Dowsett-Lemaire & Chapman, For. Fl. Malawi: 307 (2001). Type: Tanzania, Songea Dist., R. Mkukira, *Milne-Redhead & Taylor* 8272 (K holotypc, BR).

Shrub or small tree to 8 m (18 m in East Africa) tall. Branchlets brown, longitudinally ridged when dry, sparsely pubescent, glabrescent. Leaves paripinnate; stipules not seen, very early caducous; petiole 4.5–7.2(9) cm long, sparsely to densely pubescent; rachis 16–22.5 cm long, sparsely pubescent; leaflets opposite, in 6–7 pairs, 4.5–11.5 × 2.4–5.2 cm, the distal pair the largest, ovate to ovate-elliptic, slightly asymmetric, apex bluntly acute, base cuneate, sometimes asymmetric, glabrous above, pubescent beneath particularly on the main veins; venation raised on both surfaces. Inflorescences terminal, paniculate; axis puberulous; bracts c. 8 × 1.5 mm, very narrowly triangular, densely puberulous; bracteoles similar, slightly smaller; bracts and bracteoles persisting until anthesis; pedicel (2)3–3.5 cm long at anthesis. Calyx lobes 5, 8–10 × 4–5 mm, broadly elliptic, densely puberulous outside, shortly puberulous inside. Petals yellow, c. 25 mm in diameter, subcircular, claw c. 3 mm long. Stamens 10, filaments glabrous, those of 3 stamens c. 25 mm long, of 4 stamens 6–8 mm long, of 3 stamens c. 4 mm long; anthers of 7 longer stamens 5 mm long, those of the 3 shortest stamens smaller. Ovary c. 35 mm long and 1–2 mm in diameter, filiform, densely sericeous-pubescent, multiovulate; stipe c. 4 mm long; stigma terminal, recessed. Pod (according to Brenan) blackish brown, 32–42 cm long, 1–1.5 cm in diameter, subcylindrical, attenuate at apex, somewhat so at base, sparsely pubescent to subglabrous, transversely but not longitudinally partitioned within. Seeds brown, c. 7 × 6 × 3 mm.

Malawi. C: Bua R. drift on Kasungu–Nkhota Kota (Kota Kota) Road, fl. 13.i.1959, *Robson & Jackson* 1141 (K).

Also in Tanzania. 'Fringing forest of *Parkia filicoidea*'; 850 m.

Conservation notes: Only a single record from the Flora area, and little-recorded elsewhere. Critically Endangered in the Flora area, but either Vulnerable B1 or Data Deficient on a global scale.

This description has been partially prepared from Tanzanian material, particularly the type (*Milne-Redhead & Taylor* 8272 (K)).

The Malawi specimen is less densely hairy on all its parts and the inflorescence is smaller with fewer flowers than in most Tanzanian specimens. This remains a poorly-known taxon and any division into infraspecific taxa would be premature. More material is needed.

5. **Cassia afrofistula** Brenan in Kew Bull. **13**: 236 (1958); in F.T.E.A., Legum.-Caesalp.: 62, fig.11/1–5 (1967). —Lock, Leg. Afr. Check-list: 27 (1989). —Beentje, Kenya Trees, Shrubs & Lianas: 239 (1994). —M. Coates Palgrave, Trees Sthn. Africa: 340 (2002). Type from Tanzania, Tanga Dist., Sawa, *Faulkner* 1956 (K holotype, BR).

Cassia sieberiana sensu Brenan, Check-list For. Trees Shrubs Tang. Terr.: 97 (1949) in part, non DC.

Shrub or small tree, 1–4(6) m tall. Branchlets brown, longitudinally ridged, glabrous. Leaves paripinnate; stipules not seen, probably very early caducous; petiole 1.9–8.5 cm long, glabrous; rachis 9–19 cm long, glabrous; petiolules 2–4 mm long; leaflets opposite to subopposite, in 5–6(8) pairs, 2.5–7.2 × 1.6–3.4 cm, ovate to broadly ovate, somewhat asymmetric, apex acute to rounded, base asymmetrically cuneate, glabrous above, glabrous to sparsely pubescent beneath; venation slightly prominent on both surfaces. Inflorescences of axillary and terminal racemes, held erect or horizontally, often branched; axis glabrous or pubescent; bracts c. 6 × 1 mm, linear, puberulous; bracteoles c. 2 × 0.5 mm, linear, puberulous; bracts and bracteoles falling well before anthesis; pedicels 2.5–4 cm long. Calyx lobes 5, 9–11 × 4–6 mm, ovate to elliptic, sparsely appressed-pubescent outside, glabrous inside. Petals yellow, 16–20 × 10–14 mm including a 2 mm long claw, obovate-elliptic. Stamens 10; filaments glabrous, anthers pubescent, 3 stamens with filaments c. 16 mm long, sigmoidally curved in the lower third, abruptly swollen above the middle, anthers c. 4 mm long; 4 with straight filaments 6–7 mm long, anthers c. 4 mm long; 3 with filaments either straight or helically coiled towards the apex, anthers c. 1.5 mm long. Ovary 16–24 × 1–1.5 mm, linear, pubescent, multiovulate; stigmatic chamber lateral, fringed with hairs. Pod blackish, 40–44 × 1–1.2 cm, cylindrical, terete, many-seeded; pod wall easily crushed between finger and thumb. Seed dark brown, shiny, c. 6 × 6 × 3 mm, almost circular, flattened, enclosed in pulp.

Var. **afrofistula**. FIGURE 3.2.**35**/1–5.

Leaflets very slightly pubescent towards the base beneath, otherwise glabrous. Inflorescence axis glabrous or sparsely appressed-pubescent.

Mozambique. Z: Maganja da Costa, road to Namacurra, fr. 13.ix.1949, *Andrada* 1931 (LISC). MS: Muanza Dist., Chiniziua, near R. Macalaua, fl. & fr. 9.v.1957, *Gomes e Sousa* 4383 (COI, K, LISC). GI: Manjacaze Dist., Chedinguele to Zavala road, near Boa Paz lighthouse, fl. 11.xii.1944, *Mendonça* 3386 (K, LISC).

Also in Kenya and Tanzania; cultivated elsewhere in the tropics. Mostly in coastal bushland and thicket; stated to be a fixer of dunes (*Barbosa & Lemos* 8501); 0–300 m.

Conservation notes: A widespread taxon, in a specific but extensive habitat; Lower Risk, Least Concern.

It is surprising that this variety, which is widespread in the coastal regions of East Africa, occurs further south in Mozambique than var. *patentipila*, which is not recorded for East Africa.

A cultivated specimen from Zimbabwe, Harare, Marlborough Nursery, fl. 29.i.1981, *Biegel* 5820 (K, SRGH), appears to be this taxon.

Var. **patentipila** Brenan in Kew Bull. **13**: 238 (1958); in F.T.E.A., Legum.-Caesalp.: 58, fig.11/6–10 (1967). Type: Mozambique, Zambézia, Namagoa, fl. xii.1946, *Faulkner* Pretoria 343 (K holotype, COI). FIGURE 3.2.**35**/6–10.

Leaflets sparsely or very sparsely crisped-pubescent, particularly towards the base and on the midrib. Inflorescence axis spreading-pubescent.

Mozambique. N: Pemba (Porto Amélia), by road to Maringanha lighthouse, fl. & imm.fr. 21.iii.1960, *Gomes e Sousa* 4543 (K). Z: near Mocuba, fl. 12.xii.1943, *Torre* 4798 (K, LISC).

Fig. 3.2.**35**. CASSIA AFROFISTULA var. AFROFISTULA. 1, part of leaf (× ²⁄₃); 2, part of lower surface of leaflet (× 10); 3, part of inflorescence rachis with base of pedicel (× 6); 4, part of inflorescence (× ²⁄₃); 5, flower (× 2), 1–5 from *Faulkner* 343. C. AFROFISTULA var. PATENTIPILA. 6, part of lower surface of leaflet (× 10); 7, part of inflorescence rachis with base of pedicel (× 6); 8, pod (× ¹⁄₂); 9, 9, transverse section of pod (× 1); 10, seed (× 2), 6–10 from *Faulkner* 343. Drawn by Margaret Stones. From F.T.E.A.

Not known elsewhere (but see below). Coastal woodland, bushland and thicket; 0–100 m.

Conservation notes: Endemic to the Flora area, where it is apparently not infrequent in a specific but extensive habitat; Lower Risk, Least Concern.

A specimen from Malawi (N: Chitipa Dist., 72 km (45 miles) down Nthalire Road from Nyika, 41.5 km (26 miles) S of Chisenga, 1750 m, fr. 18.iv.1975, *Pawek* 9356 (K, MO, SRGH)) appears to be this taxon. The material at Kew consists of a single leaf and a pod. The indumentum of the leaflets is consistent with this variety, but the locality is far from the main centre of distribution and at a much higher altitude than other collections. Further material, preferably in flower, would be most welcome.

The distinctions between *C. afrofistula* and the introduced *C. fistula* are discussed above (p. 164).

28. MEZONEURON Desf.

Mezoneuron Desf. in Mém. Mus. Hist. Nat. **4**: 245, fig.10, 11 (1818) as *Mezonevron.*
Caesalpinia subgen. *Mezoneuron* (Desf.) Herend. & Zarucchi in Ann. Missouri Bot. Gard. **77**: 854 (1990). —Pedley in Austrobaileya **5**: 97–102 (1997).

Climbing shrubs, usually armed with prickles on stem and leaves but rarely (not in the Flora area) unarmed. Leaves bipinnate; petiole and rachis without specialised glands; leaflets opposite or alternate (not in the Flora area); stipules very small, caducous. Flowers in terminal and axillary racemes or panicles; bracts inconspicuous. Flowers hermaphrodite, with a usually conspicuous hypanthium. Sepals 5, imbricate, the lower curving upwards over the others. Petals 5, subequal or the upper one somewhat modified. Stamens 10, fertile, alternately long and short; filaments pubescent or villous in the lower part or glabrous (not in the Flora area). Ovary sessile or shortly stipitate; stigma small, oblique, ciliolate or glabrous. Pods oblong-ellipsoid, thickened along the upper margin, or flat with a broad longitudinal wing along the upper margin, chartaceous to coriaceous, indehiscent, 1–9 seeded. Seeds compressed, without endosperm.

A genus of about 26 species from tropical Asia to Australia and Polynesia, with one species in Madagascar and 2 in continental Africa. The genus differs from *Caesalpinia* mainly in its characteristic indehiscent pods.

Pinnae 13–19 pairs per leaf; leaflets 14–25 pairs per pinna, 6–9 mm long; pods oblong-ellipsoid, thickened along the upper suture .... **1.** *welwitschianum*
Pinnae 4–10 pairs per leaf; leaflets 6–8 pairs per pinna, 10–28 mm long; pods flat, winged along the upper margin .... **2.** *angolense*

1. **Mezoneuron welwitschianum** Oliv. in F.T.A. **2**: 261 (1871). —Baker, Legum. Trop. Afr.: 613 (1930). —Wilczek in F.C.B. **3**: 260, fig.20 (1952). —Torre & Hillcoat in C.F.A. **2**: 171 (1956). —White, F.F.N.R.: 126, figs. 20K, 26 (1962). Type from Angola (Cuanza Norte).

Caesalpinia welwitschiana (Oliv.) Brenan in Kew Bull. **17**: 203 (1963); in F.T.E.A., Legum.-Caesalp.: 35 (1967). —Lock, Leg. Afr. Check-list: 23 (1989).

Liane recorded up to 18 m high in the F.Z. area, or to 45 m in the Congo. Stems glabrous to brown-pubescent, armed with downwardly hooked prickles up to 7 mm long. Leaves: petiole and rachis together 18–36 cm, pubescent to tomentose, armed with downwardly hooked prickles up to 4 mm long and usually, but not exclusively, in pairs at the insertions of the pinnae; pinnae 13–19 pairs; pinna rachis 3–7(8 in F.C.B.) cm; leaflets 14–25 pairs per pinna, (4)6–9 × (1.5)2–3 mm, narrowly oblong, asymmetrical at the base, rounded at the apex, minutely pubescent beneath, pubescent or usually glabrous above. Racemes 8–36 cm long, terminal and usually also axillary; bracts minute, caducous; pedicels 1–2.3 cm. Lateral sepals 7–10 mm long;

all sepals glabrous. Petals 11–13 × 7–8 mm, yellow but at least sometimes with red veins. Stamens 12–15 mm long, only slightly exceeding the petals; filaments densely villous in the basal part but the distal part glabrous; anthers glabrous. Ovary glabrous. Pods 2.5–6 × 2.2–3.1 cm (in F.T.E.A.), ± oblong-elliptic, compressed, unarmed, glabrous, the upper margin thickened but not winged, indehiscent, 1–3-seeded; seeds 1.2–1.6(2) cm diameter, subglobose, olive-brown.

Zambia. W: Kabompo Dist., below boma, on sandbank in Kabompo R., fl. 23.xi.1952, *Holmes* 1011 (K); Kabompo, fl. 23.ix.1964, *Fanshawe* 8924 (K, NDO).

From Cameroon to Uganda, W Tanzania (Kigoma Dist.) and Angola; known from the Flora Zambesiaca area only from the two specimens cited. Riverine, elsewhere in lowland rainforest, river banks; c. 800–1200 m.

The specimen *Holmes* 1011 is labelled: "Scrambler up to top of tall trees 55 ft high. Leaves glabrous, dull on both surfaces; main rachis and rachis of pinna with orange-buff hairs. Flowers — sepals glabrous, a little mealy on the outside, green-yellow with green on margins; inside yellow-green with margins red-purple. Petals rich yellow streaked with blood-red sap, glabrous near base. Stamens — anther black, upper filament yellow, lower with dense long white and blood-red hairs. Stigma yellow merging into red at the base. Ovary red."

Brenan (Kew Bull. 17, 1963) transferred this species to *Caesalpinia* because the fruits are not flattened and winged as in the core group of species in this genus, and drew attention to its similarity to other Asiatic species included in *Caesalpinia* at that time, notably *C. tortuosa* Roxb. These Asiatic species are now also referred to *Mezoneuron* in current concepts, see Lewis, Schrire, Mackinder & Lock, Legumes of the World (2005).

2. **Mezoneuron angolense** Oliv. in F.T.A. **2**: 261 (1871). —Harms in Engler, Pflanzenw. Afrikas **3**(1): 513 (1915). —Baker, Legum. Trop. Afr. **3**: 613 (1930). —Brenan, Check-list For. Trees Shrubs Tang. Terr.: 105 (1949); in F.T.E.A., Legum.-Caesalp.: 40, fig.6 (1967). —Wilczek in F.C.B. **3**: 259 (1952). —Torre in Mendonça, Contrib. Conhec. Fl. Moçamb. **2**: 66 (1954). —Torre & Hillcoat in C.F.A. **2**: 170 (1956). White, F.F.N.R.: 125 (1962). —Lock, Leg. Afr. Check-list: 25 (1989). Types: Angola, Pungo Andongo, *Welwitsch* 606 (LISU lectotype), and Golungo Alto, between Sange and Camilungo, *Welwitsch* 607 (LISU syntype). FIGURE 3.2.**36**.

Caesalpinia angolensis (Oliv.) Herend. & Zarucchi in Ann. Missouri Bot. Gard. **77**: 854 (1990).

Liane recorded up to 14 m high in the Flora Zambesiaca area (up to 20 m high and stems 8 cm diameter in F.C.B. area). Stems with sparse hairs or subglabrous, armed with scattered, downwardly hooked prickles up to 11 mm long, which on older stems become enlarged and raised on subconical-cylindrical corky bosses up to 2 cm long and 1.5 cm diameter (see *Milne-Redhead* 4551 in carpological collection at Kew). Leaves: petiole and rachis together (6)10–30 cm, armed on the lower side with downwardly hooked prickles often in pairs particularly at the insertions of the pinnae; pinnae 4–10 pairs; pinna rachis 3–8 cm, sometimes with small prickles at the insertions of the lower leaflets; leaflets (4)6–8(9) pairs per pinna, (6)10–28 × (4)6–15(19) mm, elliptic-oblong or broadly so, or the terminal usually slightly larger and broadly obovate, rounded at the apex, glabrous except usually on the midrib beneath and often also the basal part of the lower surface. Flowers in terminal and lateral racemes 5–30(40) cm long; bracts linear-triangular exceeding the very young buds and recurving, but falling very early and seldom seen in herbarium specimens; pedicels (4)6–11 mm. Lateral sepals 5–7(8) mm long; all sepals puberulous to pubescent. Petals 7–9 mm long, the lamina suborbicular with a short claw, yellow or greenish-yellow. Longest stamens 13–17 mm; filaments villous in the lower part; anthers 1–1.5 mm long. Ovary glabrous or tomentose [see note below]. Pods 6.5–14.5 × (2)2.5–4.3 cm, including the wing on the upper margin 0.5–1.2 cm broad, elliptic-oblong to narrowly elliptic, with the persistent receptacular disc forming a collar round its base; seeds normally 3–4(?5) but sometimes only 1–2 developing, 8–9 × 6–7 mm (in F.T.E.A.), much compressed.

Fig. 3.2.**36**. MEZONEURON ANGOLENSE. 1, part of flowering branch (× $^{2}/_{3}$), from *Purseglove* 911; 2, flower (× 2); 3, flower, longitudinal section (× 2); 4, stamen (× 2), 2–4 from *Purseglove* 2644; 5, part of infructescence with pod (× $^{1}/_{2}$); 6, part of valve of opened pod, to show seed (× $^{2}/_{3}$), 5 & 6 from *Purseglove* 911. Drawn by Lura Ripley. From F.T.E.A.

Zambia. W: Mwinilunga Dist., by R. Matonchi, st. 12.ii.1938, *Milne-Redhead* 4551 (K). **Mozambique.** N: Mueda Dist., between Mueda and Chomba, fr. 22.ix.1948, *Barbosa* 2230 (K, LISC). Z: region of Maganja da Costa, fl. 30.vii.1943, *Torre* 5736 (BR, LISC). MS: Muanza Dist., Cheringoma, R. Chiniziua (Chiniciua), fr. 22.x.1949, *Pedro & Pedrógão* 8860 (PRE).

Also in Liberia, Cameroon, Congo, Uganda, Kenya, Tanzania (including Zanzibar) and Angola. In lowland rainforest, riverine and swamp forest, coastal evergreen bush or open *Parinari* woodland (*Correia* 93); sea level to 1450 m.

In most of the range of this species the ovary is densely tomentellous, but in specimens from Tanzania and Mozambique in which this character can be seen the ovary is ± glabrous. Further investigation of the distribution of plants with glabrous ovaries is required when more material is available.

Plants of this species with a glabrous ovary are very similar to plants from tropical Asia and Indonesia currently referred to *M. enneaphyllum* (Roxb.) Benth. and *M. latisiliquum* (Cav.) Merrill. When a full revision of the whole genus is undertaken it may perhaps prove necessary to sink *M. angolense* and also *M. hildebrandtii* Vatke from Madagascar into one of these species, but in view of the present very inadequate understanding of the Asian species no such decision has been taken here.

29. PTEROLOBIUM Wight & Arn.

Pterolobium R. Br. [in Salt, Voy. Abyss., append.: lxiv (1814) nom. nudum] ex Wight & Arn., Prodr. Fl. Ind. Or. **1**: 238 (1834) nom. conserv. —Vidal & Thol in Bull. Mus. Nat. Hist. Nat., sér. 3, Bot. 15, **227**: 1–29 (1974).

Shrubs, usually climbing, armed with prickles on stem and leaves. Leaves bipinnate; petiole and rachis without specialised glands; leaflets opposite. Flowers smallish, in ± dense many-flowered terminal and lateral racemes often grouped into panicles; bracts inconspicuous. Flowers hermaphrodite, with a small receptacular disc. Sepals 5, imbricate, the lower one curving upwards over the others. Petals 5, subequal. Stamens 10, all fertile, alternately longer and shorter; filaments pubescent in the lower part. Ovary subsessile; stigma transversely flattened, not peltate; ovule 1, attached towards distal end of ovary. Pods resembling the samara of *Acer* species with a basal, ± broadly elliptic, sometimes slightly inflated, seed-containing portion, extending obliquely on the upper margin into a broad wing usually 2–4 times as long as the basal part, the style sometimes persistent for a short time at the apex; seed solitary, compressed, without endosperm.

10 species in Asia and Malesia and a single one in Arabia and Africa.

The valid publication of the generic name has been discussed by Brummitt in Taxon **17**: 598 (1968).

Pterolobium stellatum (Forssk.) Brenan in Mem. New York Bot. Gard. **8**: 425 (1954); in F.T.E.A., Legum.-Caesalp.: 42, fig.7 (1967). —Roti-Michelozzi in Webbia **13**: 181, fig.7 (1957). —White, F.F.N.R.: 128, fig.20 (1962). —Breitenbach, Indig. Trees S. Africa **3**: 350 (1965). —Ross in F.S.A. **16**(2): 115, fig.23 (1977). —Polhill & Thulin in Fl. Ethiopia **3**: 54, fig.95.2.1 (1989). —White, Dowsett-Lemaire & Chapman, Evergr. For. Fl. Malawi: 309, fig.108 (2001). —M. Coates Palgrave, Trees Sthn. Africa: 343 (2002). Type: Yemen, Kurma, *Forsskål* s.n. (C lectotype). FIGURE 3.2.**37**.

Mimosa stellata Forssk., Fl. Aegypt.-Arab.: cxxiii, 177 (1775). —Vahl, Symb. Bot. **1**: 81 (1790).

Cantuffa exosa J.F. Gmel. in Linnaeus, Syst. Nat., ed.13, **2**: 677 (1791). Type: Bruce, Travels 5: app.49 (1790).

Pterolobium lacerans R. Br. in Salt, Voy. Abyss., append.: lxiv (1814) nom. illegit. —Oliver in F.T.A. **2**: 264 (1871). —Harms in Engler, Pflanzenw. Afrikas **3**(1): 503, fig.268 (1915). —Eyles in Trans. Roy. Soc. S. Afr. **5**: 368 (1916). Type as for *Pterolobium stellatum.*

Pterolobium exosum (J.F. Gmel.) Baker f., Legum. Trop. Afr. **3**: 621 (1930). —Burtt Davy, Fl. Pl. Ferns Transvaal: 330 (1932). —Steedman, Trees, Shrubs & Lianes S. Rhod.: 21 (1933). —Brenan, Check-list For. Trees Shrubs Tang. Terr.: 106 (1949). —Suessenguth & Merxmüller in Proc. & Trans. Rhod. Sci. Ass. **43**: 17 (1951). —Wilczek in F.C.B. **3**: 256 (1952). —Burtt Davy & Hoyle Check List For. Trees Shrubs Nyasaland Prot.: 38 (1958).

A scrambling or climbing shrub up to 20 m high. Stems densely puberulous when young, and armed with downwardly hooked prickles up to 6 mm long in pairs at the insertions of the leaves and occasionally singly elsewhere, the prickles on older stems becoming raised on corky bosses as in *Mezoneuron angolense* (in photograph of *Wild* s.n. from Zimbabwe in SRGH). Leaves: petiole and rachis together 5–17(25) cm long, armed on the lower side with downwardly hooked prickles in pairs at the insertions of the pinnae and occasionally singly between the pinnae, and on the upper side usually also with straight, erect or forwardly pointed prickles singly at the insertions of the pinnae; pinnae (5)7–13 pairs; pinna rachis (1.5)2.5–5(6) cm; leaflets 7–16 pairs per pinna, (4)6–10(12) × 2–3.5(5 in F.T.E.A.), oblong, rounded to slightly emarginate at the apex; lower surface pubescent, upper surface glabrous or sometimes pubescent. Flowers sweetly scented, small, in dense terminal and lateral racemes up to 18 cm long, aggregated into pseudopanicles up to 35 cm long; bracts subulate, falling very early; pedicels 3–5(6) mm. Sepals 2–3 × 0.75–1.4 mm, usually ± reflexed in flower, pubescent, greenish. Petals about equalling the sepals, erect, oblanceolate-oblong, glabrous, cream coloured. Stamens all exserted, the longer ones almost twice as long as the petals; anthers c. 0.5 mm long. Ovary densely pubescent. Pods (including the wing) 3–5(6) cm long, the wing up to 1.7 cm broad, the seed-containing basal part up to 2 × 1 cm, pubescent or at maturity glabrescent, a striking scarlet or wine-red while maturing, ultimately brown; seed 9–11 × 5–6.5 × 3.4 mm, ellipsoid.

Zambia. N: Mbala Dist., on track to Iruma village, fl. 18.iv.1959, *Richards* 11275 (K, SRGH). C: Lusaka Dist., c. 13 km NW of Lusaka, fr. 1.vi.1958, *Best* 134 (K, SRGH). S: Choma Dist., Mapanza West, near R. Munyeke, fl. & fr. 14.v.1953, *E.A.Robinson* 222, 222b (K). **Zimbabwe.** N: Makonde Dist., Whindale Farm, fl. 20.iv.1962, *Jacobsen* 1681 (SRGH). W: Bulawayo Dist., old Gwanda road c. 48 km from Bulawayo, fr. v.1956, *Judge* 1/56 (K, SRGH). C: Marondera (Marandellas), by cave on granite, fl. 5.iv.1950, *Wild* 3250 (K, SRGH). E: Mutare Dist., E fringe of Murahwa's (Murakwa's) Hill, Commonage, fl. 6.iii.1963, *Chase* 7959 (SRGH). S: Gutu Dist., Gutu Reserve, fl. & fr. iii.1961, *Davies* 2914 (SRGH). **Malawi.** N: Mzimba Dist., Kabumba R., 22 km S of Engucwini, on road from Rumphi to Kafukule, fr. 6.vii.1970, *Brummitt* 11837 (K). C: Nkhotakota Dist., Ntchisi Mt, fr. 29.vii.1946, *Brass* 17023 (K, SRGH). S: Mulanje Dist., Machemba Hill, W side, fr. 1.ix.1984, *Patel & Morris* 1560 (K, MAL). **Malawi** S or **Mozambique** Z: Chikwawa (Shibisa) to Tshinsunze (Tshinmuze), fr. ix.1859, *Kirk* (K). **Mozambique.** MS: Mossurize Dist., near Missão Católica de Espungabera (Spungabera), fr. 9.vi.1942, *Torre* 4272 (K, LISC, LMA). GI: Zavala Dist., Quissico (Quisica), fr. 1908, *Sim* 21160 (PRE).

Also in South Africa (Mpumalanga) and northwards through East Africa to eastern Congo to the Sudan, Eritrea and Yemen. In riverine forest, upland evergreen forest, dambos, open bush and rocky places; up to 1750 m.

This species has become infamous for the vicious character of its thorns. It was first described from Africa by Bruce, Travels **5**: 49 (1790) under the Abyssinian vernacular name *Kantuffa*, and his comments bear repetition today: "This thorn, like many men we meet daily in society, has got itself into a degree of reputation and respect from the noxious qualities and powers of doing ill which it possesses, and the constant exertion of these powers.....The soldier screens himself by a goat's, leopard or lion's skin, thrown over his shoulder, of which it has no hold. As his head is bare he always cuts his hair short before he goes to battle, lest his enemy should take advantage of

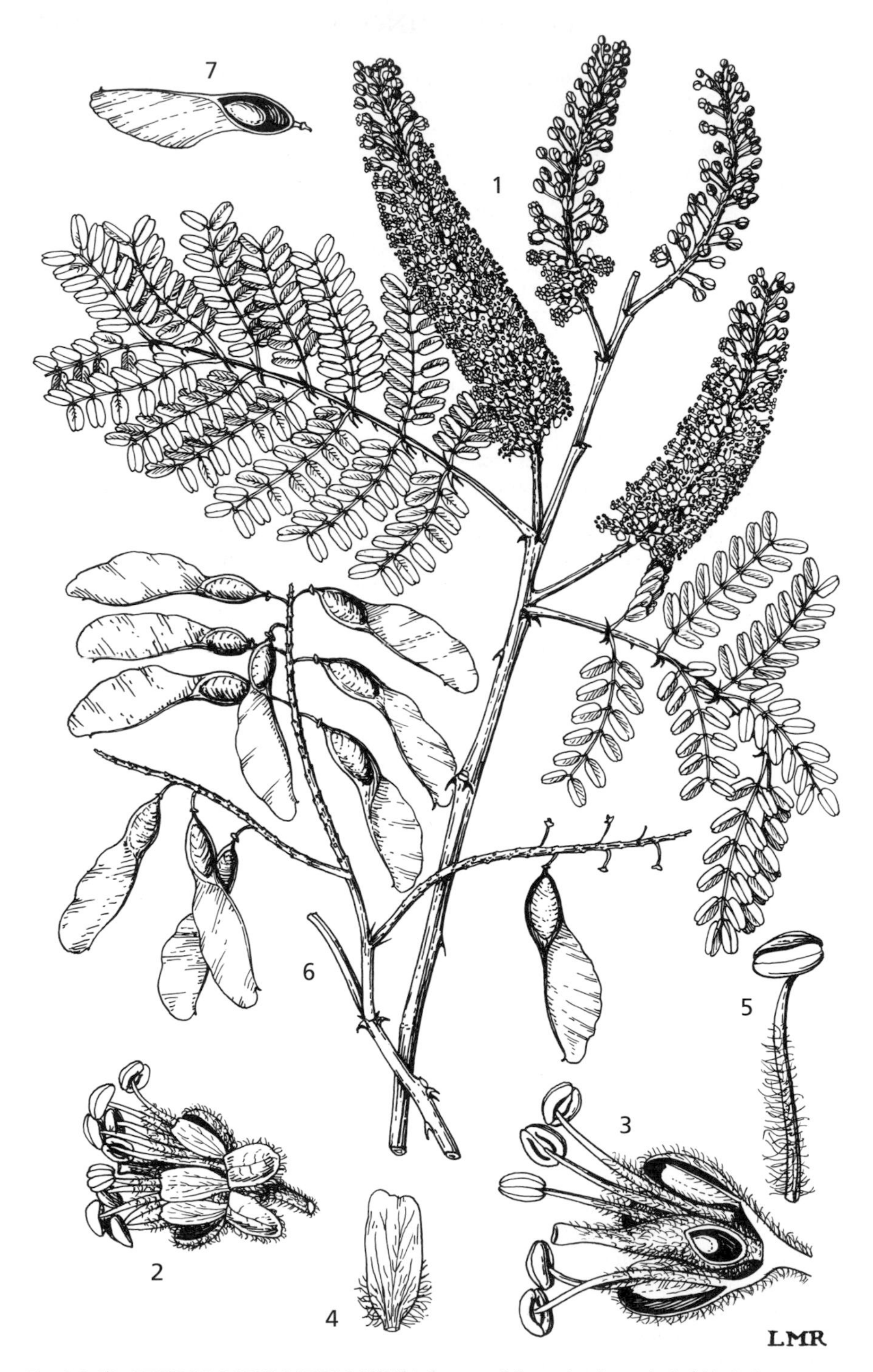

Fig. 3.2.**37**. PTEROLOBIUM STELLATUM. 1, part of flowering branch (× $^{2}/_{3}$); 2, flower (× 4); 3, flower, longitudinal section (× 6); 4, petal (× 9); 5, stamen (× 9), 1–5 from *Richards* 11275; 6, part of branchlet with pods (× $^{2}/_{3}$), from *Eggeling* 3400; 7, pod, with seed bearing part opened (× $^{2}/_{3}$), from *Sandwith* 25. Drawn by Lura Ripley. From F.T.E.A.

it; but the women, wearing their hair long, and the great men....being always cloaked it never fails to incommode them....If their cloak is fine muslin the least motion against it puts it all in rags; but if it is a thick, soft cloth....it buries its thorns, great and small, so deep in it that the wearer must either dismount and appear naked, which to principal people is a great disgrace, or else much time will be spent before he can disengage himself from its thorns. In the time when one is thus employed it rarely fails to lay hold of you by the hair, and that again brings on another operation....but much more painful than the other."

30. CAESALPINIA L.

Caesalpinia L., Sp. Pl.: 380 (1753); Gen. Pl., ed. 5: 178 (1754) as *Caesalpina.*

Shrubs, erect or usually scrambling or climbing, or sometimes trees, usually armed with spines or prickles but sometimes unarmed. Leaves bipinnate with paired pinnae; leaflets opposite, glandular or eglandular; without specialised glands on petiole and rachis, though sometimes with a general glandular indumentum; stipules minute to conspicuous and leafy. Flowers in terminal or lateral racemes or panicles, these rarely much reduced to single or few flowers; bracts usually caducous. Hypanthium shallow, ± obconic. Sepals 5, imbricate to sometimes almost valvate, the lowermost curving up to overlap the other 4. Petals 5, subequal except for the uppermost one which is usually modified in shape and size. Stamens 10, alternately longer and shorter; filaments pubescent to villous and often glandular in the lower part; anthers dorsifixed, dehiscing by longitudinal slits. Ovary ± sessile or shortly stipitate, glabrous or with various indumentum; stigma truncate or oblique, ciliolate or glabrous. Pods very variable, usually compressed, not winged, dehiscent or not, sometimes spiny; seeds variable, endosperm present or absent.

About 25 species, 10 in Africa and Arabia, 1 in Madagascar and c. 14 in the New World, mostly in the Caribbean and Central America. Many other species formerly included in this genus have now been dispersed to other segregate genera, see Lewis in Lewis, Schrire, Mackinder & Lock, Legumes of the World (2005).

Caesalpinia sappan L., native of India and Malesia, is widely cultivated in tropical regions and is known in the Flora area from a single collection — **Mozambique.** M: Maputo (Lourenço Marques), st. 1908, *Sim* 21085 (PRE). It is a shrub or small tree with sparse prickles on the stem, leaves up to c. 40 cm long with up to 12 pairs of pinnae, each pinna with up to 20 pairs of ± rhombic coriaceous leaflets up to c. 20 × 9 mm and shiny on the upper surface and emarginate at the apex; racemes often aggregated to form a pseudopanicle; with yellow petals c. 13 mm long, about equalled by the stamens; and with smooth, ± obtriangular pods up to 10 × 4 cm.

1. Longest stamens 4–7 cm long; pedicels 2.5–7 cm long .... **1.** *pulcherrima*
– Longest stamens 0.8–2 cm long; pedicels 0.4–2.7 cm long (longer elsewhere) . . 2
2. Midrib arising from lower corner of leaflet at the base, running ± diagonally, the leaflet thus very asymmetric at the base; cultivated straggling shrub or small tree .... *sappan*
– Midrib at base of leaflet arising subcentrally, the base thus ± symmetric 3
3. Leaflets 1–2.3 cm long, without gland dots; petiole and rachis of leaf with numerous curved prickles as well as erect prickles on upper side at insertion of leaflets; petals yellow or yellowish white .... **4.** *decapetala*
– Leaflets 0.3–1.1 cm long, with numerous gland dots particularly on lower surface; petiole and rachis unarmed apart from subulate stipellae at insertion of leaflets; petals pink to red 4

4. Pinnae 4–12 pairs per leaf; leaflets 8–18 pairs per pinna, 1.3–6 × 0.6–2 mm; lower sepal without a conspicuous rostrate beak but occasionally a small rudimentary beak ± 1 mm long present . . . . **2.** *rubra*
– Pinnae 3–7 pairs per leaf; leaflets 6–11 pairs per pinna, 3–11 × 2–4.5 mm; lower sepal with a very conspicuous apical, upwardly curved rostrate beak 2.5–4 mm long . . . . **3.** *rostrata*

1. **Caesalpinia pulcherrima** (L.) Sw., Observ. Bot. Pl. Ind. Occ.: 166 (1791). —Oliver in F.T.A. **2**: 262 (1871). —Baker, Legum. Trop. Afr. **3**: 616 (1930). —Brenan, Check-list For. Trees Shrubs Tang. Terr.: 95 (1949); in F.T.E.A., Legum.-Caesalp.: 31 (1967). —Wilczek in F.C.B. **3**: 254 (1952). —Torre & Hillcoat in C.F.A. 2: 172 (1956). —Roti-Michelozzi in Webbia **13**: 214 (1957). —White, F.F.N.R.: 119 (1962). —Ross in F.S.A. **16**(2): 129 (1977). —Polhill & Thulin in Fl. Ethiopia **3**: 51 (1989). —Beentje, Kenya Trees, Shrubs & Lianas: 238 (1994). Type: India, *Herb. Linnaeus* 529.1 (LINN syntype).

Poinciana pulcherrima L., Sp. Pl.: 380 (1753). —Sims in Bot. Mag. **25**: fig.995 (1807). —Howes in Kew Bull. **1**: 78 (1947).

Shrub 1–6 m high, occasionally ± scrambling. All parts except the stamen filament glabrous; stems smooth, unarmed or sometimes with sparse ± spreading prickles up to 6 mm long. Leaves: petiole and rachis together 6–30 cm, unarmed or occasionally with paired prickles (?modified stipellae) at the insertions of the pinnae; pinnae 3–10(14) pairs; pinna rachis 1.8–9 cm long; leaflets 5–12(20) pairs per pinna, (5)9–20(28) × 4–12(15 in F.T.E.A.) mm, oblong to oblong-elliptic, rounded to emarginate at the apex, asymmetrical at the base, glabrous. Racemes terminal or terminal and axillary; bracts ± linear-lanceolate, up to 7.5 mm long, falling at an early stage; pedicels (1.8)2.5–7(10.5) cm, lengthening as the buds mature and so often forming a rounded corymb. Lateral sepals 7–14 mm long, lower sepal up to 17 mm long. Lateral and lower petals (12)15–26 × 10–20 mm, the upper one smaller, all erose-undulate at the distal margin, red, red and yellow, orange or (var. *flava*) yellow. Longer stamens 4–7 cm, long-exserted; filaments red, hairy near the base. Ovary glabrous. Pod 6–12 × 1.4–2.5 cm, obliquely oblong to oblanceolate, compressed, smooth, brownish- or purplish-black, dehiscing down both sides; seeds 9–10 × 7–8 × 3 mm, obovate or obtriangular, compressed, brownish-black.

Zambia. C: Lusaka Dist., Mt. Makulu Res. Station, near Chilanga, cultiv., fl. & fr. 18.ii.1960, *Angus* 2145 (SRGH). E: Chipata (Fort Jameson), fl. iii.1950, *Gilges* 30 (SRGH). **Malawi.** C: Salima Dist., Lake Nyasa Hotel, Salima, stabilised dunes, fl. 19.v.1957, *McQueen* 54 (K, SRGH). S: Kundwelo Village, Phalombe (Palombe) Plain, fl. 29.vii.1956, *Newman & Whitmore* 280 (SRGH). **Mozambique.** N: Mogovolas, Posto Agrícola, fl. v.1934, *Ribeiro* 43 (LISC). GI: Xai-Xai (Vila de João Belo), Tavane, fl. 7.x.1945, *Pedro* 202 (PRE, SRGH). M: Maputo (Lourenço Marques), Jardim Tunduru (Vasco da Gama), fl. & fr. 15.iv.1947, *Pedrógão* 177 (PRE).

Probably native to tropical America (though often alleged to be Asiatic) but cultivated in most parts of the tropics and frequently becoming naturalised — see Brenan in F.T.E.A. Cultivated as a hedge plant (Howes, 1947) or as an ornamental, commonly known as the Pride of Barbados. Some of the specimens cited above are certainly from cultivation, and observations on the ability of this species to become naturalised in the F.Z. area are required. It is apparently less able to establish itself than is *C. decapetala.*

2. **Caesalpinia rubra** (Engl.) Brenan in Kew Bull. **17**: 202 (1963). Type: Namibia, Karibib Dist., Usakos, *Marloth* 1432 (B holotype; BOL, PRE). FIGURE 3.2.**38.**

Hoffmannseggia rubra Engl., Bot. Jahrb. Syst. **10**: 25 (1889). —Harms in Engler, Pflanzenw. Afrikas **3**(1): 503 (1915). —Dinter in Repert. Spec. Nov. Regni Veg. **18**: 425 (1922). —Baker, Legum. Trop. Afr.: 619 (1930). —Miller in J. S. Afr. Bot. **18**: 32 (1952). —Schreiber in Merxmüller, Prodr. Fl. SW Afrika, fam. 59: 8 (1967).

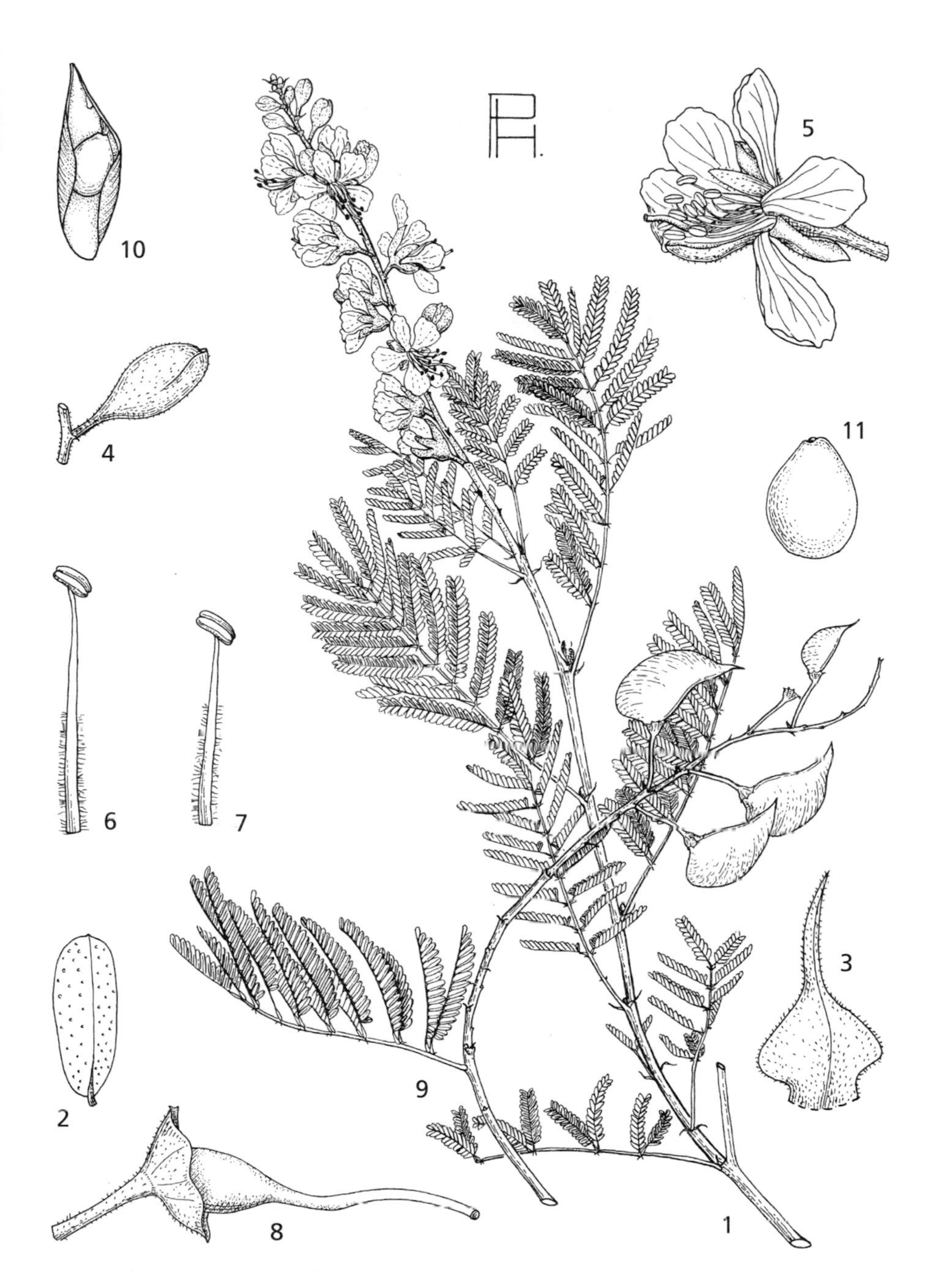

Fig. 3.2.**38**. CAESALPINIA RUBRA. 1, flowering branchlet (× ²/₃); 2, leaflet, showing gland dots (× 6); 3, bract (× 8); 4, flower bud (× 2); 5, flower (× 2); 6, 7, stamens (× 4); 8, gynoecium (× 4), 1–8 from *De Winter* 3033; 9, fruiting branchlet (× ²/₃); 10, valve of dehisced pod, showing seed (× 1); 11, seed (× 2), 9–11 from *De Winter* 3164. Drawn by Pat Halliday. Reproduced with permission from Flora of Southern Africa.

Shrub up to 1.7 m high. Stems densely grey-puberulent to -pubescent when young, armed with usually sparse, spreading or down-curved prickles up to 7 mm long, perhaps occasionally unarmed. Leaves: petiole and rachis together 1–9 cm, unarmed but with subulate stipellae at the insertions of the pinnae; pinnae 4–12 pairs; pinna rachis (0.4)0.7–3.5 cm, unarmed; leaflets 8–18 pairs per pinna, 1.3–6 × 0.6–2 mm, asymmetrically oblong, rounded at the apex, glabrous or with the midrib puberulent and margins ciliate, with numerous gland dots throughout. Racemes (1)3–18 cm long, generally terminal; bracts up to 3 × 2 mm long excluding the terminal arista c. 2 mm long, broadly ovate, acuminate-aristate at the apex, densely grey-puberulent, falling before opening of the subtended flower; pedicels 4–14 mm. Lateral sepals 4–8 mm long; lower sepal occasionally with a rudimentary beak on its dorsal surface at the extreme end of the flower; all sepals densely puberulent when young, sometimes glabrescent. Lateral and lower petals 7–12 mm long; all petals red or pink. Longest stamens 8–12 mm. Ovary glabrous, blackish. Pods 2.0–2.4 × 1.2–1.4 cm, obliquely pyriform, the point of attachment being sublateral, the apex acute, brown, dehiscing down both margins; seeds c. 8 × 7 × 2 mm, obovate, compressed.

Botswana. SE: Central Dist., 3 km E of Letlhakane (Lothlekane), young bud, 23.iii.1965, *Wild & Drummond* 7251 (K, SRGH); Central Dist., near Serowe, fr. 13.viii.1968, *Buerger & Boshoff* 90/37 (K); Kalahari Desert, Letlhakane (T'Klakane) Pits, fl. 24.x.1899, *E.J. Lugard* 305 (K).

Also in Namibia; known in the Flora area only from the 3 specimens cited. Rocky places, river beds; 750–1050 m.

3. **Caesalpinia rostrata** N.E. Br. in Hooker's Icon. Pl. 28: fig.2702 (1901). —Harms in Engler, Pflanzenw. Afrikas **3**(1): 510 (1915). —Torre in Mendonça, Contrib. Conhec. Fl. Moçamb. **2**: 67 (1954). —Ross in F.S.A. **16**(2): 127 (1977). —Germishuizen, Coetzer & Condy in Fl. Pl. Africa **57**: 70, pl. 2173 (2001). Type from cultivation in Durban Botanic Garden (South Africa), raised from seed obtained from "Delagoa Bay", Maputo (Lourenço Marques), *Wood* 7943 (K holotype; BOL, NH, PRE).

Scrambling shrub 2–4 m high. Stems puberulous or appressed-pubescent when young, armed with scattered straight spreading prickles up to 10 mm long. Leaves: petiole and rachis together (0.8)1.5–9.5 cm, unarmed but with subulate stipellae at the insertions of the pinnae; pinnae (1)3–7 pairs; pinna rachis 1–5.5 cm, unarmed; leaflets (4)6–11 pairs per pinna, (2)3–11 × (1.5)2–4.5 mm, oblong to oblong-elliptic, asymmetrical at the base, rounded to truncate at the apex, glabrous or the midrib puberulent beneath, with numerous dark gland dots particularly noticeable on the lower surface. Racemes 2–13 cm long, terminal or axillary; bracts up to c. 7 × 8 mm excluding the terminal arista c. 2 mm long, very broadly elliptic to suborbicular and concave, truncate-aristate at the apex, somewhat scarious, brown, pubescent, falling before opening of their subtended flower; pedicels 3–5 mm. Lateral sepals c. 5–7 mm long; lower sepal strongly hooded over the others and with a conspicuous distinct curved beak 2.5–4 mm long on its dorsal surface at the extreme end of the flower; all sepals pubescent. Lateral and lower petals c. 12 mm long, broadly obovate, the upper one somewhat shorter and much narrower, ± spathulate, all red or pink. Longest stamens 10–12 mm. Ovary glabrous, black. Pods 2.7–3.2 × 1.6–1.9 cm, broadly oblong, smooth, brown, dehiscing down both margins; seeds not seen.

Mozambique. M: Magude, banks of Incomati R., Delagoa Plantation, fl. 29.xi.1944, *Mendonça* 3123 (LISC); Magude, Xrangagine, fr. ix.1946, *Pimenta* s.n. (LISC, LMA).

Apparently confined to S Mozambique (only 2 collections known) and adjacent South Africa (Mpumalanga). Recorded from river banks, but further information required.

An interesting but little-known species. The gland-dotted leaflets and the characters of the bracts and the pods indicate that its affinities are clearly with *C. rubra* (Engl.) Brenan as described above and *C. trothae* Harms from Tanzania to

Ethiopia. It is interesting to note that the most unusual beak on the lower sepal, so characteristic of this species, is occasionally also very slightly developed in *C. rubra* (see *Dinter* 6840 and *de Winter* 3033 at K).

The species was first described from plants grown at Durban Botanic Garden in South Africa from seed collected at "Delagoa Bay" (Maputo) by James Wilson. Specimens from cultivation at Durban are dated 1895 and 1900 (K) but J.H. Ross informs me that it is no longer in cultivation there and there is no record of how long it remained. The exact location of the original collection of the seed is unknown. It is now known in the wild only from the 2 collections from S Mozambique cited above and a single specimen from the adjacent part of South Africa, Crocodile Bridge area, Komati River gorge through the Lebombos, Kruger Nat. Park, fr. 9.xi.1954, *van der Schiff* 3999 (K, PRE). The inclusion of Zimbabwe in the distribution given by Torre (1954) is an error.

4. **Caesalpinia decapetala** (Roth) Alston in Trimen, Handb. Fl. Ceylon **6** (suppl.): 89 (1931). —Brenan, Check-list For. Trees Shrubs Tang. Terr.: 94 (1949); in Mem. New York Bot. Gard. **8**: 425 (1954); in F.T.E.A., Legum.-Caesalp.: 36 (1967). —Wilczek in F.C.B. **3**: 253 (1952). —Wild, Common Rhod. Weeds: fig.51 (1955). —Torre & Hillcoat in C.F.A. **2**: 172 (1956). —White, F.F.N.R.: 118, fig.20H (1962). —Ross in F.S.A. **16**(2): 125 (1977). —Polhill & Thulin in Fl. Ethiopia **3**: 53, fig.95.1.2 (1989). —White, Dowsett-Lemaire & Chapman, Evergr. For. Fl. Malawi: 306 (2001). —M. Coates Palgrave, Trees Sthn. Africa: 344 (2002). —Schmidt, Lötter & McCleland, Trees Shrubs Mpumalanga & Kruger Nat. Park: 184 (2004). Type: India, *Heyne* s.n. (K). FIGURE 3.2.**39**.

Caesalpinia sepiaria Roxb. [Hort. Bengal.: 32 (1814), nom. nud.] Fl. Ind. **2**: 360 (1832). —Baker, Legum. Trop. Afr.: 615 (1930). —Howes in Kew Bull. **1**: 63 (1947). Type: India, *Roxburgh* in *Wallich* 5834a (K).

Scrambling or climbing shrub, occasionally small tree, up to 8(10 in F.T.E.A.) m high. Stems in most African plants densely brown-pubescent or (commonly in Asia) subglabrous when young, armed with usually ± straight spreading prickles 1–8(10) mm long. Leaves: petiole and rachis together (8)12–35(47) cm, armed on the lower side with downwardly hooked prickles up to 6 mm long often in pairs especially at the insertions of the pinnae, and on the upper side usually by solitary upwardly curved prickles at the insertions of the pinnae; pinnae 4–11 pairs; pinna rachis 2.5–9 cm long, unarmed or rarely with short (1 mm) downwardly curved prickles at insertions of leaflets; leaflets 8–13 pairs per pinna, (0.7)1–2.3(2.7) × (0.2)0.3–0.8(1.3 or 1.7 in F.T.E.A.) cm, oblong to narrowly oblong or somewhat oblanceolate, ± rounded at the apex, the lower surface pubescent, upper surface pubescent or glabrous; stipules 4–20 mm long, asymmetrically ovate, acuminate. Racemes 10–30 cm long, terminal; bracts up to 9 × 2.5 mm, ± lanceolate, falling before opening of their subtended flower; pedicels 1.5–4(4.6 in fruit) cm (but in the Flora area not seen more than 2.7 cm), strongly ascending, but the flowers borne at right angles so that they are deflexed. Lateral sepals 7–11 mm long. Petals (8)10–15(17) mm long or the upper one somewhat smaller, their margins ± entire, yellow to yellowish-white. Longest stamens (9)12–20 mm. Ovary pubescent. Pods 6–9.5 × (2.0)2.3–2.7(3.5) cm excluding the beak, ± oblong and often slightly curved, compressed, unarmed, glabrous at maturity, brown, the epicarp often flaking off and exposing the pronounced reticulate venation of the mesocarp, the whole pod tardily dehiscing along the upper margin; seeds 9–10 mm long and 7 mm diameter, ± ovoid, mottled brown and blackish or all black.

Zambia. N: Mbala Dist., Fwambo area, Fisa River Gorge, fl. 3.ix.1956, *Richards* 6087 (K). C: Kabwe (Broken Hill), fl. xi.1928, *van Hoepen* 1263 (PRE). E: Lundazi Boma, fl. & imm.fr. 27.iv.1952, *White* 2478 (K). S: Mazabuka Dist., at Munyenga near the Magoye R., fl. 24.v.1963, *van Rensburg* 2204 (K, SRGH). **Zimbabwe.** W: Hwange Dist., Victoria Falls, in rainforest, fl. & fr. 4.x.1964, *Whellan* 2192 (SRGH). C: Harare Dist., Avondale, fl. vi.1931, *Rattray* 5078 (SRGH). E: Mutare Dist., Upper Park R., fl. 28.vi.1952, *Chase* 4564 (BM, K, LISC, SRGH). **Malawi.** N: Misuku Hills, Iponjolo, fr.

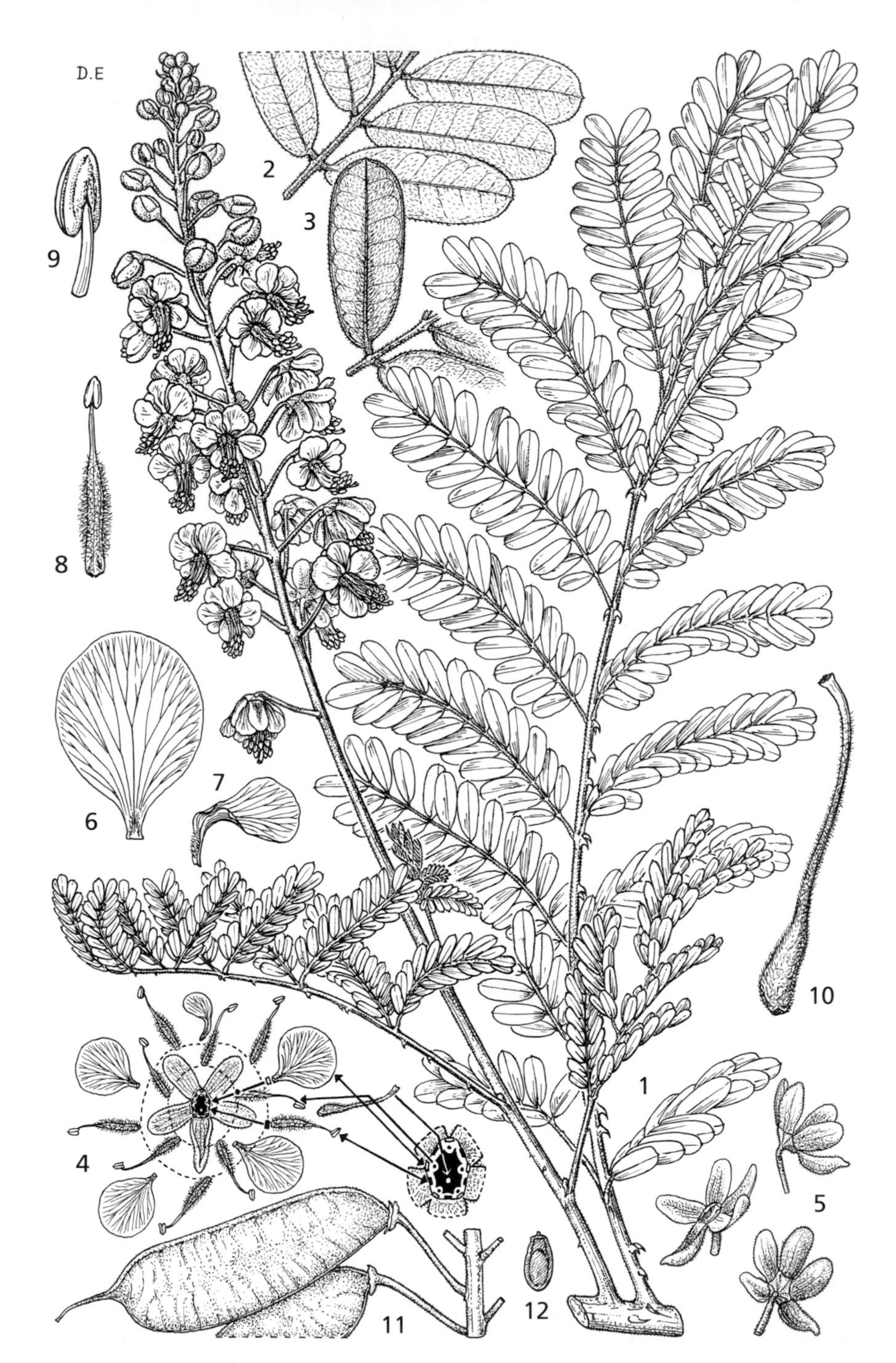

Fig. 3.2.**39**. CAESALPINIA DECAPETALA. 1, flowering branchlet (× ²⁄₃), from *Rutherford-Smith* 11062; 2, 3, leaflets, viewed from above and beneath (× 2), from *White* 2478; 4, flower, with parts separated (× ²⁄₃), and centre of flower enlarged (× 1¹⁄₂); 5, calyx, three views (× ²⁄₃); 6, upper petal (× 2); 7, lateral petal (× 2); 8, stamen (× 2); 9, anther (× 6); 10, gynoecium (× 3), 4–10 from *Chase* 4564; 11, pods (× ²⁄₃); 12, seed (× 1¹⁄₂), 11 & 12 from *Myre* 2528. Drawn by Derek Erasmus.

2.i.1977, *Pawek* 121193 (K, MAL). C: Dedza Mt. Forest, fl. 23.vi.1969, *Salubeni* 11358 (K, MAL, SRGH). S: Thyolo (Cholo) Mt., fl. & imm.fr. 26.ix.1946, *Brass* 17818 (K, SRGH). **Mozambique.** T: Angónia, near Posto Zootécnico, fl. & fr. 28.viii.1946, *Gouveia & Pedro* 1861 (PRE). GI: Manjacaze Dist., between Chidenguele (Chidenguel) and Régulo do Banhine, st. 13.viii.1947, *Pedro & Pedrógão* 1721 (PRE).

Native of tropical Asia, but widely cultivated in tropical countries, particularly as a hedge plant (see Howes, 1947), popularly known as the Mauritius Thorn or Mysore Thorn; now completely naturalised in many places, especially in high rainfall areas. Recorded from the Flora area from montane forest edges and associated bushland, riverine forest, thickets, river banks and open places; up to 1750 m.

The earliest-collected specimens seen from the Flora Zambesica area are from Malawi, Kondowe to Karonga, by *Whyte* in 1896 and Shire Highlands, *Adamson* s.n., without date but apparently between 1878 and 1912 (K). The earliest from Zimbabwe is dated 1914, from Harare (Salisbury), *Craster* 24 (K), and from Zambia 1928 (see above). M. Coates Palgrave (2002) remarks that the species is becoming a serious weed in South Africa.

As noted by Brenan in F.T.E.A., the species shows a much wider range of variation in Asia where it is native than it does in Africa where it is introduced. In the Flora area nearly all plants have densely pubescent stems when young but in India plants with subglabrous stems are common. Pod and pedicel lengths also show conspicuously less variation in the Flora area than in Asia.

The specific epithet is unfortunate since the plant has 5, not 10, petals.

31. POMARIA Cav.

Pomaria Cav., Icon. **5**: 2, fig.402 (1799). —Simpson in Lundellia **1**: 47 (1998). — Simpson & Lewis in Kew Bull. **58**: 175–184 (2003).

Hoffmannseggia sensu auct. afr. non Cav. —Brummitt & Ross in Kew Bull. **29**: 417 (1974). — Ross in F.S.A. **16**(2): 117 (1977).

Herbs or low shrubs, sometimes almost acaulescent; all parts of plant except petals and stamens with numerous dark, scattered glands; stems sometimes with plumose setae. Leaves bipinnate, imparipinnate; leaflets opposite, usually small; stipules fringed, serrate or laciniate; stipels when present small and setaceous. Inflorescence a terminal or leaf-opposed raceme, elsewhere sometimes axillary; bracts small, fringed, deciduous; bracteoles lacking. Flowers hermaphrodite, pedicellate, pink to red (elsewhere sometimes yellow with red markings). Calyx tube very short; lobes 5, unequal, the lower one larger than the others, projecting between the lateral petals and cupping the fertile parts, all shed before the fruit matures. Petals 5, imbricate, unequal, the upper one modified, ± funnel-shaped, the others in two lateral pairs. Stamens 10; filaments alternately longer and shorter, variously hairy, villous basally in the Flora area; anthers uniform, opening by longitudinal slits. Ovary subsessile, few–many-ovulate; style often incurved and clavate apically; stigma terminal. Pods oblong to ovate, straight or falcate, compressed, longitudiunally dehiscent, the valves often glandular. Seeds compressed.

16 species, 9 in North America, 4 in South America and 3 in southern Africa. The genus has recently been segregated from *Hoffmannseggia. Pomaria* was revised by Simpson (1998) and *Hoffmannseggia* (a wholly American genus of c. 24 species) by Simpson in Lundellia **2**: 14–54 (1999). The relationship of *Caesalpinia, Hoffmannseggia* and *Pomaria* is discussed by Simpson & Lewis (2003).

Herb with decumbent to ascending stems or rarely a subshrub; young stems sparsely appressed to densely spreading pubescent, plumose setae usually conspicuous; terminal pinna 1.5–10.5 cm long, typically considerably longer than the lateral pinnae, with 5–12 pairs of leaflets; pods straight or slightly curved . . . **1.** *burchellii*

Low much-branched shrub; young stems densely appressed-pubescent with whitish hairs, plumose setae few or absent; terminal pinna 0.4–2.4 cm long, not clearly longer than the lateral pinnae, with 4–6 pairs of leaflets; pods distinctly curved to sickle-shaped .... **2.** *lactea*

1. **Pomaria burchelli** (DC.) B.B. Simpson & G.P. Lewis in Kew Bull. **58**: 182 (2003). Type: South Africa, Northern Cape Province, Vryburg Dist., "Chooi Desert between Desert Stn. and Giraffe Stn.", *Burchell* 2345 (G-DC holotype; K).

Shortly erect or decumbent-ascending herb or rarely a subshrub with stems 12–45 cm long arising from a woody rootstock, at least in subsp. *burchellii*, bearing fusiform tubers up to 15 cm long. Stems with sparsely appressed to densely spreading hairs usually interspersed with conspicuous longer pale plumose setae up to 2 mm long; dark glands present on stems, leaves, inflorescences, calyces and pods. Leaves: petiole and rachis pubescent like the stem, with plumose setae aggregated at insertion of pinnae and leaflets; petiole 0.5–2.7 cm long; rachis (excluding terminal pinna) 0–6.5 cm long; pinnae 1–4 pairs; rachides of lateral pinnae 0.7–2.4 cm long, with 2–6 pairs of leaflets; rachides of terminal pinna 1.5–10.5 cm long, with 4–12 pairs of leaflets; leaflets 4–11× 2–5(7) mm, usually oblong to oblong-elliiptic, glabrous to appressed-pubescent; stipules up to 8 × 1.5 mm, lanceolate with marginal plumose setae, pubescent like the stem. Inflorescence (including peduncle) up to 12.5 cm long; pedicels up to 3(6) mm long in flower; bracts linear-lanceolate with plumose setae, deciduous before the buds open. Calyx lobes unequal, the lowest up to 6 × 3.5 mm, the upper ones up to 6 × 2 mm. Petals unequal, up to 6 × 2.5 mm, recorded as red, orange, salmon, pink or purple. Stamens 10, filaments alternately longer and shorter, up to 6 mm long. Ovary subsessile, pubescent. Pods 1.8–3.2 × 0.8–1 cm, obliquely oblong, straight or slightly curved, buff with conspicuous dark glands, with or without short pubescence, densely covered with plumose setae, 3–4-seeded. Seeds 5–6.5 × 3.5–6 mm.

Subsp. **burchellii**. FIGURE 3.2.**40**.

Melanosticta burchellii DC., Mem. Leg.: 475, fig.69 (1826). —Harvey in F.C. **2**: 270 (1862).

Caesalpinia burchellii Sprengel in Linnaeus, Syst. Veg., ed.16, 4, Cur. Post.: 169 (1827) nom. illeg. Type as for *Melanosticta burchellii.*

Hoffmannseggia burchellii (DC.) Oliv. in F.T.A. **2**: 263 (1871). —Schinz in Mém. Herb. Boiss. **1**: 123 (1900). —Dinter in Repert. Spec. Nov. Regni Veg. **18**: 424 (1922). —Baker, Legum. Trop. Afr.: 618 (1930). —Leistner in Mem. Bot. Surv. S. Afr. **38**: 124 (1967). —Schreiber in Prodr. Fl. SW Afrika, fam. 59: 16 (1967). —Brummitt & Ross in Kew Bull. **29**: 418 (1974). —Ross in F.S.A. **16**(2): 118 (1977). —Lock, Leg. Afr. Check-list: 24 (1989).

Stems short, up to 30 cm long, often subacaulescent, ± ascending. Leaves: pinnae 1–2(3) pairs; terminal pinna with 5–12 pairs of leaflets, that of the larger leaves of each plant with at least 7 pairs of leaflets.

Botswana. N: 110 km W of Nokaneng, fr. 112.iii.1965, *Wild & Drummond* 6913 (K, LISC, SRGH). SW: 320 km W of Gaborone, fr. 20.x.1975, *Mott* 766 (K, UBLS). SE: 20 km NW of Molepolole, 2.x.1954, *Codd* 8928 (K, PRE). **Zimbabwe.** W: Nyamandhlovu Dist., Nyamandhlovu Pasture Res. Station, *Corby* 1523 (SRGH).

Also in Namibia and South Africa (N Cape Province). Mixed woodland and wooded grassland, dry riverbeds, on Kalahari sands; 900–1200 m.

Fig. 3.2.**40**. POMARIA BURCHELLII subsp. BURCHELLII. 1, habit (× 2/3); 2, 3, leaflets from above and beneath (× 4), 1–3 from *Wild & Drummond* 6913; 4, flower (× 3); 5, calyx (× 3); 6–8, calyx lobes (× 6); 9–11, upper, upper lateral and lower lateral petals respectively (× 6); 12, 13, stamens (× 4); 14, gynoecium (× 4), from *Galala* 72; 15, pod (× 3), with enlargement of a hair; 16, part of valve of pod showing seed (× 3), 15 & 16 from *Wild & Drummond* 6913. Drawn by Derek Erasmus.

15
16
12
13
8
7
2
11
6
10
3
1
9
5
4
14
D

Subsp. **rubro-violacea** (Baker f.) Brummitt, comb. nov. Type: Zimbabwe, Henry's Pan, *Holub* s.n. (Z holotype).

Hoffmannseggia rubro-violacea Baker f. in Vierteljahrsschr. Naturf. Ges. Zürich **70**: 216 (1925); Legum. Trop. Afr.: 619 (1930).

Hoffmannseggia burchellii subsp. *rubro-violacea* (Baker f.) Brummitt & Ross in Kew Bull. **29**: 419 (1974). —Lock, Leg. Afr. Check-list: 24 (1989).

Stems decumbent, up to 45 cm long. Leaves: pinnae (2)3–4 pairs; terminal pinna with 4–6 pairs of leaflets.

Botswana. N: Nata R. c. 11 km from mouth near Madsiara Drift, 21.iv.1957, *Drummond & Seagrief* 5158 (K, PRE, SRGH). **Zimbabwe.** W: Henry's Pan, *Holub* s.n. (Z).

Also in South Africa (Limpopo Province). Dry river course on sand; c. 900 m.

Ross (1977) suggests the subspecies may also differ by the leaflets being sensitive, folding up when touched in subsp. *rubro-violacea* but not in subsp. *burchellii.* The characteristic root tubers seen in subsp. *burchellii* have not been seen in subsp. *rubro-violacea.*

2. **Pomaria lactea** (Schinz) B.B. Simpson & G.P. Lewis in Kew Bull. **58**: 181 (2003). Lectotype, chosen by Brummitt & Ross in Kew Bull. **29** (1974): Namibia, Keetmanshoop, *Fenchel* 172 (Z lectotype).

Hoffmannseggia sandersonii var. *lactea* Schinz in Mém. Herb. Boiss. **1**: 124 (1900).

Hoffmannseggia lactea (Schinz) Schinz in Bull. Herb. Boiss., sér. 2, **1**: 875 (1901). —Dinter in Repert. Spec. Nov. Regni Veg. **18**: 425 (1922). —Baker, Legum. Trop. Afr.: 618 (1930). —Schreiber in Prodr. Fl. SW Afrika, fam. 59: 16 (1967). —Brummitt & Ross in Kew Bull. **29**: 420 (1974). —Ross in F.S.A. **16**(2): 119 (1977).

Hoffmannseggia pearsonii E.P. Phillips in Bull. Misc. Inform., Kew **1911**: 262 (1911). Lectotype, chosen by Brummitt & Ross (1974): Namibia, Keetmanshoop Dist., Aub riverbed at Gobas, *Pearson* 3747 (K lectotype; BM).

Low much-branched shrub up to 50 cm high. Stems densely appressed-pubescent with whitish hairs, plumose setae very few or absent; dark glands present and very conspicuous on stems, leaves, inflorescences, calyces and pods. Leaves: petiole and rachis glabrous to appressed or spreading pubescent, setae, if present, not plumose; petiole 0.3–2.3 cm long, often persisting as a rigid subspine after the fall of the rest of the leaf; rachis (excluding terminal pinna) 0.3–3.8 cm; pinnae (1)2–5 pairs; rachides of lateral pinnae (0.3)0.6–1.8 cm long, with 3–7 pairs of leaflets; rachides of terminal pinna 0.4–2.4 cm long, with 4–6 pairs of leaflets; leaflets 3–8 × 1–3 mm, oblong to elliptic-oblong, glabrous; stipules up to 6 × 2 mm, laciniate, often rigid and persistent. Inflorescence (including peduncle) up to 21 cm long; pedicels up to 3 mm long in flower; bracts up to 5 × 2 mm, laciniate, deciduous before the buds open. Calyx with a very short tube; lobes 5, unequal, the lower lobe up to 7 × 3 mm, the upper ones up to 6 × 1 mm. Petals unequal, up to 8 × 3 mm, recorded as red, dark red, brownish-pink and purple-orange. Stamens 10; filaments alternately longer and shorter, up to 5 mm long. Ovary subsessile, up to 4 × 1 mm. Pods 2.4–3.4 × 0.7–0.9 cm, distinctly curved or falcate, pale buff with conspicuous dark glands, with a short dense indumentum interspersed with plumose setae, 3–5-seeded. Seeds 5–5.5 × 4.5–5 mm.

Botswana. SW: Kgalagadi, 11.5 km from Middlepits on road to Bokspits, 19.iii.1991, *Cook et al.* 86 (K).

Also in Namibia and South Africa (N Cape Province). Probably in mixed woodland and wooded grassland on sandy soils; c. 1000 m.

32. **ERYTHROSTEMON** Klotzsch

Erythrostemon Klotzsch, in Link, Klotzsch & Otto, Icon. Pl. Rar. Horti Berol. **2**: 97, fig.39 (1844).

Shrubs or perennial herbs, unarmed. Leaves bipinnate, generally with a terminal pinna, the pinnae opposite; leaflets opposite or elsewhere sometimes alternate, often with punctate glands in a submarginal row; stipules often fimbriate-glandular and subpersistent. Flowers in terminal and leaf-opposed racemes. Flowers hermaphrodite; pedicels articulated or not. Hypanthium resembling a short calyx tube. Sepals 5, imbricate, the lowest larger and curving up to overlap the other 4, pectinate or fimbriate at the apex, abscissing before the fruit matures. Petals 5, subequal except the innermost, which is modified to a greater or lesser extent (lesser in the Flora area). Stamens 10, long exserted in the Flora area; filaments densely hairy near the base. Ovary sessile to shortly stipitate, hairy and with stipitate glands in the Flora area; stigma terminal. Pod straight or curved, elastically dehiscent, with stalked, swollen headed glands in the Flora area. Seeds laterally compressed, often with fracture lines.

12–13 species from southern USA to Argentina and Chile. Formerly included in *Caesalpinia.*

Erythrostemon gilliesii Klotzsch in Link, Klotzsch & Otto, Icon. Pl. Rar. Horti Berol. **2**: 97, fig.39 (1844). Type from Argentina. FIGURE 3.2.**41**.

Poinciana gilliesii Hook., Bot. Misc. **1**: 129, t.34 (1830); —Bot. Mag.: t.4006 (1843).

Caesalpinia gilliesii (Hook.) Benth. in Martius, Fl. Bras. **15**(2): 71 (1870). —Baker, Legum. Trop. Afr.: 616 (1930). —Burtt Davy, Fl. Pl. Ferns Transvaal: 328 (1932). —Brenan, Check-list For. Trees Shrubs Tang. Terr.: 95 (1949); in F.T.E.A., Legum.-Caesalp.: 29 (1967). —Roti-Michelozzi in Webbia **13**: 215 (1957). —Ross in F.S.A. **16**(2): 129 (1977). —Lewis, *Caesalpinia*: 161, fig.31 (1988). —Polhill & Thulin in Fl. Ethiopia **3**: 53 (1989).

Shrub or small tree 1–3(5) m high. Young stems and inflorescences pubescent and fairly densely covered with stipitate, peltate, usually blackish or brown glands, without prickles. Leaves: petiole and rachis together 8–20 cm, glabrous, occasionally with a few glands; pinnae 7–15 on each side of the rachis, opposite, subopposite or alternate, and often with a solitary terminal pinna; pinna rachis 1–4 cm; leaflets 7–10(12) per pinna, 2–8(11) × 1–2.5(3.5) mm, oblong-elliptic, glabrous, usually with black gland dots just inside the margins. Racemes 6–16(24) cm long, all terminal; bracts up to 28 × 8 mm, lanceolate to oblanceolate, ± pectinate, conspicuous and concealing the young buds but falling before anthesis; pedicels 1.5–3.5 cm, with stipitate glands. Lateral sepals 13–21 mm long; all sepals pubescent and usually with numerous stipitate glands, the lower one usually pectinate or erose at the apex. Petals (17)23–32 mm long, yellow. Stamens 5–10.5 cm long, much exceeding the petals, the filaments bright red. Ovary tomentose and with shortly stipitate glands; style somewhat exceeding the stamens. Pods 6–10 × 1.7–2.1 cm, ± obliquely oblong, straight or somewhat curved upwards, compressed, light brown, pubescent and usually conspicuously spotted with dark glands when young but glabrescent and smooth later, dehiscing down both margins; seeds 5–8, c. 9 × 7 × 2.5 mm, ± broadly obovate, compressed, brown flecked with purplish-black.

Zimbabwe. C: Harare, Marlborough Nurseries, cultiv., fl.& imm.fr. 2.xii.1971, *Biegel* 3665 (K, SRGH). **Mozambique.** GI: Guijá, Brigada Técnica do Limpopo, fl. 9.x.1958, *Barbosa* 8349 (COI, K, LISC). M: Maputo (Lourenço Marques), Polana Hotel, fl. & fr. x.1944, *Pimenta* 17319 (LISC).

Native of South America (Argentina and possibly Uruguay) and widely introduced as an ornamental shrub in tropical, subtropical and warm-temperate countries. The plants cited from Mozambique were no doubt originally planted; information is required on whether it is naturalized.

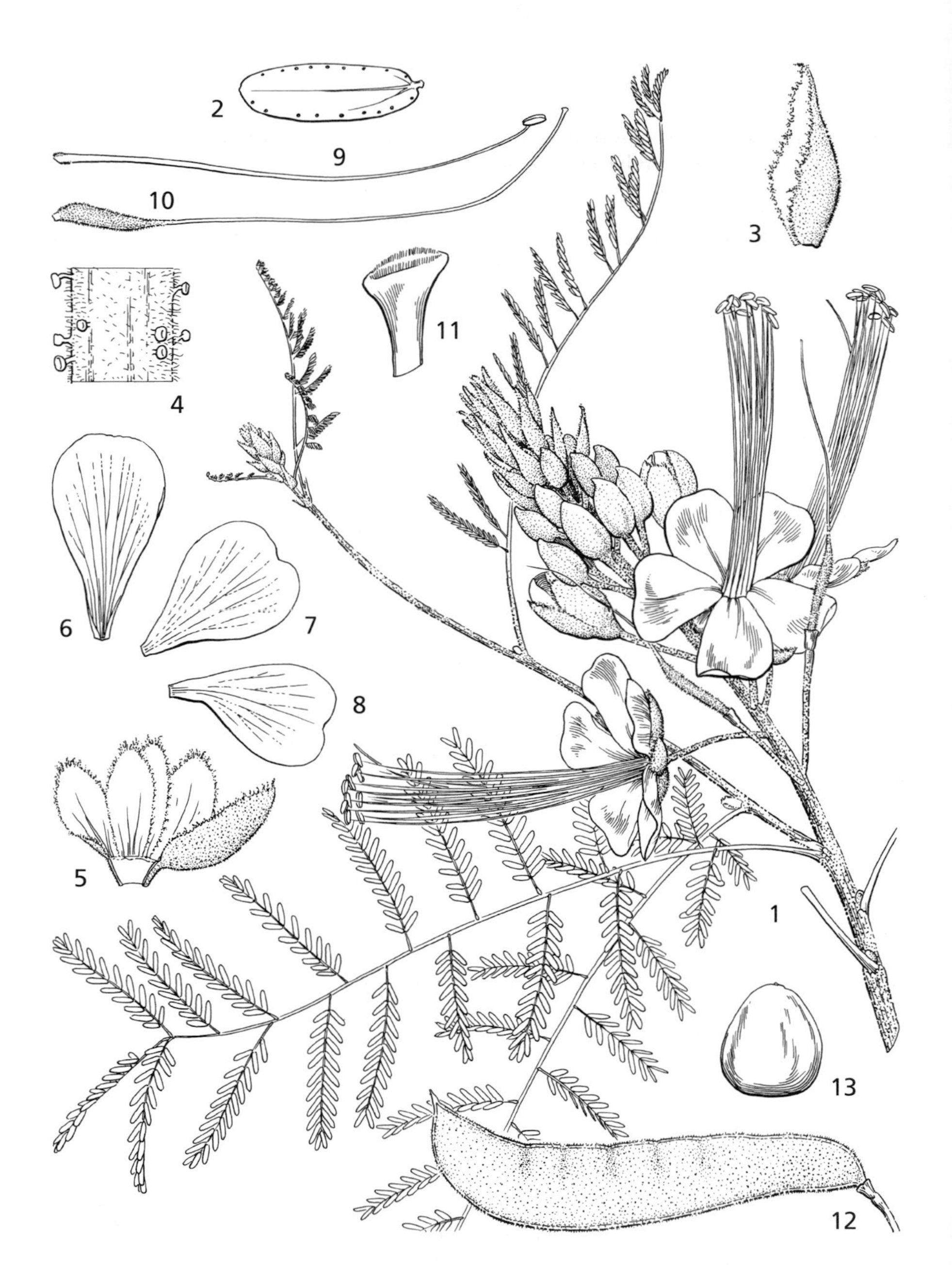

Fig. 3.2.**41**. ERYTHROSTEMON GILLIESII. 1, flowering brnchlet (× ²/₃), from *Venturi* 5365; 2, leaflet undersurface showing submarginal glands (× 4), from *Kiesling et al.* 4891; 3, bract (× 1¹/₂); 4, glands on pedicel (× 10); 5, calyx opened out (× 1); 6, 7, 8, upper, upper lateral and lower lateral petal respectively (× 1); 9, stamen (× 1); 10, gynoecium (× 1); 11, stigma (× 15), 3–11 from Cult. Kew 213-69.01878 (source Palermo Bot. Gard., Sicily); 12, pod (× ²/₃), from *Kiesling et al.* 4891; 13, seed (× 2), from *Lewis* 1417. Drawn by Eleanor Catherine. From Kew Bulletin.

33. GUILANDINA L.

Guilandina L., Sp. Pl.: 381 (1753).; Gen. Pl., ed. 5: 179 (1754).
Caesalpinia subgen. *Guilandina* (L.) Gillis & Proctor in J. Arnold Arb. **55**: 426 (1974).

Shrubs or small bushy trees, spreading or scrambling, usually covered with straight or hooked prickles. Leaves bipinnate, with paired pinnae, eglandular; leaflets opposite; stipules minute to conspicuous and leafy. Flowers in terminal and supraxillary racemes or panicles; bracts caducous or persistent. Flowers unisexual, the carpellate flowers seemingly bisexual but anthers sterile, the staminate flowers with a rudimentary ovary; pedicels jointed. Hypanthium shallow. Sepals 5, almost equal, the lowest one somewhat boat-shaped. Petals 5, subequal, the uppermost somewhat modified. Stamens 10, not exceeding the petals; filaments pubescent. Ovary often spiny; stigma terminal. Pod compressed, thick, rounded, rigidly chartaceous, often spiny. Seeds usually 2–4, hard, globose, ovoid or oblong-ellipsoid, with transverse fracture lines.

Probably about 8 species, pantropical, two widespread as coastal plants with drift seeds, formerly included in *Caesalpinia* though generally recognised as one of the more distinctive of the putative segregates.

Guilandina bonduc L., Sp. Pl.: 381 (1753). —Harvey in F.C. 2: 269 (1862). Type: Sri Lanka, *Herb. Hermann* 3, fig.35 (BM lectotype). FIGURE 3.2.**42**.

Caesalpinia bonduc (L.) Roxb., Fl. Ind. **2**: 362 (1832). —Dandy & Exell in J. Bot. **76**: 179 (1938). —Brenan, Check-list For. Trees Shrubs Tang. Terr.: 94 (1949); in F.T.E.A., Legum.-Caesalp.: 37 (1967). —Wilczek in F.C.B. **3**: 250 (1952). —Keay in F.W.T.A., ed.2, **1**(2): 481, fig.154 (1958). —Torre & Hillcoat in C.F.A. **2**: 171 (1956). —Roti-Michelozzi in Webbia **13**: 204 (1957). —Ross in F.S.A. **16**(2): 124 (1977). —M. Coates Palgrave, Trees Sthn. Africa: 344 (2002).

Guilandina bonducella L., Sp. Pl., ed.2, **1**: 545 (1762). Type as above.

Caesalpinia bonducella (L.) Fleming in Asiat. Res. **11**: 159 (1810). —Oliver in F.T.A. **2**: 262 (1871). —Taubert in Engler, Pflanzenw. Ost-Afr. **C**: 202 (1895).

Caesalpinia crista L., Sp. Pl.: 380 (1753) sensu auctt. in part but excluding the type. —Hutchinson & Dalziel, F.W.T.A.: 348, fig.135A (1928). —Baker, Legum. Trop. Afr.: 614 (1930).

Shrub or sometimes small bushy tree, spreading or sometimes scrambling, up to 5 m high. Stems brown pubescent at least when young, and copiously armed with spreading or slightly deflexed prickles of varying length and stoutness. Leaves: petiole and rachis together up to 50 cm, armed on the lower side with reflexed prickles often in pairs particularly at the insertions of the pinnae, and occasionally also with solitary ± straight prickles on the upper side especially at the insertions of the pinnae, or the petiole often with numerous prickles all round; pinnae 3–10 pairs; pinna rachis (4)6–18(22) cm, usually with paired reflexed prickles at the insertions of the leaflets but occasionally unarmed; leaflets (4)6–9(10) pairs per pinna, (0.8)1.3–4(5) × (0.5)0.8–2.2(2.5) cm, asymmetrically ovate or elliptic to ovate-oblong, obtuse or subacute or sometimes somewhat acuminate at the apex, both surfaces usually pubescent with brownish hairs when young but often glabrescent except on midrib and margins; stipules conspicuous, leafy, compound, with usually 3 ovate to oblate and usually retuse lobes up to 2.5 × 3.6 cm. Racemes up to 35 cm long, terminal and lateral (axillary or supra-axillary) and often branched; bracts up to 14 mm long, linear-lanceolate, much exceeding the subtended buds and reflexing as the buds develop, falling at about anthesis; pedicels (4)6–9(20 in fruit) mm. Lateral sepals 4–7 mm long; all sepals brown-pubescent. Petals 8–10 mm long, oblanceolate-oblong, the upper one broader and stouter than the others, yellow to greenish-yellow. Stamens 4.5–7 mm. Ovary covered with stiffish hairs. Pod 4.5–8 × 3.5–5.3 cm, broadly elliptic-oblong, fairly densely covered by stiff spines up to 9 mm long, apparently rather tardily dehiscing along the lower margin and perhaps sometimes the upper as well; seeds usually 1–2 per pod, subglobose, 1.5–2 cm diameter, leaden-grey, hard, the testa showing a number of parallel, regular, encircling lines looking like fine cracks.

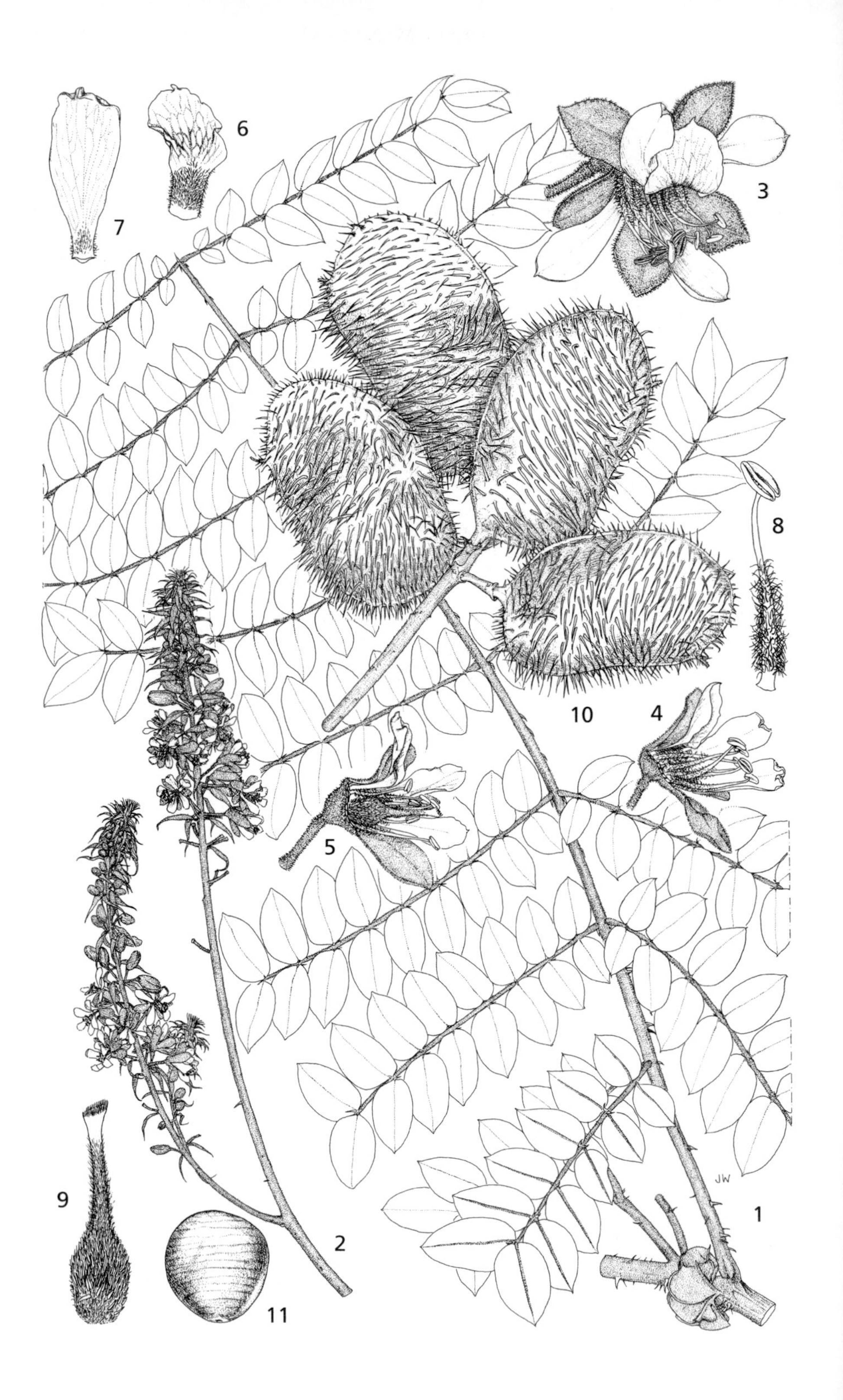

6
7
3
8
10
4
5
JW
1
9
2
11

Mozambique. N: Pemba Dist., coast of Pemba (Porto Amélia) at side of way to the lighthouse of Maringanha, 12 km S of town, fl. 21.iii.1960, *Gomes e Sousa* 4539 (COI, K, PRE, SRGH). Z: Pebane, by the lighthouse, fl. & fr. 8.iii.1966, *Torre & Correia* 15065 (LISC). MS: Búzi Dist., Sofala (Nova Sofala), fr. 21.vi.1961, *Leach & Wild* 11121 (K, SRGH). GI: Xai-Xai, Chipenhe, Chiconela, fl. & fr. 1.iv.1959, *Barbosa & Lemos* 8450 (COI, K, LISC, PRE). M: near Maputo (Lourenço Marques), fl. & fr. v.1946, *Pimenta* 17329 (LISC).

Widespread on tropical coasts of the Old and New World. Coastal scrub and open places, particularly on sandy shores and dunes.

The early nomenclature of this species was very confused and is discussed in detail by Dandy & Exell (1938).

34. **BUSSEA** Harms

Bussea Harms in Bot. Jahrb. Syst. **33**: 159 (1902).

Trees or occasionally shrubs, unarmed. Young shoots and inflorescences with conspicuous brown indumentum. Leaves usually mostly opposite or subopposite but frequently some not so, bipinnate, without specialised glands on petiole and rachis; pinnae up to 8 pairs; leaflets in 3–15 pairs or sometimes (not in the Flora area) alternate; stipules small, subulate. Flowers in fairly dense racemes which are aggregated to form terminal panicles or rarely are single and terminal; bracts 2–8 mm long, often concave. Receptacle forming a concave disc which breaks away from the pedicel as a ring after flowering. Sepals imbricate, with scarious erose margins. Petals unequal, the upper one smaller than the rest, all ± spathulate with a brown tomentum on the outside of the claw and extending up the middle of the blade. Stamens 10; filaments glabrous in the upper part but the broader basal part conspicuously brown-villous; anthers dorsifixed, dehiscing by longitudinal slits. Ovary subsessile, curving upwards towards the end, brown-tomentose with 2–3 ovules; style pubescent in the lower part gradually becoming glabrous distally; stigma peltate. Pods 7–12 cm long, linear-oblanceolate, woody, erect, brown-tomentose especially when young, dehiscing into 2 recurving valves, each valve having a conspicuous longitudinal median groove or rarely being concave; seeds 1–3, up to 2.2 cm long, oblong, compressed, elongated longitudinally in the pod, with a short funicle at one corner; endosperm absent.

A genus of 5 species in Africa and two in Madagascar.

Leaflets (3)5–10(11) pairs per pinna, each 0.4–1.4(1.9) cm broad . . . **1.** *massaiensis*
Leaflets 9–13 pairs per pinna, each 0.4–0.55 cm broad . . . . **2.** *xylocarpa*

1. **Bussea massaiensis** (Taub.) Harms in Bot. Jahrb. Syst. **33**(1): 159 (1902); in Engler, Pflanzenw. Afrikas **3**(1): fig.274 (1915). —Baker, Legum. Trop. Afr.: 617 (1930). —Brenan, Check-list For. Trees Shrubs Tang. Terr.: 94 (1949); in F.T.E.A., Legum.-Caesalp.: 26 (1967). —Verdcourt in Kew Bull. **12**: 350 (1957). Type: Tanzania, Dodoma Dist., Saranda, *Fischer* 226 (B† holotype).

Peltophorum massaiense Taub. in Engler, Pflanzenw. Ost-Afr. **C**: 202 (1895).

Fig. 3.2.**42**. GUILANDINA BONDUC. 1, leaf (× $^1/_2$), from *Gomes e Sousa* 4539; 2, inflorescence (× $^1/_2$), from *Faulkner* Kew ser. 2418; 3, flower (× 3), from *Gomes e Sousa* 4539; 4, section of male flower (× 2), from *Faulkner* Kew ser. 2418; 5, section of hermaphrodite flower (× 2), from *Gomes e Sousa* 4447; 6, upper petal (× 3); 7, lower petal (× 3); 8, stamen (× 5), 6–8 from *Faulkner* Kew ser. 2418; 9, ovary (× 5), from *Gomes e Sousa* 4447; 10, pods (× $^1/_2$), from *Mogg* 31450; 11, seed (× 1), from *Faulkner* Kew ser. 931. Drawn by Juliet Williamson.

Shrub or tree up to 14 m high with smooth grey bark. Young stems brown-pubescent but quickly glabrescent, unarmed. Leaves: petiole and rachis together 2–14 cm, densely brown-pubescent when young but almost glabrescent; pinnae (1)2–5 pairs; pinna rachis 2–12 cm; leaflets (3)5–10(11) pairs per pinna, (0.5)1–3.4 × (0.3)0.4–1.4(1.9) cm, asymmetrically elliptic to elliptic-oblong, obtuse to emarginate at the apex, strongly asymmetrical at the base; lower surface pubescent and upper surface glabrous or pubescent when young, but both often glabrescent. Flowers in fairly dense racemes grouped together to form a terminal panicle, the terminal raceme up to 15 cm long, the others shorter, the axes densely brown-pubescent; bracts up to 8 × 4 mm, lanceolate or ovate to elliptic, falling at or slightly before anthesis; pedicels 4–10 mm long. Sepals 5–8 mm long, densely brown-pubescent except on scarious margins. Petals unequal, the lower and lateral ones 1.2–2.5 × 0.9–1.4 cm, the upper one 0.9–1.0 × 0.7–0.8 cm, all yellow, distal margins usually erose. Stamen filaments 7–9 mm long. Ovary brown-tomentose. Pods 7–12 × 1.2–2.2 cm, broadest in the distal third, erect, shortly brown-tomentose especially in the median groove of each valve; seeds 1–3, 1.6–2.2 × 0.9–1.2 cm, oblong, elongated longitudinally.

Subsp. **rhodesica** Brenan in Kew Bull. **17**: 197 (1963). —Lock, Leg. Afr. Checklist: 20 (1989). Type: Zambia, Northern Prov., Kangiri, *Fanshawe* 4894 (K holotype; LISC, SRGH). FIGURE 3.2.**43**.

Bussea massaiensis sensu White, F.F.N.R.: 118 (1962).

Leaves: petiole and rachis together 5–14 cm, with (3)4–5 pairs of pinnae; pinna rachis (3)5–12 cm; leaflets (5)7–10(11) pairs per pinna. Larger petals 1.2–1.5 × 0.9–1.2 cm.

Zambia. N: Mporokoso Dist., Sumbu–Malango road, fl. 24.ix.1956, *Richards* 6266 (K).

Subspecies known only from the northern part of Northern Province of Zambia; an important and sometimes dominant constituent of 'mateshi' thicket, in which it is often the tallest species; 900–1250 m.

Subsp. *massaiensis* differs in its leaves (petiole and rachis together 2–9(14) cm, with (1)2(3) pairs of pinnae; pinna rachis 3–8(9) cm; leaflets (3)5–7(8) pairs per pinna) and usually larger lower petals (1.5–2.5 × 1.0–1.4 cm). It occurs in thickets and deciduous woodland in central Tanzania.

2. **Bussea xylocarpa** (Sprague) Sprague & Craib in Bull. Misc. Inform., Kew **1909**: 351 (1909). —Baker, Legum. Trop. Afr.: 617 (1930). —Torre in Mendonça, Contrib. Conhec. Fl. Moçamb. **2**: 67 (1954). —Verdcourt in Kew Bull. **12**: 350 (1957). —M. Coates Palgrave, Trees Sthn. Africa: 345 (2002). Type: Mozambique, Lupata, Zambesi, *Kirk* (K holotype).

Calliandra xylocarpa Sprague in Bull. Misc. Inform., Kew **1908**: 290 (1908).
?*Bussea massaiensis* sensu Torre in Mendonça, Contrib. Conhec. Fl. Moçamb. **2**: 67 (1954).

Leaves: petiole and rachis 4.5–14 cm, pubescent; pinnae 3–7 pairs; pinna rachis 4–8 cm; leaflets 9–13 pairs per pinna, 10–19 × 4–5.5 mm, oblong, rounded to emarginate at the apex, strongly asymmetrical at the base; lower surface finely pubescent, upper surface glabrous at least at fruiting. Pods 7–9 × 1.5–1.7 cm; seeds c. 1–7 × 1 cm.

Mozambique. T: Lupata, Zambezi R., fr. vii.1859, *Kirk* (K).

Known with certainty only from the original collection, from which the above description is drawn, and which has mature fruit but no flowers. There is, however, an interesting further collection of this genus from considerably further south in Mozambique, GI: Panda Dist., Mawayela (Maueele), 24°20'S, 34°08'E in a population of *Androstachys johnstonii*, 100 m, fl. xii.1936, *Gomes e Sousa* 1917 (K, SRGH), which is at a much younger stage and has flowers but no fruit. The leaves suggest that it might be referable to *B. xylocarpa*, but alternatively it might represent a hitherto undescribed species. The specimen is described as follows: Leaves — petiole and

Fig. 3.2.**43**. BUSSEA MASSAIENSIS subsp. RHODESICA. 1, flowering branchlet (× $^2/_3$); 2, detail of upper surface of leaflet (× 4); 3, same of lower surface (× 4); 4, flower (× $1^1/_2$); 5, sepals (× $1^1/_2$); 6, rear view of calyx (× $1^1/_2$); 7, petals (× 3); 8, 9, stamens (× 3); 10, gynoecium and view of stigma from above (× 3), 1–10 from *Richards* 13702; 11, pods (× $^2/_3$); 12, seed (× $^2/_3$), 11 & 12 from *Phipps & Vesey-FitzGerald* 3210. Drawn by Derek Erasmus.

rachis together 8–16 cm, densely and conspicuously rusty-brown tomentose; pinnae 5–8 pairs; pinna rachis 7–10 cm long; leaflets 11–15 pairs per pinna, 9–21 × 4–8 mm, oblong or ovate-oblong, rounded to emarginate at the apex, strongly asymmetrical at the base; lower surface very finely pubescent or subglabrous, upper surface glabrous and ± glossy. Inflorescence axes and pedicels densely rusty-brown tomentose; pedicels 8 mm. Sepals 7–8 mm long, densely brown-pubescent except on scarious margins. Lower petals 17–20 × 12–14 mm, yellow, erose. Stamen filaments c. 9 mm.

The relationship of these two Mozambique collections to each other and to *B. massaiensis* can only be evaluated when more material is available.

35. PELTOPHORUM (Vogel) Benth.

Peltophorum (Vogel) Benth. in Hooker's J. Bot. Kew Gard. Misc. **2**: 75 (1840) nom. conserv.

Small to medium trees, unarmed. Young shoots and inflorescences covered with brown indumentum and sometimes scattered stalked, swollen-headed glands. Leaves alternate, bipinnate, without specialised glands; pinnae up to 21 pairs; leaflets up to 32 pairs per pinna, usually oblong; stipules caducous. Flowers in terminal and lateral racemes often aggregated to form a pseudopanicle; bracts usually linear-lanceolate, caducous. Receptacle comprising a lower narrow part simulating the pedicel and an upper broad shallow disc. Sepals appearing separate but actually joined by a very short tube [see note], imbricate with scarious margins, usually reflexed in flower. Petals somewhat unequal, the upper one shorter than the others, all ± spathulate with brown pubescence on the outside of the claw and extending up the middle of the blade. Stamens 10, not exceeding the petals; filaments glabrous in the upper part but the broader basal part conspicuously brown-villous; anthers dorsifixed, dehiscing by longitudinal slits. Ovary subsessile, brown-pubescent or -tomentose; style pubescent in lower part, gradually becoming glabrous distally; stigma peltate. Pods 4–12 cm long, oblong to narrowly elliptic, not conspicuously pubescent, strongly compressed, with distinct wings 3–6 mm broad along each margin, indehiscent; seeds 1 to few, oblong or discoid, strongly compressed, elongated longitudinally within the pod.

A pantropical genus of 5–7 species, only one native in Africa.

Barneby in Brittonia **48**: 178 (1996) notes that in contrast to previous generic descriptions *Peltophorum* does have a very short calyx tube. After anthesis the calyx lobes fall and the tube separates from the hypanthium by a circumscissile fracture, persisting as a loose ring around the base of the fruit.

1. Leaflets rounded and usually emarginate at the apex, not mucronate, 4–8 mm broad; sepals 6–8 mm long (introduced species) . . . . **3.** *pterocarpum*
– Leaflets rounded to subacute at the apex and distinctly mucronate, 1.5–3 mm broad; sepals 4–6 mm long (native or introduced) 2
2. Stems with very fine, closely appressed puberulence or pubescence; stipules minute or absent (introduced) . . . . **2.** *dubium*
– Stems with ± spreading puberulence or pubescence; stipules linear-subulate with up to 7 alternate lateral branches, but quickly caducous (native) . **1.** *africanum*

1. **Peltophorum africanum** Sond. in Linnaea 23: 35 (1850). —Oliver in F.T.A. **2**: 260 (1871). —Sim, For. Fl. Port. E. Africa: 47, fig.49 (1909). —Harms in Engler, Pflanzenw. Afrikas **3**(1): 512 (1915). —Eyles in Trans. Roy. Soc. S. Afr. **5**: 368 (1916). —Baker, Legum. Trop. Afr.: 611 (1930). —Gomes e Sousa in Bol. Soc. Estud. Colón. Moçamb.: 70 (1936) (as *Poltophorum* sphalm.). —Brenan, Check-list For. Trees Shrubs Tang. Terr.: 105 (1949); in F.T.E.A., Legum.-Caesalp.: 17

(1967). —Suessenguth & Merxmüller, Proc. & Trans. Rhod. Sci. Assoc. **43**: 17 (1951). —Pardy in Rhod. Agric. J. **49**(4): 218 (1952). —Wilczek in F.C.B. **3**: 262 (1952). —Torre in Mendonça, Contrib. Conhec. Fl. Moçamb. **2**: 65 (1954). —Torre & Hillcoat in C.F.A. **2**: 169 (1956). —O. Coates Palgrave, Trees Central Africa: 111 (1957). —White, F.F.N.R.: 126 (1962). —Breitenbach, Indig. Trees S. Africa **3**: 352 (1965). —Gomes e Sousa, Dendrol. Moçamb. Estudo Geral **1**: 245, fig.48 (1966). —Schreiber in Merxmüller, Prodr. Fl. SW Afrika, fam. 59: 17 (1967). —Palmer & Pitman, Trees Sthn. Africa **2**: 887 (1972). —Ross in F.S.A. **16**(2): 132, fig.26 (1977). —M. Coates Palgrave, Trees Sthn. Africa: 345 (2002). Type: South Africa, Gauteng, N slopes of Magalisberg at Crocodile R., *Zeyher* 554 (BM, K, OXF, P). FIGURE 3.2.**44**.

Small tree 3–9(14) m high, with a spreading crown. Young stems with rusty-brown to greyish, ± spreading tomentum or puberulence, sometimes also with stalked, swollen-headed glands, particularly on saplings [see note]. Leaves: petiole and rachis together 4–28 cm, densely brown-pubescent to -tomentose, sometimes with glands; pinnae 4–13(17) pairs; pinna rachis 1.5–8(10.5) cm; leaflets 6–28 pairs per pinna, (2)4–9(12) ×(1)1.5–3(4) mm, oblong or occasionally ovate-oblong, asymmetrical at the base, rounded to subacute at the apex, and mucronate; both surfaces appressed-pubescent; stipules up to 1.4 cm long, linear-subulate with up to 7 linear, alternate, lateral appendages to 6 mm long, the whole usually very quickly caducous. Flowers in fairly dense, terminal and axillary racemes up to 24 cm long, usually grouped at the ends of branches to form a pseudopanicle, the axes densely brown pubescent; bracts up to 7 × 1 mm, linear-lanceolate, caducous; pedicels 3–8(10 in fruit) mm excluding the receptacle. Sepals (4)5–6(7) mm long, brown-pubescent except on the scarious margins, ± reflexed in flowering. Petals 10–14(17) mm long, obtriangular-spathulate with a short claw, or the upper one somewhat shorter and with a broader and stouter claw, the distal margins all ± undulate, all yellow. Stamens 8–13 mm, shorter than the petals; anthers 1.5–2.75 mm long. Ovary appressed brown-pubescent, often ± velutinous. Pods 4–9 × 1.4–2.6 cm, elliptic to narrowly elliptic, the wings down each margin 3–6 mm broad, the whole densely puberulent, or glabrescent at maturity, indehiscent; seeds 1 or 2, 9–12 × 5–8 × 1 mm, ± oblong, elongated longitudinally in the pod, with the hilum at one end.

Caprivi Strip. Katima Mulilo area, fr. 24.xii.1958, *Killick & Leistner* 3069 (K). **Botswana.** N: Ngamiland, c. 10 km E of Goha Hills, in floodplain E of Falinyanti R., fr. 26.iv.1944, *Kokot* 20 (PRE). SE: northern Kalahari Desert, fl. i.1899, *E.J. Lugard* 241 (K); Central Dist., E of Mahalapye, fl. 12.xi.1958, *de Beer* 792 (K, SRGH). **Zambia.** B: Zambezi (Balovale), fl. xii.1953, *Gilges* 319 (K, PRE, SRGH). N: Mpika, 2 km W of Katete-Luangwa confluence, st. 4.v.1965, *B.L. Mitchell* 2836 (K). W: Solwezi Dist., Mutanda Bridge, fr. 28.vi.1930, *Milne-Redhead* 611 (K). C: c. 10 km E of Lusaka, fl. & fr. 11.xi.1955, *King* 202 (K). E: Petauke Dist., road to Kaulu Dam, fl. 3.xii.1958, *Robson* 810 (K, LISC, SRGH). S: Namwala Dist., between Jacob's Gate and Mumbwa, fl. 27.xi.1962, *van Rensburg* 1008 (K, SRGH). **Zimbabwe.** N: Mt Darwin Dist., near Mukumbura (Mkumburu) R. on road to Tete, fr. 23.i.1960, *Phipps* 2408 (K, SRGH). W: Matobo Dist., Farm Besna Kobila, fl. xi.1953, *Miller* 1970 (K, LISC, SRGH). C: Makoni Dist., Rusape, fl. & fr. xi.1952, *Dehn* 312 (SRGH). E: Chimanimani Dist., Nyahode (Nyhodi) Valley, c. 29 km S of Chimanimani (Melsetter), fl. 25.i.1951, *Crook* 356 (K, SRGH). S: Masvingo Dist., Mutirikwi Recr. Park (Kyle Game Res.), fr. 11.ii.1965, *Blizard* 103 (SRGH). **Malawi.** C: near Nambuma R., fr. 30.vii.1936, *Abdullah* 6084 (K). S: Zomba Dist., Magomero, fl. xii.1957, *Emtage* in GHS 86215 (SRGH). **Mozambique.** T: between Muchena and Massamba 13.9 km from Muchena, fr. 7.vii.1949, *Barbosa & Carvalho* 3476 (K, LISC). MS: between Chimoio (Vila Pery) and Manica (Macequece), fl. 4.xi.1943, *Torre* 6077 (K, LISC, LMU). GI: Inhambane, adjoining route Mavume–Aluise R. near 50 km from Mavume, fr. ii.1939, *Gomes e Sousa* 2203 (COI, K, LISC). M: Maputo Dist., Inhaca Is., Ponta Vermelha, fl. 11.xii.1959, *Lemos & Balsinhas* 2b (COI, K, LISC, PRE, SRGH).

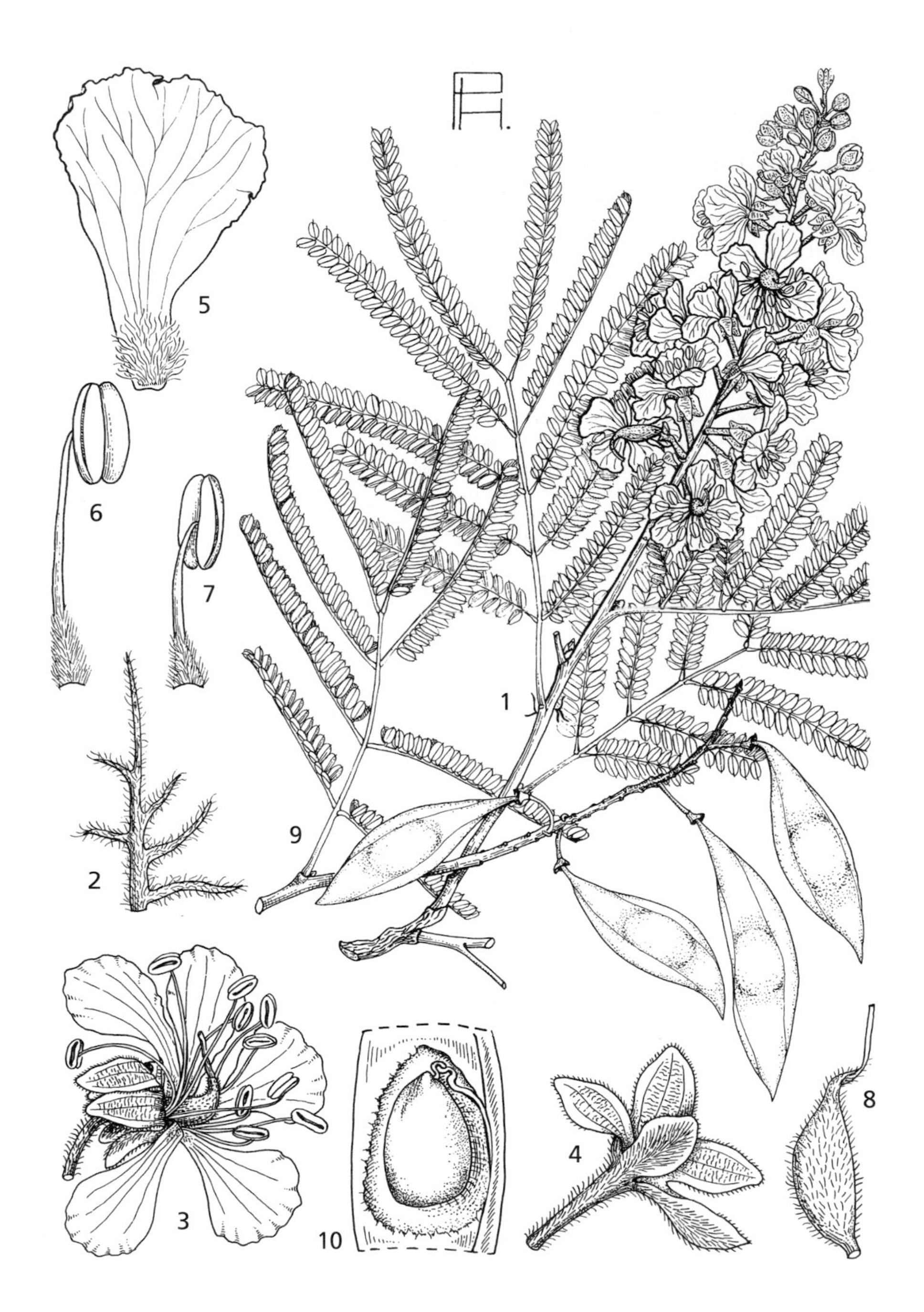

Fig. 3.2.**44**. PELTOPHORUM AFRICANUM. 1, flowering branchlet (× ²⁄₃); 2, stipule (× 4); 3, flower (× 2); 4, calyx, with rest of flower removed (× 2); 5, petal (× 4); 6, 7, stamens (× 4); 8, gynoecium (× 4), 1–8 from *De Winter* 3630; 9, fruiting twig (× ²⁄₃); 10, part of pod showing seed (× 2), 9 & 10 from *De Winter* 3865. Drawn by Pat Halliday. Reproduced with permission from Flora of Southern Africa.

Also in Congo (Katanga), Angola, Namibia, Swaziland and South Africa (North-West and Limpopo Provinces, Mpumalanga, N KwaZulu-Natal); occasionally cultivated as an ornamental tree in other parts of Africa. Usually in open woodland or wooded grassland, or on rocky outcrops or river banks, often on sandy soil; up to 1400 m.

Mitchell 2836, a sapling from the Mpika Dist. and the only record for N Zambia, seems anomalous at first sight as it is covered with stalked, swollen-headed glands 0.5–0.8 mm long. But such glands are to be found to a lesser extent on other material and on *P. dubium*; see note by Barneby in Brittonia **48**: 178 (1996).

The wood is valued for timber.

2. **Peltophorum dubium** (Spreng.) Taub. in Engler & Prantl, Nat. Pflanzenfam. III. **3**: 176 (1892). —Britton in Ann. New York Acad. Sci. **7**: 88 (1893). Type from South America.

Caesalpinia dubia Spreng. in Linnaeus, Syst. Veg. **2**: 343 (1825).

Peltophorum africanum var. *speciosum* Burtt Davy in Bull. Misc. Inform., Kew **1921**: 50 (1921) excl. spec. Maurit. —Baker, Legum. Trop. Afr.: 612 (1930). Type: Zimbabwe, vicinity of Matopos, 1917, *Dowsett* under *Burtt Davy* 16426 (BOL chosen here as lectotype, K).

Tree up to 20 m high. Young stems with fine, dense, rusty-brown to greyish, closely appressed puberulence or pubescence, also with small, stalked, swollen-headed glands. Leaves: petiole and rachis together 4–30(50) cm; pinnae (7)10–21 pairs; pinna rachis 3–11 cm; leaflets 10–32 pairs per pinna, 5–12 × 1.5–3(4) mm, oblong, asymmetrical at the base, rounded to subacute at the apex and mucronate; lower surface pubescent, upper surface glabrous or rarely pubescent; stipules simple and inconspicuous or absent. Inflorescence as in *P. africanum*; bracts occasionally persisting to flowering; pedicels 3–8 mm excluding the receptacle. Sepals 4–5(6) mm long. Petals 12–17 mm long. Ovary and pods as in *P. africanum*.

Zimbabwe. W: Rhodes Matopos Nat. Park, no date, *Dowsett* s.n. (K).

Native of tropical South America and apparently only rarely introduced in Africa.

This species is very similar to the native *P. africanum* from which it may be distinguished by the different indumentum on the young stems and the lack of branched stipules (which, however, fall early in *P. africanum* and are often not seen in flowering or fruiting specimens).

3. **Peltophorum pterocarpum** (DC.) K. Heyne, Nutt. Pl. Ned.-Ind. **2**: 755 (1927). —Brenan, Check-list For. Trees Shrubs Tang. Terr.: 106 (1949); in F.T.E.A., Legum.-Caesalp.: 17 (1967). —Wilczek in F.C.B. **3**: 263 (1952). —Keay in F.W.T.A., ed.2, **1**(2): 442 (1958). —Ross in F.S.A. **16**(2): 133 (1977). Type from Indonesia (Timor).

Inga pterocarpa DC., Prodr. **2**: 441 (1825).

Caesalpinia ferruginea Decne., Descr. Herb. Tim.: 134 (1834).

Peltophorum ferrugineum (Decne.) Benth., Fl. Austral. **2**: 279 (1864). —Baker, Legum. Trop. Afr.: 612 (1930).

Peltophorum africanum var. *speciosum* Burtt Davy in Bull. Misc. Inform., Kew **1921**: 50 (1921), in part regarding Mauritius but not Zimbabwe material.

Tree up to 17 m high. Young stems shortly brown-tomentose, glabrescent. Leaves: petiole and rachis together up to 35 cm, brown-tomentose when young; pinnae 7–14 pairs; pinna rachis 4–13 cm; leaflets (5)9–17(20) pairs per pinna, (5)8–21 × (2)4–8(9) mm, ± oblong or oblong-rhombic, very asymmetrical at the base, rounded and usually emarginate (never mucronate) at the apex; both surfaces pubescent to glabrous; stipules simple and inconspicuous or apparently absent. Inflorescence like that of *P. africanum* but usually longer and with more branches; pedicels 4–7(10 in fruit) mm, approximately as long as the calyx. Sepals (5)6–8(10) mm long. Petals 13–23 mm long. Ovary as in *P. africanum*. Pods (4)5–12 × 1.6–3.2 cm, narrowly elliptic to oblong; seeds 1–3(?4).

Mozambique. GI: Inhambane, fl. xii.1936, *Gomes e Sousa* 1917 (COI); Inhambane, in town, fl. x.1939, *Gomes e Sousa* s.n. (K, PRE). M: Maputo (Lourenço Marques), fl. 22.i.1946, *Gomes e Sousa* s.n. (COI).

Native of SE Asia to Australia but widely planted as an ornamental tree in tropical regions. The notes with *Gomes e Sousa* 1916 suggest that it was growing in natural vegetation with *Androstachys johnstonii*, but this may be an error.

This attractive ornamental species has been widely known under several different names. A further species from tropical Asia, *P. dasyrachis* (Miq.) Baker, is also grown in Africa but is not yet known from the Flora area. It is similar to *P. pterocarpum* but is easily distinguished by its much longer pedicels.

36. PARKINSONIA L.

Parkinsonia L., Sp. Pl.: 375 (1753); Gen. Pl., ed.5: 177 (1754).

Shrubs or small trees, not climbing, armed with spines (modified branches or leaf rachis) or unarmed, not glandular. Leaves bipinnate, sometimes (in both F.Z. species) with the pinna rachis ± phyllodial and the leaflets very reduced or absent; stipules very small to conspicuous and spinescent. Flowers in long or short axillary racemes; bracts caducous. Receptacle comprising a disc and a lower elongate part up to 10 mm long simulating the pedicel from which it is demarcated by a distinct joint. Sepals 5, valvate to narrowly imbricate (perhaps varying between flowers on the same plant), subglabrous. Petals 5, subequal, the upper one usually with a distinct claw. Stamens 10, not or slightly exceeding the petals; filaments alternately longer and short, densely pubescent in the lower part; anthers dorsifixed dehiscing by longitudinal slits. Ovary shortly stipitate, glabrous to pubescent, with up to 8 ovules; style sometimes twisted; stigma truncate, ciliolate or glabrous. Pods ± linear-elliptic to linear, flat or ± turgid, sometimes constricted between the seeds, indehiscent or the valves sometimes separating slightly from each other; seeds up to 8 but frequently only a few developing, usually longitudinally elongate in the pod with the hilum at one end, compressed, with endosperm.

The generic concept adopted here is that of Brenan in Kew Bull. **17**: 203–209 (1963). The genus thus includes about 12 species, mostly in drier areas of N and S America but also one in southern Africa and 3 in E and NE tropical Africa; one American species is also widely cultivated in many warm countries.

Pinna rachis up to 14 cm long, not flattened or winged, usually without leaflets; leaf rachis not spinescent, the spines being modified shoots .... **1.** *africana*
Pinna rachis up to 40 cm long, winged laterally and so appearing flattened, usually with numerous small leaflets; leaf rachis forming a spine 0.2–1.7 cm long .... **2.** *aculeata*

1. **Parkinsonia africana** Sond. in Linnaea **23**: 38 (1850). —Harvey in F.C. **2**: 269 (1862). —Miller in J. S. Afr. Bot. **18**: 34 (1952). —Leistner in Koedoe **2**: 163 (1959). —Brenan in Kew Bull. **17**: 206 (1963). —Schreiber in Merxmüller, Prodr. Fl. SW Afrika, fam. 59: 116 (1967). —Palmer & Pitman, Trees Sthn. Africa **2**: 887 (1972). —Ross in F.S.A. **16**(2): 109, fig.21 (1977). —M. Coates Palgrave, Trees Sthn. Africa: 342 (2002). Type: South Africa, Western Cape Province, Calvinia Dist., Springbokkeel, *Burke & Zeyher* 557 (BM, K, OXF, PRE). FIGURE 3.2.**45**.

Shrub or small tree up to 7 m high. Stems subglabrous or occasionally finely pubescent when young, armed with stout spines which are modified lateral shoots and may often be branched and bear leaves. Leaves reduced to one or two pairs of sessile, green pinna rachides (2)5–14 cm long, slightly grooved on the upper surface, with up to 15 nodes but no leaflets, or occasionally with very small linear or oblanceolate leaflets up to 1.5 mm long. Flowers in fairly lax racemes

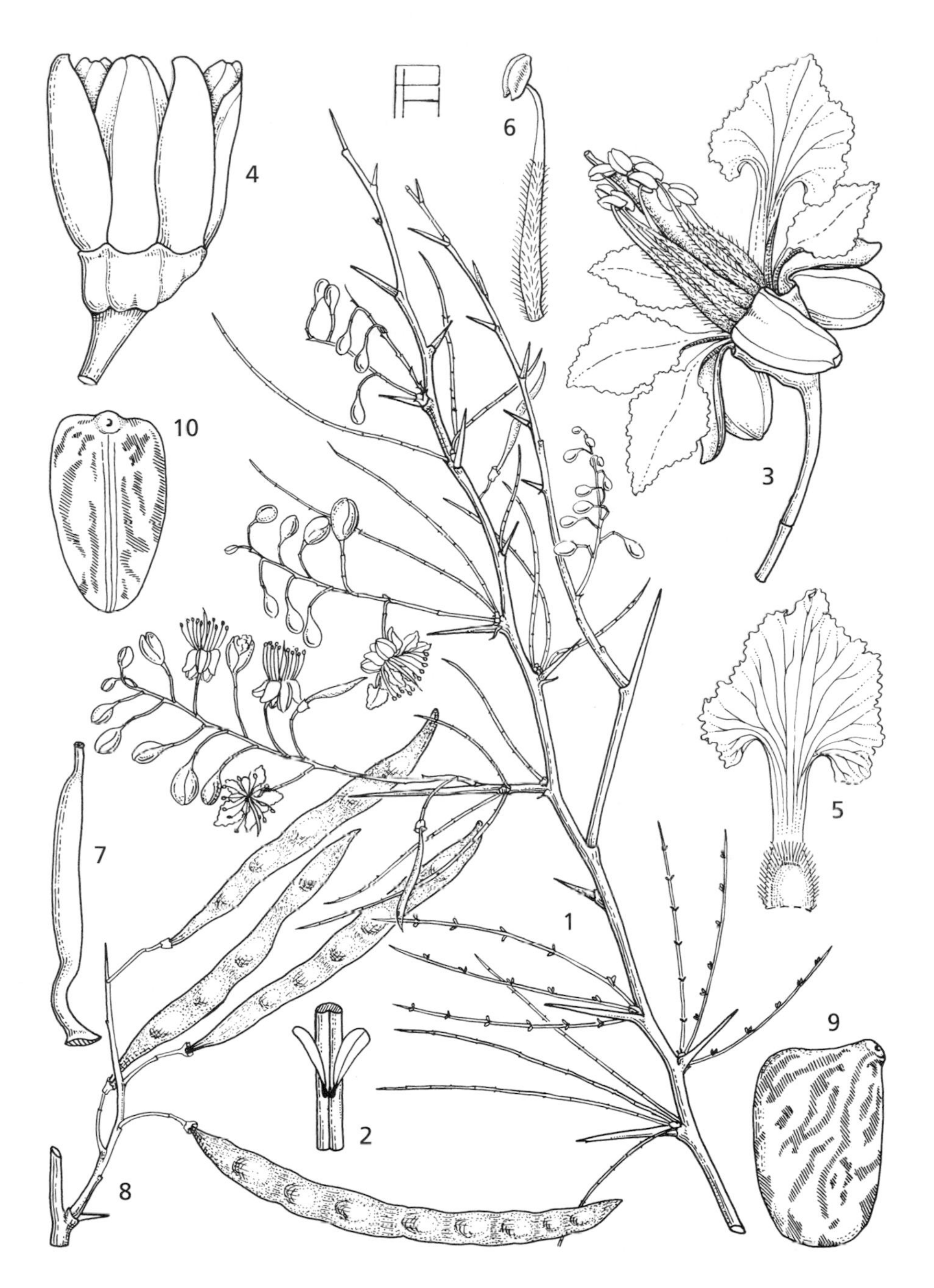

Fig. 3.2.45. PARKINSONIA AFRICANA. 1, flowering branch (× ⅔), from *Keet* 1659; 2, part of pinna rachis with two leaflets (× 6), from *Gerstner* 6284; 3, flower (× 3); 4, sepals and hypanthium (× 4); 5, upper petal (× 4); 6, stamen (× 4); 7, gynoecium (× 4), 3–7 from *Keet* 1659; 8, fruiting twig (× ⅔); 9, 10, seed, side and hilar views (× 4), 8–10 from *Pearson* 9293. Drawn by Pat Halliday. Reproduced with permission from Flora of Southern Africa.

up to 16 cm long; bracts up to 1.5 mm long, ± ovate, caducous; pedicels 3–10 mm long excluding the very elongate lower part of the hypanthium. Receptacle comprising a disc up to 3 mm radius which usually breaks away in fruit to form a small collar, and a lower elongate part up to 10 mm long simulating the pedicel, shorter or longer than the pedicel proper. Sepals 5–8 mm long, ± oblong, narrowly imbricate, becoming reflexed in full flower, with short sparse hairs or subglabrous. Petals 7–12 mm long, the standard with a distinct claw and ± reniform blade, all yellow. Stamens about equalling or slightly exceeding the petals; anthers 1–1.7 mm long. Ovary glabrous or with sparse whitish hairs, particularly on the margins. Pods 5–12 × 0.6–0.8 cm, linear, flattened, straight or curved, often somewhat constricted between the seeds, glabrous, light grey-brown; seeds up to 8 per pod but often fewer (frequently only 1 or 2 developing), 7–9.5 × 5–6 × 2–3 mm, obovate to oblong, elongated longitudinally in the pod with the hilum at one end, light brown with dark markings.

Probably occurring in Botswana SW (shown on map in M. Coates Palgrave, Trees Sthn. Africa, 2002), but no specimens seen from the Flora area; recorded by Miller (1952), as collected by Gerstner on the border of South Africa (Northern Cape Province) and Botswana near Rietfontein, by Leistner (1959), from the Kalahari Gemsbok Nat. Park just outside Botswana, and as probably occurring in SW Botswana by Ross (1977).

Also in South Africa (Northern Cape Province) and Namibia. Sandy plains and semi-desert, especially near water courses.

The seeds are said to be used for making an excellent coffee.

2. **Parkinsonia aculeata** L., Sp. Pl.: 375 (1753). —Oliver in F.T.A. **2**: 267 (1871). —Harms in Engler, Pflanzenw. Afrikas **3**(1): 500, fig.267 (1915). —Baker, Legum. Trop. Afr.: 625 (1930). —Brenan, Check-list For. Trees Shrubs Tang. Terr.: 105 (1949); in Kew Bull. **17**: 206 (1963); in F.T.E.A., Legum.-Caesalp.: 43 (1967). —Wilczek in F.C.B. **3**: 248 (1952). —Torre in Mendonça, Contrib. Conhec. Fl. Moçamb. **2**: 68 (1954). —Torre & Hillcoat in C.F.A. **2**: 174 (1956). —Roti-Michelozzi in Webbia **13**: 179 (1957). —Keay in F.W.T.A., ed.2, **1**(2): 483, fig.154B (1958). —Ross in F.S.A. **16**(2): 110 (1977). —Polhill & Thulin in Fl. Ethiopia **3**: 56 (1989). —Beentje, Kenya Trees, Shrubs & Lianas: 246 (1994). Type from South America.

Shrub or small tree up to 8(13) m high. Stems subglabrous or shortly appressed-pubescent when young, armed with short spines derived from modification of the leaf rachis or sometimes stipules as well. Leaves with 1–2(3) pairs of very long narrow pinnae inserted very close together near the base of the spinescent rachis, so looking like 2 or 4(6) simple pinnate leaves borne on a short spine; spines (rachis) 0.2–1.7 cm long; pinnae up to 40 cm long, their rachis broadly winged and so appearing flattened, up to 2.5 mm broad; leaflets up to 6(9) × 2(3) mm, oblong or oblanceolate, opposite or alternate, up to 80 along each margin of the rachis, or sometimes absent or caducous; stipules often spinescent like the rachis. Flowers in fairly lax racemes up to 20 cm long; bracts up to 2 mm long, caducous; pedicels 7–16 mm long excluding the elongate lower part of the hypanthium. Receptacle forming a narrow disc about 1 mm broad and an elongate lower part up to 3(4) mm long, shorter than the pedicel proper. Sepals 5–7 mm long, oblong, narrowly imbricate, becoming reflexed in full flower, with few short hairs or subglabrous. Petals 8–13 mm long, the standard with a distinct claw, all yellow. Stamens 4–9 mm long; anthers 1.3–2 mm long. Ovary clothed with longish white hairs, glabrescent. Pods (2)3–12(15) × 0.5–0.8 cm, linear, compressed, usually ± straight, constricted between the seeds, glabrous, light grey-brown; seeds up to 8 but commonly only 1–4 developing, c. 9 × 6 × 2.5 mm, elliptic-oblong, light brown with dark markings.

Zimbabwe. C: Harare, Kia-Ora Nursery, cultiv., fl. x.1969, *L.H. Smith* 4 (K, SRGH). W: Matsheumhlope R. near Bulawayo Golf Course, fl. 26.ix.1972, *Simon* 2288 (K, LISC, SRGH). E: Chipinge, cult., fl. 28.x.1950, *Crook* 340 (SRGH). **Mozambique.** T: Changara

Dist., near Boroma, fl. & fr. 26.vi.1941, *Torre* 2924 (BR, LISC). GI: along the Limpopo, c. 244 m, fl. & fr. no date, *Rowland-Jones* 52 (K, PRE, SRGH). M: near Namaacha, slopes of Libombo Mts, fl. & fr. 27.viii.1948, *Myre & Carvalho* 160 (K, SRGH).

Native of southern, central and SW North America, but now widely cultivated as an ornamental tree or hedge plants in many tropical and subtropical countries. Often becoming naturalised and sometimes a troublesome weed. In the Flora area known from numerous localities in Mozambique but elsewhere only the few specimens cited from Zimbabwe. Known in America as Paloverde or Jerusalem Thorn.

37. DELONIX Raf.

Delonix Raf., Fl. Tellur. **2**: 92 (1836).

Unarmed trees. Leaves bipinnate with 2–23 pairs of leaflets, without specialised glands on petiole and rachis; leaflets opposite in (4)6–32 pairs, eglandular; stipules small, subulate and caducous or pinnately compound with up to 6 pairs of pinnae and often persistent. Racemes axillary, subcorymbose, with up to c. 20 fairly large flowers, often in groups and so appearing compound; bracts ± ovate, caducous. Flowers with a distinct hypanthium, the lower part of which is narrower and simulates the pedicel but is clearly demarcated from it by a joint, the upper part being expanded into a shallowly concave disc or a conspicuous cup, persisting into fruiting. Sepals 5, valvate, somewhat leathery. Petals 5, subequal except that the upper one usually differs slightly in size, shape and colour, all with a long claw, the broad part suborbicular to reniform. Stamens 10, all fertile, often conspicuously exserted, eglandular; anthers dorsifixed, dehiscing by longitudinal slits. Ovary ± sessile or shortly stipitate, glabrous to sericeous-tomentose, with numerous ovules, tapered above into a long style approximately equalling the stamens and somewhat broadened at the apex to form a flattish, ciliate stigma. Pods 13–70 cm long, linear-oblong to linear, flattened, coriaceous or woody, dehiscing down both sides; seeds numerous, transversely elongate in the pod, compressed or subcylindrical, with endosperm.

A genus of 11 species, 9 in Madagascar, one from Africa to India and one in the Horn of Africa.

Petals less than 4 cm long, yellow, orange or white; stamen filaments much exceeding the petals; pods less than 25 cm long and 3.6 cm wide; stipules subulate, caducous . . . . **1.** *elata*
Petals usually more than 4 cm long, scarlet red or occasionally orange to yellow; stamen filaments shorter than the petals; pods usually more than 30 cm long and 3.6 cm wide; stipules pinnately compound, often persistent . . . . **2.** *regia*

1. **Delonix elata** (L.) Gamble, Fl. Madras **1**(3): 396 (1919). —Baker, Legum. Trop. Afr.: 624 (1930). —Brenan, Check-list For. Trees Shrubs Tang. Terr.: 102 (1949); in F.T.E.A., Legum.-Caesalp.: 23 (1967). —Roti-Michelozzi in Webbia **13**: 195 (1957). —White, F.F.N.R.: 433 (1962). —Ross in F.S.A. **16**(2): 10 (1977). —Polhill & Thulin in Fl. Ethiopia **3**: 50, fig.95.1.1 (1989). —Beentje, Kenya Trees, Shrubs & Lianas: 241 (1994). Types: India, *Herb. Linnaeus* 529.3, 529.4 (LINN syntypes). FIGURE 3.2.**46**.

Poinciana elata L., Cent. Pl. **2**: 16 (1756). —Oliver in F.T.A. **2**: 266 (1871). —Harms in Engler, Pflanzenw. Afrikas **3**(1): 507, fig.271 (1915).

Tree 2.5–15 m high with rounded spreading crown and rather smooth, buff or grey bark. Young stems thinly appressed-pubescent to densely spreading-pubescent, later glabrescent. Leaves: petiole and rachis together (4)8–20 cm long; pinnae 2–12 pairs; pinna rachis (0.5)1–9 cm long; leaflets (8)11–25 pairs, 4–17 × (1)1.5–4(4) mm, oblong, ± rounded at the apex,

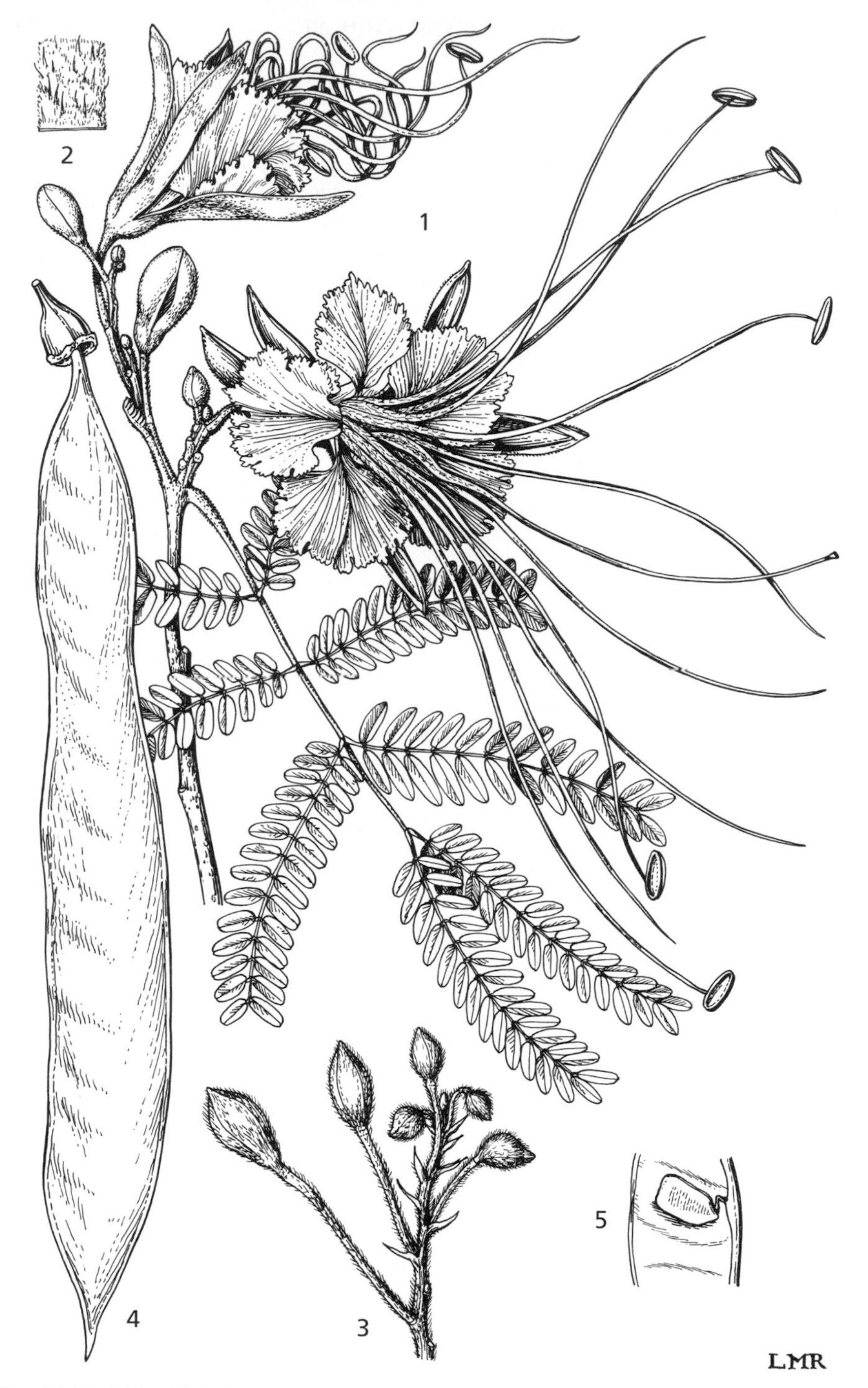

Fig. 3.2.**46**. DELONIX ELATA. 1, part of flowering branchlet (× ²/₃), from *Newbould* 3446, *Wilson* 189 & *Padwa* 294; 2, part of outside of sepal to show indumentum (× 10), from *Newbould* 3446; 3, young inflorescence with spreading indumentum to show variation (× ²/₃), from *B.D. Burtt* 794; 4, pod (× ²/₃), from *Newbold* 3446; 5, seed attached to part of valve of pod (× ²/₃), from *Wilson* 189. Drawn by Lura Ripley. From F.T.E.A.

asymmetrical at the base, finely appressed-pubescent on both surfaces, sometimes glabrescent above; stipules subulate, up to 3 mm long, caducous. Racemes short, the axis up to 5 cm long and ± densely appressed- or spreading-pubescent, each with about 5–20 flowers; bracts up to 3.5 × 2.5 mm, but usually falling at a very early stage, ovate, densely pubescent; pedicels 1.5–3.5(5) cm long. Hypanthium including a narrow basal part 1–4 mm long abruptly expanded above into a conspicuous campanulate part 3–7(8) mm long, the whole on the outside finely pubescent with appressed grey hairs to densely brown-tomentose. Sepals 1.2–3.3(4) cm long, lanceolate to oblong, tapering above to an ± acute apex, similar to the hypanthium in pubescence. Petals 1.6–3.2(3.8) cm long, the upper one smaller than the others and yellow to orange, the others white but fading to yellow or orange, all ± lacerate and crisped at the upper margin. Stamen filaments 3–11(14) cm long, bright red, much exceeding the petals. Ovary shortly stipitate, densely pubescent, often ± sericeous. Pods 13–26 × 2.1–3.7 cm; seeds 8–19, c. 15 × 8 × 3 mm, compressed.

Zambia. N: Mbala Dist., near L. Tanganyika, Cipanta village, fl. i.1955, *Lawton* 144 (K).

From Tanzania north to Egypt and in Arabia and India. Deciduous thickets and bushland.

D. elata is doubtfully native in the Flora area, being known only from the specimen cited. This is labelled by the collector "presumably an exotic". White (F.F.N.R., 1962) gives it as "possibly planted, possibly indigenous", while Brenan (F.T.E.A., Legum.-Caesalp.: 25 (1967)), states it is almost certainly an introduction in Zambia. Our plant corresponds with the variant from NW and central Tanzania discussed by Brenan (1967); this differs from the more northerly typical *D. elata* in having a dense, spreading tomentum on the inflorescence and flower buds, generally more numerous pinnae (mostly 10–11), smaller flowers and fewer seeds per pod (8–9). The nearest known localities in Tanzania are in Tabora Dist., *Hughes* 158 (K) and Iringa Dist., Ruaha Nat. Park, *Greenway & Kanuri* 14049 (K).

2. **Delonix regia** (Hook.) Raf., Fl. Tellur. **2**: 92 (1836). —Baker, Legum. Trop. Afr.: 623 (1930). —Brenan, Check-list For. Trees Shrubs Tang. Terr.: 102 (1949); in F.T.E.A., Legum.-Caesalp.: 23 (1967). —Wilczek in F.C.B. **3**: 248 (1952). —Torre in Mendonça, Contrib. Conhec. Fl. Moçamb. **2**: 67 (1954). —Torre & Hillcoat in C.F.A. **2**: 173 (1956). —Roti-Michelozzi in Webbia **13**: 194 (1957). —Keay in F.W.T.A., ed.2, **1**(2): 481 (1958). —White, F.F.N.R.: 122 (1962). —Ross in F.S.A. **16**(2): 10 (1977). —Polhill & Thulin in Fl. Ethiopia **3**: 50 (1989). Type cultivated in England, originally from Madagascar.

Poinciana regia Hook. in Bot. Mag. **56**: fig.2884 (1829). —Oliver in F.T.A. **2**: 266 (1871). —Harms in Engler, Pflanzenw. Afrikas **3**(1): 507, fig.271 (1915).

Tree 3–15(18) m high. Young stems subglabrous to thinly spreading-pubescent. Leaves: petiole and rachis together (11)14–35 cm long; pinnae 9–18(23) pairs; pinna rachis (4)6–12(15) cm long; leaflets 10–32 pairs per pinna, 4–11(17) × 2–4(5) mm, oblong, obtuse to rounded at the apex, asymmetrical at the base, finely appressed-pubescent on both surfaces or rarely subglabrous; stipules pinnately compound, the rachis up to 1.4 cm long, with 2–6 pairs of pinnae, each oblong or oblanceolate up to 9 × 3 mm. Racemes with the axis up to 12 cm excluding the peduncle, subglabrous to sparsely pubescent, each with (3)6–14 flowers; bracts up to 9 × 5 mm, ovate or elliptic, acute at the apex, usually falling shortly before anthesis; pedicels 3.5–9(10.5) cm long. Hypanthium 2–8 mm long, the narrow basal part gradually expanded above into a shallowly concave disc but the upper part not campanulate, the whole glabrous or sparsely pubescent. Sepals 1.8–3 cm long, lanceolate to oblong, glabrous to very sparsely pubescent. Petals (3.5)4.2–6.5(7.4) cm long, the lower 4 subequal but the upper one slightly longer than the others with the claw broader and less clearly defined, all scarlet red or sometimes (especially the upper one and the claws of the others) orange or ± yellow, the distal margins undulate. Stamen filaments (2.4)3.5–4.5(5) cm long, shorter than the petals. Ovary ± sessile, appressed-pubescent with longish but usually sparse hairs. Pods (20)30–70 × 3.6–5.8 cm, strongly woody at maturity, with seeds sunk in regular transverse cavities in the woody endocarp; seeds up to 50 or more, c. 2.0 × 0.6 × 0.4 cm, linear-ellipsoid, not compressed, light brown mottled with darker brown.

Zimbabwe. C: Harare, cultivated in a garden, fl. 30.xi.1969, *Biegel* 3221 (K). E: Border E of Mutare (Umtali), fl. 10.xii.1913, *Peter* 47820 (K). **Mozambique.** T: Tete, fl. xi.1858, *Kirk* (K). M: Maputo (Lourenço Marques), in the avenues, fl. & fr. x.1945, *Pimenta* 15302 (LISC, PRE, SRGH).

Extensively planted as an ornamental tree (the Flamboyant or Flame Tree) and possibly naturalized in places. Recorded by White (F.F.N.R., 1962) from Livingstone, Luwingu, Ndola and elsewhere. A rare native of Madagascar now widely grown in many tropical and subtropical parts of the world.

The specimen from Tete cited above is annotated by Kirk "This species '*P. [Poinciana] regia*' is cultivated in the public square of Moçambique and has now been extended to Quilimane and as far as Tete it grows freely but has not spread beyond the towns and there only where planted, the pod is 6–8 inches long, thick hard and black".

38. ERYTHROPHLEUM R.Br.

Erythrophleum R. Br. in Denham, Clapperton & Oudney, Travels N. & C. Afr., J. Excurs.: 235 (1826). *

Unarmed trees. Leaves bipinnate with 2–5 pairs of pinnae, without specialised glands on petiole and rachis; leaflets up to 17 per pinna, alternate, eglandular, petiolulate; stipules very small and quickly falling. Inflorescences of dense spike-like racemes of small ± contiguous flowers (similar in appearance to those of many genera of Mimosoideae), usually ± aggregated into panicles; bracts very small, falling as or before the flowers open. Flowers hermaphrodite. Sepals 5, connate in their lower part or almost free to the base, slightly imbricate but open from an early stage. Petals 5, equal, free, ± imbricate, pubescent, oblong to oblanceolate-spathulate. Stamens 10, often alternately longer and shorter; filaments glabrous or hairy; anthers dorsifixed, dehiscing by longitudinal slits; connective not projecting beyond the anther. Ovary shortly stipitate, pubescent to tomentose, with several ovules, tapering into a narrowly conical style; stigma minute, punctiform, cup-shaped and minutely ciliate. Pods stipitate, ± oblong, straight or curved, coriaceous to woody, dehiscing along one or both margins; seeds up to 11, compressed, not areolate, with endosperm.

10 species in the tropical regions of Africa, Madagascar, Asia, Malesia and Australia.

The genus contains powerful alkaloids. The bark is used as a fish poison, an insecticide in grain stores, a local medicine and formerly for ordeal trials, see Palmer & Pitman, Trees Sthn. Africa: 835 (1972) and M. Coates Palgrave, Trees Sthn. Africa: 315–316 (2002).

1. Leaflets obtuse to emarginate, not acuminate; inflorescence axis and peduncle densely pubescent to subtomentose with spreading grey-brown hairs; stamen filaments hairy; pods dehiscing simultaneously along both margins, ± symmetrical at the base . . . . **3.** *africanum*
– Leaflets bluntly acuminate (i.e. with a short drip-tip); inflorescence axis and peduncle puberulent to pubescent with short rusty-brown hairs; stamen filaments hairy or glabrous; pods dehiscing along one or both margins, the stipe usually inserted markedly to one side 2

* For discussion concerning publication of the generic name see Brenan in Taxon **9**: 193–194 (1960) and Rickett in Taxon **13**: 181 (1964).

2. Stamen filaments glabrous; pedicel 0.1–1 mm long; pods dehiscing along the upper margin only; leaflet midrib usually pubescent beneath .... **1.** *suaveolens*
– Stamen filaments hairy; pedicel 2–3 mm long; pods dehiscing along both margins; leaflet midrib glabrous beneath .... **2.** *lasianthum*

1. **Erythrophleum suaveolens** (Guill. & Perr.) Brenan in Taxon **9**: 194 (1960); in F.T.E.A., Legum.-Caesalp.: 18, fig.1 (1967). —Boughey in J. S. Afr. Bot. **30**: 159 (1964). —Beentje, Kenya Trees, Shrubs & Lianas: 243 (1994). —White, Dowsett-Lemaire & Chapman, Evergr. For. Fl. Malawi: 307, fig.108 (2001). —M. Coates Palgrave, Trees Sthn. Africa: 315 (2002). Type: Gambia ('Senegal'), Albreda on R. Gambia, *Perrottet* s.n. (P holotype). FIGURE 3.2.**47**.

Fillaea suaveolens Guill. & Perr. in Guillemin, Perrottet & A. Richard, Fl. Seneg. Tent.: fig.55 (?June–July 1932) & 242 (Oct. 1832).

Erythrophleum guineense G. Don, Gen. Syst. **2**: 424 (Oct.1832). —Oliver in F.T.A. **2**: 320 (1871). —Sim, For. Fl. Port. E. Afr.: 52 (1909). —Greenway in Bull. Misc. Inform., Kew **1928**: 203 (1928). —Baker, Legum. Trop. Afr.: 779 (1930). —Brenan, Check-list For. Trees Shrubs Tang. Terr.: 103 (1949). —Wilczek in F.C.B. **3**: 243, fig.18 (1952). —Keay in F.W.T.A., ed.2, **1**(2): 483, fig.154D (1958). —Topham in Burtt Davy & Hoyle, Check List For. Trees Shrubs Nyasaland Prot.: 37 (1958). —White, F.F.N.R.: 124, 20 (1962). —Breitenbach, Indig. Trees S. Africa **3**: 318 (1965). —Gomes e Sousa, Dendrol. Moçamb. Estudo Geral **1**: 242, fig.46 (1966). Type: West Africa ("Sierra Leone and other parts of Guinea"), *G.Don* s.n. (BM holotype).

Erythrophleum ordale Bolle in Peters, Naturw. Reise Mossambique **6**(1): 10 (1861). Type: Mozambique, Mocuba, Boror, *Peters* (B† holotype).

Tree up to 30 m high; bark dark brown, very rough. Young branches and leaf rachides glabrous to shortly pubescent. Leaves: petiole and rachis together (6)11–35 cm long; pinnae 2–4(5) pairs; pinna rachis 8–20 cm long; leaflets (6)9–13(15) per pinna, (3)4.5–7(9) × (1.5)2.2–4.5(5.3) cm, obliquely ovate to elliptic, the base asymmetrical (except the terminal leaflet), the apex bluntly acuminate, i.e. drawn out into a short 'drip-tip', the surfaces glabrous except on the midrib beneath which is usually strongly pubescent; petiolules 2–4.5(5) mm long. Racemes 2.5–8(11) cm long including a peduncle 0.7–1.5 cm long, the axis and peduncle densely puberulous to pubescent with short, crisped, appressed, rusty-brown hairs. Flowers 6–8(9) mm long from base of pedicel to anthers, cream or lemon-yellow to greenish-yellow; pedicel 0.6–1 mm long at anthesis. Calyx 1–2(2.5) mm long, fused for almost half its length, rusty-brown pubescent like the inflorescence axis. Petals 2–3.5 mm long, greyish-brown pubescent. Stamen filaments glabrous. Pods (7)10–14(17) × 3–4.5(5.3) cm excluding the stipe, oblong or somewhat curved, rounded at both ends, often somewhat woody, dehiscing along one margin only; stipe (1)1.5–2(2.5) cm long, inserted ± asymmetrically; seeds (4)6–9 × 10–12 × 4–7 mm, oblong-ellipsoid, on a funicle 7–10 mm long.

Zambia. N: Chinsali Dist., Mbesuma area, Chambeshi R. levee, fl. & fr. 20.x.1961, *Astle* 992 (K, SRGH). W: Mwinilunga Dist., banks of Zambezi, fr. 27.ix.1952, *Holmes* 917 (K). C: Kabwe Dist., Musafwa's, fr. viii.1934, *Trapnell* 1532 (K). S: Namwala Dist., Lunga R., Kafue Nat. Park, fl. xii.1962, *Mitchell* 20/40 (SRGH). **Zimbabwe.** E: Chipinge Dist., Nyagadze R., 3 km S of Chikore Mission, fr. iii.1962, *Goldsmith* 92/62 (K, LISC, SRGH). S: Masvingo Dist., Masvingo Communal Land (Victoria Reserve), fr. 13.ii.1956, *Furness* 19/56 (SRGH). **Malawi.** N: Nkhata Bay, old graveyard, st. 14.vi.1954, *Jackson* 1340 (K). S: Zomba, fl. 1901, *Sharpe* 184 (K). **Mozambique.** N: Mandimba, between Belém and Luambala, fl. & fr. 24.x.1948, *Andrada* 1442 (COI, LISC, LMA). Z: between Quelimane and Marral, fr. 17.ix.1941, *Torre* 3446 (LISC). MS: Gorongosa Dist., 35 km from Gorongosa (Vila Paiva) on road to R. Vundúzi, fl. & fr. 26.x.1965, *Torre & Pereira* 12607 (LISC).

From West Africa (to Senegal) and the Sudan to East Africa and southwards to the Congo and the Flora Zambesiaca area. Riverine, lakeshore and lowland forests; 500–1200(1350) m.

Fig. 3.2.**47**. ERYTHROPHLEUM SUAVEOLENS. 1, flowering branchlet (× $^2/_3$), from *Elliot* 1493; 2, lower surface of leaflet (× 1), from *Verdcourt* 2129; 3, flower (× 6); 4, stamen (× 6); 5, ovary, showing insertion in hypanthium (× 6); 6, stigma (× 8), 3–6 from *Elliot* 1493; 7, pod (× $^2/_3$), from *Verdcourt* 2129; 8, end-on view of dehisced pod (× $^1/_2$), from *Paulo* 149; 9, seed (× $^2/_3$), from *Lebrun* 4224 and *Eyles* 7917. Drawn by Lura Ripley. From F.T.E.A.

Also grown in a plantation of native species in Mozambique, Maputo Prov., Marracuene, Projecto Foz, Unidade Michafutene, fl. & fr. 15.ix.1989, *Groenendijk & Dungo* 2172 (L, LMU).

2. **Erythrophleum lasianthum** Corbishley in Bull. Misc. Inform., Kew **1922**: 27 (1922). —Torre in Mendonça, Contrib. Conhec. Fl. Moçamb. **2**: 86 (1954). —Breitenbach, Indig. Trees S. Africa **3**: 319 (1965). —Palmer & Pitman, Trees Sthn. Africa: 833 (1972). —Ross in F.S.A. **16**(2): 12, fig.1 (1977). —Pooley, Trees Natal: 144 (1993). —M. Coates Palgrave, Trees Sthn. Africa: 315 (2002). Type: South Africa, KwaZulu-Natal, Ingwavume, *Dist. Magistrate* as PRE 1228 (K holotype, FHO, PRE).

Erythrophleum guineense var. *swaziense* Burtt Davy, Fl. Pl. Ferns Transvaal: 330 (1932). —Breitenbach, Indig. Trees S. Africa **3**: 318 (1965). —Gomes e Sousa, Dendr. Moçamb., Estudo Geral. **1**: 244 (1966). Type: Swaziland, no locality *B.Nicholson* s.n. (K holotype; PRE).

Erythrophleum guineense sensu Henkel, Woody Pl. Natal Zululand: 236 (1934), non G. Don.

Erythrophleum suaveolens sensu auct. non (Guill. & Perr.) Brenan. —Compton, Annot. Check List Fl. Swaziland: 46 (1966). —sensu B. & M. de Winter & Killick, 66 Transvaal Trees: 163 (1966).

Tree up to 14 m high. Young branches and leaf rachides glabrous to shortly and thinly puberulent or pubescent. Leaves: petiole and rachis together 6–20 cm long; pinnae 2–4 pairs; pinna rachis 8–18 cm long; leaflets (7)10–13 per pinna, (2)2.5–5(6) × 1.3–3.5 cm, similar in shape and texture to those of *E. suaveolens*, both surfaces completely glabrous; leaflet petiolules 3–4 mm long. Racemes 5–10 cm long including a peduncle 0.8–1.8 cm long, the axis and peduncle densely puberulous to pubescent with appressed or shortly spreading rusty-brown hairs. Flowers (7)9–11 mm from base of pedicel to anthers, cream to greenish-yellow; pedicel 2–3 mm long at anthesis. Calyx 2–3 mm long, fused for third to half its length, rusty-brown pubescent. Petals 3–4 mm long, pubescent particularly at their margins. Stamen filaments with soft, irregular, spreading hairs. Pods (7)11–15 × 2.9–4.2 cm excluding the stipe, sometimes curved, rounded to pointed at the apex, scarcely woody, dehiscing simultaneously along both margins; stipe 1–2 cm long, ± asymmetrically placed; seeds probably potentially 8–10 × 10–12 × 4–6 mm, ± ellipsoid.

Mozambique. GI: between Vilankulo (Vilanculos) and Nova Mambone, fl. 1.ix.1942, *Mendonça* 85 (K, LISC). M: Matutuíne Dist., Picado do Seco, near Matutuíne (Bela Vista)–Catuane road, fl. 6.ix.1948, *Gomes e Sousa* 3831 (COI, K, LISC, PRE, SRGH).

Also in South Africa (Mpumalanga and KwaZulu Natal) and Swaziland. Open woodland and riverine forest at low altitudes.

This species seems to combine the characters of *E. suaveolens* and *E. africanum*, having the leaflet shape and texture, inflorescence indumentum and pod shape of the former but the pubescent filaments and pod dehiscence of the latter. It is also closely allied to *E. couminga* Baill. from Madagascar.

3. **Erythrophleum africanum** (Benth.) Harms in Repert. Spec. Nov. Regni Veg. **12**: 298 (1913). —De Wild., Contrib. Fl. Katanga, Suppl. 1: 21 (1927). —Greenway in Bull. Misc. Inform., Kew **1928**: 202 (1928). —Baker, Legum. Trop. Afr.: 777 (1930). —Brenan, Check-list For. Trees Shrubs Tang. Terr.: 103 (1949); in F.T.E.A., Legum.-Caesalp.: 20 (1967). —Wilczek in F.C.B. **3**: 244 (1952). —Pardy in Rhod. Agric. J. **52**(6): 513 (1955). —Torre & Hillcoat in C.F.A. **2**: 252 (1956). —Keay in F.W.T.A., ed.2, **1**: 484 (1958). —Topham in Burtt Davy & Hoyle, Check List For. Trees Shrubs Nyasaland Prot.: 37 (1958). —White, F.F.N.R.: 124, fig.20B (1962). —Schreiber in Merxmüller, Prodr. Fl. SW Afrika, fam. 59: 14 (1967). —

Palmer & Pitman, Trees Sthn. Africa **2**: 831 (1972). —Ross in F.S.A. **16**(2): 11 (1997). —M. Coates Palgrave, Trees Sthn. Africa: 314 (2002). Types: Angola, Huila, Mumpula, *Welwitsch* 591 (BM), and Pungo Andongo, Calundo, *Welwitsch* 573 (BM, K).

Gleditsia (*Gleditschia*) *africana* Benth. in Trans. Linn. Soc., London **25**: 304 (1865). —Oliver in F.T.A. 2: 265 (1871).

Erythrophleum pubistamineum Hennings in Gartenflora **38**(2): 39, fig.8 (1889). —Eyles in Trans. Roy. Soc. S. Afr. **5**(4): 364 (1916). Type: Angola, Malange, *Mechow* 15 (B† holotype).

Erythrophleum pubistamineum var. *parvifolium* Schinz in Mem. Herb. Boissier **1**: 119 (1900). Type: Angola, Omupanda in UuKuanjama, *Wulfhorst* 31 (Z holotype).

Erythrophleum africanum var. *micrantherum* De Wild., Contrib. Fl. Katanga, Suppl. 1: 23 (1927). —Baker, Legum. Trop. Afr.: 778 (1930). Type from Congo.

Erythrophleum africanum var. *angustifolium* De Wild., Contrib. Fl. Katanga, Suppl. 1: 29 (1927). —Baker, Legum. Trop. Afr.: 778 (1930). Type from Congo.

Erythrophleum africanum forma *glabrissimum* and forma *subglabrum* De Wild., Contrib. Fl. Katanga, Suppl. 1: 28 (1927). —Baker, Legum. Trop. Afr.: 778 (1930). Type from Congo.

Erythrophleum africanum var. *stenocarpum* Harms in Notizbl. Gart. Berlin **13**: 414 (1936). —Brenan, Check-list For. Trees Shrubs Tang. Terr.: 103 (1949); see also obs. in F.T.E.A., Legum.-Caesalp.: 20 (1967). Type: Tanzania, Lindi Dist., L. Lutamba, *Schlieben* 6536 (B† holotype, BM).

Tree 4–14(20) m high. Young branches and leaf rachides usually densely pubescent with grey-brown hairs, sometimes glabrescent, rarely subglabrous when young. Leaves: petiole and rachis together (5)7–15(23) cm long; pinnae (2)3–4(5) pairs (in the Flora area the great majority of leaves with 3 pairs); pinna rachis (4.5)7–14(19) cm long; leaflets (8)10–15(17) per pinna, 1.5–6(8) × 0.9–3.3(4.6) cm, ± rhombic and asymmetrical, obtuse to emarginate at the apex, not acuminate, the surfaces glabrous to densely pubescent; leaflet petiolules 1–2(3.5) mm long. Racemes 4–11 cm long including a peduncle 0.5–2 cm long, the axis and peduncle densely pubescent to subtomentose with spreading grey-brown hairs. Flowers (7)8–10 mm long from base of pedicel to anthers, cream to greenish-yellow; pedicel 0.5–1 mm long at anthesis. Calyx 1.5–2.5 mm long, the lobes free almost to the base, grey-brown pubescent. Petals (2)2.5–3.5(4) mm long, pubescent. Stamen filaments with soft, irregular, spreading hairs, or very rarely subglabrous. Pods (5)8–17 ×(2)2.4–4(4.9) cm excluding the stipe, rounded or pointed at the apex, not woody, dehiscing simultaneously along both margins; stipe 1–1.5 cm long, inserted medially; seeds potentially 6 but frequently only (1)2–5 developing to maturity, 10–14 × 8–12 × 3–4 mm, ± lenticular.

Botswana. N: Okavango Delta, 3.1 km S of Shakawe on main road to Sepopa, fr. 24.iv.1975, *P.A. Smith* 1346 (K, LISC, SRGH). **Zambia.** B: Mabete Is., opposite Nangweshi, fl. & fr. 22.vii.1952, *Codd* 7146 (COI, K, SRGH). N: Mbala (Abercorn), in grounds of T.M.V.I., fl. 19.ix.1956, *Richards* 6227 (K). W: Solwezi, near guest house, fr. 6.vi.1930, *Milne-Redhead* 428 (K). C: c. 14 km N of Lusaka on road to Kabwe (Broken Hill), fl. without date, *Cole* 41 (K). E: Luangwa Valley, Mfuwe, fl. ix.1967, *Astle* 5108 (K). S: Livingstone Dist., Victoria Falls, fr. 7.vii.1930, *Pole Evans* 2768 (K, PRE). **Zimbabwe.** N: Sebungwe, Mabunda R., fr. 8.viii.1951, *Whellan* 517 (K, LISC, SRGH). W: Hwange Dist., Victoria Falls, near Hotel, fl. & fr. 23.xi.1949, *Wild* 3181 (K, LISC, SRGH). E: Mutare Dist., c. 48 km on Birchenough Bridge road, fr. 22.iv.1955, *Chase* 5557 (COI, K, LISC, SRGH). S: Mwenezi Dist., Gonakudzingwa, fr. 26.vi.1960, *Farrell* 220 (COI, SRGH). **Mozambique.** N: Cabo Delgado, near Montepuez, fl. 17.x.1942, *Mendonça* 916 (K, LISC). Z: Mocuba Dist., 16 km from crossing of Mocuba and Maganja da Costa roads, fr. 8.ii.1966, *Torre & Correia* 14462, with photo (LISC). T: Moatize Dist., between Zóbuè and Tete, 22.1 km from Zóbuè, fr. 21.vii.1949, *Barbosa & Carvalho* 3734 (K). MS: Barué Dist., near Catandica (Vila Gouveia), fl. & fr. 29.x.1943, *Torre* 6097 (LISC).

From West Africa (Senegal) to the Sudan and Tanzania (not Uganda and Kenya) and southwards to Namibia and the Flora area. Deciduous woodland; 250–1600 m.

A single specimen, *Gairdner* 84 from Zambia, without precise locality (K), has been seen in which the stamen filaments are ± glabrous.

Although this species was recorded from Malawi by Topham (in Check List For. Trees Shrubs Nyasaland Prot.: 37, 1958) no specimens have been seen by the author.

The wood is reported to be heavy, hard and durable.

39. BURKEA Benth.

Burkea Benth. in Hooker's Icon. Pl. **6**: fig.593/4 (1843).

Unarmed tree. Leaves bipinnate with (1)2–7 pairs of pinnae, without specialised glands on petiole and rachis; leaflets up to 15 per pinna, alternate, eglandular, petiolulate; stipules very small and quickly falling. Inflorescence of pendulous racemes of small flowers, fairly dense but less so than in *Erythrophleum*, usually in groups at ends of branches; bracts very small, persistent until the flower opens. Flowers hermaphrodite. Calyx of 5 lobes connate on their lower part, slightly imbricate but ± open from an early stage. Petals 5, equal, free, strongly reflexed about their middle after anthesis, glabrous. Stamens 10; filaments glabrous; anthers dorsifixed, dehiscing by longitudinal slits; connective shortly projecting beyond the anther. Ovary subsessile, tomentose, with 1–2 ovules; style very short, ending in a funnel-shaped stigma split down one side. Pods stipitate, elliptic to suborbicular, coriaceous, indehiscent, one-seeded. Seeds compressed, areolate on both faces, with endosperm.

A monotypic genus widespread in tropical Africa.

Burkea africana Hook., Hooker's Icon. Pl. **6**: fig.593/4 (1843). —Harvey in F.C. **2**: 271 (1862). —Oliver in F.T.A. **2**: 320 (1871). —Harms in Engler, Pflanzenw. Afrikas **3**(1): fig.237 (1915). —R.E. Fries, Wiss. Ergebn. Schwed. Rhod.-Kongo-Exped. **1**: 65 (1914). —Eyles in Trans. Roy. Soc. S. Afr. **5**: 364 (1916). —Baker, Legum. Trop. Afr.: 776 (1930); in Bol. Soc. Brot. Sér. 2, **8**: 115 (1933). —Steedman, Trees Shrubs & Lianes S. Rhod.: 20 (1933). —Pardy in Rhod. Agric. J. **49**(3): 170 (1952). —Wilczek in F.C.B. **3**: 238 (1952). —Torre & Hillcoat in C.F.A. **2**: 250 (1956). —O. Coates Palgrave, Trees Central Africa: 89 (1957). —Keay in F.W.T.A., ed.2, **1**(2): 483 (1958). —Topham in Burtt Davy & Hoyle, Check List For. Trees Shrubs Nyasaland Prot.: 36 (1958). —Palmer & Pitman, Trees Sthn. Africa: 171 (1961). —White, F.F.N.R.: 118, fig.20 (1962). —Breitenbach, Indig. Trees S. Africa **3**: 320 (1965). —Gomes e Sousa, Dendrol. Moçamb. Estudo Geral **1**: 241, fig.45 (1966). —Brenan in F.T.E.A., Legum.-Caesalp.: 21 (1967). —Killick in Fl. Pl. Afr. **38**: fig.1505 (1967). —Schreiber in Merxmüller, Prodr. Fl. SW Afrika, fam. 59: 7 (1967). —Palmer & Pitman, Trees Sthn. Africa **2**: 837 (1972). —Ross in F.S.A. **16**(2): 15, fig.2 (1977). —Van Wyk, Trees Kruger Nat. Park, ed.3: 82 (1994). —M. Coates Palgrave, Trees Sthn. Africa: 316 (2002). Type: South Africa, North West Province, Magaliesberg, *Burke* 274 (K holotype). FIGURE 3.2.**48**.

Burkea africana var. *andongensis* Oliv. in F.T.A. **2**: 320 (1871). —Hiern, Cat. Afr. Pl. Welw. **1**: 304 (1896). —Baker, Legum. Trop. Afr.: 776 (1930). Type: Angola, Cuanza Norte, Pungo Andongo, *Welwitsch* 574 (LISU holotype; BM, K).

Burkea africana var. *cordata* Oliv. in F.T.A. **2**: 320 (1871). —Hiern, Cat. Afr. Pl. Welw. **1**: 304 (1896). —Baker, Legum. Trop. Afr.: 777 (1930). Type: Angola, Huíla Dist., between Lopolo and Monhino, *Welwitsch* 589b (LISU holotype; BM).

Tree 4–20 m high, the crown often flat, sometimes rounded. Bark grey to blackish, fissured and scaly. Twigs rather stout, with conspicuous raised leaf scars, when young covered with a fine dark red tomentum. Leaves: petiole and rachis together (3)7–24(32 in East Africa) cm long, appressed-pubescent to reddish tomentose when young, often becoming glabrous at maturity;

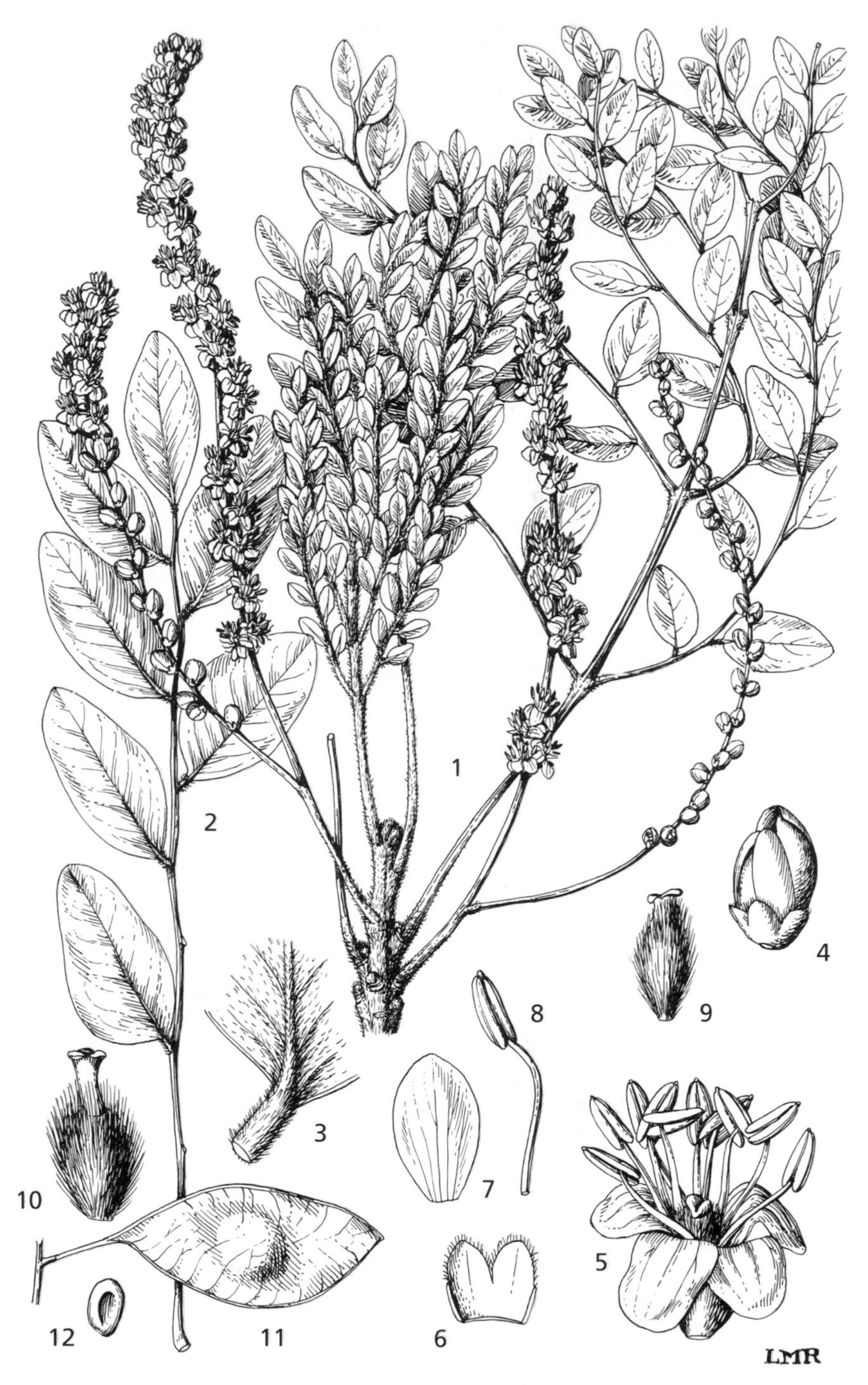

Fig. 3.2.**48**. BURKEA AFRICANA. 1, part of branchlets with inflorescences and young leaves (× $^2/_3$), from *Semsei* 1862; 2, pinna of mature leaf (× $^2/_3$); 3, basal part of mature leaflet, lower surface (× 3), 2 & 3 from *Eggeling* 5776; 4, flower bud (× 4); 5, flower (× 4); 6, two calyx segments (× 4); 7, petal (× 4); 8, stamen (× 6); 9, ovary (× 6), 4–9 from *Semsei* 1862; 10, another ovary, showing variation (× 6), from *Aylmer* 27/17; 11, pod (× $^2/_3$); 12, seed, showing areole (× $^2/_3$), 11 & 12 from *Gillman* 1543. Drawn by Lura Ripley. From F.T.E.A.

pinnae (1)2–4(7 in East Africa) pairs; pinna rachis (3)7–16(23 in East Africa) cm long; leaflets 6–12(15 in East Africa) per pinna, (1.5)2.5–5.5(7.5) × (0.7)1.2–3.6(4.2) cm, usually immature at anthesis, ± obliquely ovate to rhombic, strongly asymmetrical at the base, obtuse to emarginate at the apex, usually silvery-sericeous when very young, becoming thinly pubescent to glabrous at maturity; leaflet petiolules 2–4(5) mm long. Racemes 5–30 cm long including the peduncle, sometimes branched, pendulous and usually ± clustered at the ends of young branches, usually appearing with the young leaves; peduncle and axis pubescent to tomentose with appressed or spreading, brown or reddish hairs. Flowers 4–5.5(6.5) mm from base to anthers, sessile, white or cream or occasionally greenish, pink or brownish. Calyx 1.5–2(2.5) mm long, fused for third to two-thirds of its length, ± pubescent at least in its lower part, the teeth broadly rounded and ciliate. Petals 4–5 mm long, but strongly reflexed about their middle after anthesis, 2–2.5 mm broad, rounded to somewhat pointed at the apex. Pods (3.5)4.5–5.5(6.5) × 2–3 cm excluding the stipe, ± elliptic, slightly asymmetrical, brown; stipe 0.5–1.5(1.8) cm long; seeds 9–12 × 7–8 × 2 mm, elliptic to suborbicular, flat, brown, with areoles 7–8 × 3.5–4 mm.

Botswana. N: Ngamiland Dist., Moremi Game Reserve, fr. 17.vii.1964, *Tinley* (K, PRE). SE: Southern Dist., Kanye Hill, fl. 13.xi.1948, *Hillary & Robertson* 496 (K, PRE, SRGH). **Zambia.** B: Kalabo Boma, fr. 13.ii.1952, *White* 2060 (K). N: Mbala Dist., Ndundu, fl. 3.x.1959, *Richards* 11486 (K, LISC). W: Mwinilunga Dist., slope leading down to Muzera R. 19 km W of Kakoma, fl. 30.ix.1952, *White* 3424 (K). C: Lusaka Dist., Mt. Makulu Res. Station, near Chilanga, fl. x.1960, *Angus* 2436a (K, SRGH). E: Lundazi, fl. 17.x.1967, *Mutimushi* 2236 (K). S: Choma Dist., Mapanza, fl. & fr. 7.xi.1954, *Robinson* 942 (K). **Zimbabwe.** N: Makonde Dist., Mhangura (Mangula), fl. 1.x.1962, *Jacobsen* 1797 (PRE). W: Bulawayo, Luveve–Glenville road near Luveve, fl. 4.x.1959, *Norman* R40 (K). C: Harare Dist., Seke Dam (Prince Edward Dam) road about 10 km S of Harare, fl. 14.x.1955, *Drummond* 4895 (K, LISC, SRGH). E: Mutare Dist., Park lands, fl. 7.x.1948, *Chase* 1191 (K, LISC, SRGH). S: Masvingo Dist., E of Great Zimbabwe Ruins, fr. 1.vii.1930, *Hutchinson & Gillett* 3374 (K, SRGH). **Malawi.** N: Rumphi Dist., 9.5 km (6 miles) from Rumphi on Livingstonia road, fl. 20.x.1962, *Adlard* 513 (K, SRGH). C: Salima, lakeshore, fl. xi.1959, *Lewis* 175 (SRGH). S: Mulanje, 2000–3000 ft., fl. x.1905, *Purves* 222 (K). **Mozambique.** N: Malema Dist., Mutuáli, road to Lioma 7 km, fr. 9.iii.1953, *Gomes e Sousa* 4058 (COI, K, PRE). Z: between Gurué and Ile, 37.7 km from Gurué, fl. 19.x.1949, *Barbosa & Carvalho* 4532 (K, PRE). T: Moatize Dist., near Zóbuè, fl. 21.x.1941, *Torre* 3692 (LISC, LMA). MS: Cheringoma, mile 50, fl. 30.xi.1944, *Simão* 298 (LISC). GI: Funhalouro Dist., adjoining route between Mavume and Mocoduene, 60 km S of official residence of Mavume, fr. x.1938, *Gomes e Sousa* 2160 (COI, K, LISC, PRE).

From West Africa (to Senegal) and the Sudan, Uganda and Tanzania to Namibia and South Africa (North-West and Limpopo Provinces, Gauteng and Mpumalanga). Deciduous woodland; 40–1740 m.

INDEX TO BOTANICAL NAMES